Interstitial Fibrosis In Heart Failure

Francisco J. Villarreal
Department of Medicine
University of California, San Diego
School of Medicine
San Diego, CA 92103
fvillarr@ucsd.edu

Library of Congress Cataloging-in-Publication Data

Interstital fibrosis in heart failure / edited by Francisco J. Villarreal.
    p.; cm. – (Developments in cardiovascular medicine ; v. 253)
   Includes bibliographical references and index.
ISBN 0-387-22824-1 (alk. paper)   e-ISBN 0-387-22825-X
 1. Heart—Fibrosis. 2. Tissue remodeling. 3. Heart failure. I. Villarreal, Francisco J. II.
Series
 [DNLM: 2. Endomyocardial Fibrosis. 2. Extracellular Matrix Proteins. 3. Heart Failure,
Congestive—complications. 4. Matrix Metalloproteinases—physiology. WG 280 I617 2004]
RC682,.I535 2004
616.1'29061—dc22

                        2004052912

Printed in the United States of America.

9 8 7 6 5 4 3 2 1   SPIN 11050865

springeronline.com

Interstitial Fibrosis In Heart Failure

edited by

Francisco J. Villarreal, M.D., Ph.D.

Professor

Department of Medicine
University of California, San Diego
School of Medicine
San Diego, California

 Springer

Developments in Cardiovascular Medicine

232. A. Bayés de Luna, F. Furlanello, B.J. Maron and D.P. Zipes (eds.): *Arrhythmias and Sudden Death in Athletes.* 2000   ISBN: 0-7923-6337-X
233. J-C. Tardif and M.G. Bourassa (eds): *Antioxidants and Cardiovascular Disease.* 2000.   ISBN: 0-7923-7829-6
234. J. Candell-Riera, J. Castell-Conesa, S. Aguadé Bruiz (eds): *Myocardium at Risk and Viable Myocardium Evaluation by SPET.* 2000.ISBN: 0-7923-6724-3
235. M.H. Ellestad and E. Amsterdam (eds): Exercise Testing: New Concepts for the New Century. 2001.   ISBN: 0-7923-7378-2
236. Douglas L. Mann (ed.): The Role of Inflammatory Mediators in the Failing Heart. 2001   ISBN: 0-7923-7381-2
237. Donald M. Bers (ed.): Excitation-Contraction Coupling and Cardiac Contractile Force, Second Edition. 2001   ISBN: 0-7923-7157-7
238. Brian D. Hoit, Richard A. Walsh (eds.): Cardiovascular Physiology in the Genetically Engineered Mouse, Second Edition. 2001   ISBN 0-7923-7536-X
239. Pieter A. Doevendans, A.A.M. Wilde (eds.): Cardiovascular Genetics for Clinicians 2001   ISBN 1-4020-0097-9
240. Stephen M. Factor, Maria A.Lamberti-Abadi, Jacobo Abadi (eds.): Handbook of Pathology and Pathophysiology of Cardiovascular Disease. 2001   ISBN 0-7923-7542-4
241. Liong Bing Liem, Eugene Downar (eds): Progress in Catheter Ablation. 2001   ISBN 1-4020-0147-9
242. Pieter A. Doevendans, Stefan Kääb (eds): Cardiovascular Genomics: New Pathophysiological Concepts. 2002   ISBN 1-4020-7022-5
243. Daan Kromhout, Alessandro Menotti, Henry Blackburn (eds.): Prevention of Coronary Heart Disease: Diet, Lifestyle and Risk Factors in the Seven Countries Study. 2002   ISBN 1-4020-7123-X
244. Antonio Pacifico (ed.), Philip D. Henry, Gust H. Bardy, Martin Borggrefe, Francis E. Marchlinski, Andrea Natale, Bruce L. Wilkoff (assoc. eds): Implantable Defibrillator Therapy: A Clinical Guide. 2002   ISBN 1-4020-7143-4
245. Hein J.J. Wellens, Anton P.M. Gorgels, Pieter A. Doevendans (eds.): The ECG in Acute Myocardial Infarction and Unstable Angina: Diagnosis and Risk Stratification. 2002   ISBN 1-4020-7214-7
246. Jack Rychik, Gil Wernovsky (eds.): Hypoplastic Left Heart Syndrome. 2003   ISBN 1-4020-7319-4
247. Thomas H. Marwick: Stress Echocardiography. Its Role in the Diagnosis and Evaluation of Coronary Artery Disease 2nd Edition.   ISBN 1-4020-7369-0
248. Akira Matsumori: Cardiomyopathies and Heart Failure: Biomolecular, Infectious and Immune Mechanisms. 2003   ISBN 1-4020-7438-7
249. Ralph Shabetai: The Pericardium. 2003   ISBN 1-4020-7639-8
250. Irene D. Turpie; George A. Heckman (eds.): Aging Issues in Cardiology. 2004   ISBN 1-40207674-6
251. C.H. Peels; L.H.B. Baur (eds.): Valve Surgery at the Turn of the Millennium. 2004   ISBN 1-4020-7834-X
252. Jason X.-J. Yuan (ed.): Hypoxic Pulmonary Vasoconstriction: Cellular and Molecular Mechanisms. 2004   ISBN 1-4020-7857-9
253. Francisco J. Villarreal (ed.): Interstitial Fibrosis In Heart Failure 2004   ISBN 0-387-22824-1

Previous volumes are still available

TABLE OF CONTENTS

I. ARCHITECTURE OF THE NORMAL AND REMODELED HEART

II. DIAGNOSTIC TOOLS FOR THE IDENTIFICATION OF MYOCARDIAL FIBROSIS AND CHANGES IN VENTRICULAR MECHANICS

List of Contributors

Guillermo Torre-Amione, Texas Heart Institute at St. Luke's Episcopal Hospital, Houston, Texas, U.S.A.

Hiroshi Ashikaga, Department of Medicine, University of California San Diego, San Diego, CA, U.S.A.

Gregory L. Brower, Department of Anatomy, Physiology & Pharmacology, Auburn University, Auburn, Alabama, U.S.A.

Dale Brown, Division of Cardiology, Denver Health Medical Center and University of Colorado Health Sciences Center, Denver, CO

Lindsay Brown, Department of Physiology and Pharmacology, School of Biomedical Sciences, The University of Queensland, Australia

Laurence L. Brunton, Department of Pharmacology, University of California San Diego, San Diego, CA, U.S.A.

Vincent Chan, Department of Physiology and Pharmacology, School of Biomedical Sciences, The University of Queensland, Australia

Amanda L. Chancey, Department of Anatomy, Physiology & Pharmacology, Auburn University, Auburn, Alabama, U.S.A.

R. English Chapman, Medical University of South Carolina, Charleston, South Carolina, U.S.A.

Jack P. M. Cleutjens, Cardiovascular Research Institute Maastricht, Dept. Of Pathology, University Maastricht, Maastricht, The Netherlands

Javier Diez, Division of Cardiovascular Pathophysiology, School of Medicine, and Department of Cardiology and Cardiovascular Surgery, University Clinic, University of Navarra, Pamplona, Spain

Ian Dixon, Institute of Cardiovascular Sciences, St. Boniface General Hospital Research Centre, University of Manitoba, Winnipeg, Canada

L. Henry Edmunds, University of Pennsylvania, Philadelphia, PA, U.S.A.

Sara Epperson, Department of Pharmacology, University of California San Diego, San Diego, CA, U.S.A.

Mary F. Forman, Department of Anatomy, Physiology & Pharmacology, Auburn University, Auburn, Alabama, U.S.A.

Andrew Fenning, Department of Physiology and Pharmacology, School of Biomedical Sciences, The University of Queensland, Australia

O.H. Frazier, Texas Heart Institute at St. Luke's Episcopal Hospital, Houston, Texas, U.S.A.

Annette M. Gonzalez, Department of Pharmacology, University of California San Diego, San Diego, CA, U.S.A.

Asa Gustafsson, Department of Pharmacology, University of California San Diego, San Diego, CA, U.S.A.

Joseph H. Gorman, University of Pennsylvania, Philadelphia, PA, U.S.A.

Robert C. Gorman, University of Pennsylvania, Philadelphia, PA, U.S.A.

Ramareddy V. Guntaka, Departments of Molecular Sciences and Medicine, University of Tennessee Health Science Center, Memphis, Tennessee, U.S.A.

Jianming Hao, Institute of Cardiovascular Sciences, St. Boniface General Hospital Research Centre, University of Manitoba, Winnipeg, Canada

Mark R. Holland, Washington University in St. Louis, in St. Louis, MO, U.S.A.

Benjamin M. Jackson, University of Pennsylvania, Philadelphia, PA, U.S.A.

Lynetta J. Jobe, Department of Anatomy, Physiology & Pharmacology, Auburn University, Auburn, Alabama, U.S.A.

Joseph S. Janicki, Department of Anatomy, Physiology & Pharmacology, Auburn University, Auburn, Alabama, U.S.A.

Stephen C. Jones, Institute of Cardiovascular Sciences, St. Boniface General Hospital Research Centre, University of Manitoba, Winnipeg, Canada

Bodh I. Jugdutt, Cardiology Division of the Department of Medicine, University of Alberta, Edmonton, Alberta, Canada

Manoj M. Lalu Cardiovascular Research Group, University of Alberta, Edmonton, Alberta, Canada

Ian Legrice, The Bioengineering Institute & Department of Physiology, University of Auckland, Auckland, New Zealand

Hernando Leon, Cardiovascular Research Group, University of Alberta, Edmonton, Alberta, Canada

Carlin S. Long, Division of Cardiology, Denver Health Medical Center and University of Colorado Health Sciences Center, Denver, CO

J. Gary Meszaros, Department of Pharmacology, University of California San Diego, San Diego, CA, U.S.A.

M. Darren Mitchell, Division of Cardiology, Denver Health Medical Center and University of Colorado Health Sciences Center, Denver, CO U.S.A.

Adele Pope, The Bioengineering Institute & Department of Physiology, University of Auckland, Auckland, New Zealand

Richard Schulz, Cardiovascular Research Group, University of Alberta, Edmonton, Alberta, Canada

Francis G. Spinale, Medical University of South Carolina, Charleston, South Carolina, U.S.A.

Bruce Smaill, The Bioengineering Institute & Department of Physiology, University of Auckland, Auckland, New Zealand

Yao Sun, Division of Cardiovascular Diseases, Department of Medicine, University of Tennessee Health Science Center, Memphis, Tennessee, U.S.A.

Suresh C. Tyagi, Department of Physiology and Biophysics, University of Louisville School of Medicine, Louisville, Kentucky, U.S.A.

Francisco J. Villarreal, Department of Medicine, University of California San Diego, San Diego, CA, U.S.A.

Sonia Villegas, Department of Pharmacology, University of California San Diego, San Diego, CA, U.S.A.

Baljit S. Walia, Institute of Cardiovascular Sciences, St. Boniface General Hospital Research Centre, University of Manitoba, Winnipeg, Canada

Cynthia K. Wallace, Texas Heart Institute at St. Luke's Episcopal Hospital, Houston, Texas, U.S.A.

Karl T. Weber, Division of Cardiovascular Diseases, Department of Medicine, University of Tennessee Health Science Center, Memphis, Tennessee, U.S.A.

Samuel A. Wickline, Washington University in St. Louis, in St. Louis, MO, U.S.A.

Preface

The presence of cardiac extracellular matrix (ECM) proteins within the myocardium allows the heart to fulfill critical physiological functions. However, the excessive deposition of ECM proteins (mostly collagens) within the myocardium can contribute to the transformation of the heart whose function relies on optimal mechanics into a dysfunctional organ. The increasing prevalence of heart failure in the general population has led to greater recognition of the importance that the excess deposition of ECM in the heart has in the pathophysiology of the disease. Most cardiologists would agree that in the presence of diffuse fibrosis a true reversal of adverse cardiac remodeling may not occur if the excess deposition of ECM proteins is not either significantly reduced or eventually eliminated. The cardiac fibroblast, which is the cell that produces most of the ECM in the heart, was essentially an ignored biological entity up until the early 1990's. Since then, the number of publications related to the cellular physiology of these cells has increase substantially. This fact highlights the importance given by basic scientists to the regulation of ECM production/deposition in the heart. Over the last decade or so, greater recognition has also been given the issue of ECM "turnover". Indeed, the excess deposition of ECM proteins in the myocardium not only reflects the greater production of the proteins but also their reduced degradation. In this regard, a significant number of publications have also emerged substantiating a role for matrix metalloproteinases in regulating the turnover of ECM proteins.

The major objective of this book is to provide a timely and integrative review of the basics of cardiac ECM structure, how the process of cardiac remodeling influences its disposition, abundance and function, possible non-invasive techniques for diagnosis, as well as discuss potential drug-based or molecular therapeutic strategies that may interrupt or even reverse the course of the development of cardiac fibrosis. The book is divided into 5 sections: I. Macro and microscopic architecture of the normal and remodeled heart; II. Myocardial fibrosis: Diagnosis and alterations in ventricular mechanics; III. Pathophysiology of cardiac remodeling and fibrosis; IV. Matrix metalloproteinases in cardiac remodeling and fibrosis; V. Therapeutic prospects for myocardial fibrosis. Chapter 1 highlights the architecture of the heart with a focus on the disposition of the ECM. Chapter 2 incorporates basic molecular, structural and biochemical features of the cardiac ECM into the context of pathological cardiac remodeling. Chapter 3 highlights features of the cardiac fibroblast and how this cell responds to known pro-inflammatory mediators. Chapters 4 and 5 focus on the possibility of utilizing non-invasive methodologies to detect either an ensuing fibrotic state or the presence of fibrosis. Chapter 6 then provides an overview on how the development of adverse ventricular remodeling and fibrosis impacts the mechanical behavior of the heart.

Chapters included in section III of the book focus on systems known to be important in the pathophysiology of cardiac remodeling/fibrosis including the roles played by the renin-angiotensin system (Chapter 7), matrix metalloproteinases (Chapter 8) and mast cells (Chapter 9). The following section (IV) then takes a more focused approach in describing the role that matrix metalloproteinases play in the pathophysiology of cardiac remodeling/fibrosis in the setting of ischemia-reperfusion injury (Chapter 10), myocardial infarction (Chapter 11), and volume overload (Chapter 12). The last section (V) covers the topic of potential therapeutic approaches for cardiac fibrosis. Chapter 13 comprehensively covers aspects related to pharmacotherapy including those currently used in the pre-clinical and clinical settings. The use of ventricular assist devices has become an attractive therapeutic option for those heart failure patients who are deemed suitable candidates. This topic is covered in Chapter 14. The observation that in some patients certain degrees of "reverse" remodeling can be achieved by the use of these assist devices has also focused attention on how ECM-related changes may follow this "temporary" treatment option. The implantation and subsequent removal of assist devices provides scientists with a unique opportunity, that of sampling diseased and "treated" heart tissue. It is interesting to note that in the limited number of reports published so far many of largest changes in cardiac gene expression levels prior to and after assist device treatment are in ECM-related genes. The last 2 chapters (15 and 16) cover the topic of prospects for molecular-based targeted therapy or gene therapy of the fibrosed heart.

In summary this book attempts to cover current state of the art findings relevant to cellular and molecular processes underlying cardiac fibrosis including basic elements of structure, function, diagnosis, and treatment. My hope is that this book can be of use to both scientists and clinicians. As heart failure continues to emerge as a major threat a greater focused effort will be required to attempt to mitigate or reverse its consequences.

Acknowledgments

This book is dedicated to my mentors Drs. James Covell, Wolfgang Dillmann and Laurence Brunton for continuously supporting me in the pursuit of my academic career. I would also like to take the opportunity to thank all contributors for their excellent chapters. Finally, I would also like to express my greatest appreciation for the preparation and editorial help provided by Ms. Maria L.Pacheco and Shirley Reynolds.

I. *ARCHITECTURE OF THE NORMAL AND REMODELED HEART*

Chapter 1

The Architecture of the Heart: Myocyte Organization and the Cardiac Extracellular Matrix

Ian LeGrice, Adèle Pope, Bruce Smaill
University of Auckland, Auckland, New Zealand

1. Introduction

Bioengineering research is providing new insights into the mechanical function of the heart and the extent to which this is underpinned by myocardial architecture and the cardiac extracellular matrix [1]. The normal left ventricle ejects more than 60% of its end-diastolic cavity volume during systole and analyses of left ventricular wall motion and regional deformation using imaging modalities such as MRI, reveal that this is due to relatively uniform shortening of myocytes across the ventricular wall, slippage or shearing of adjacent groups of myocytes, and the twisting or torsion of the ventricle about its axis.

Computer modeling studies indicate that this optimal mechanical performance is critically dependent on the geometry of the cardiac chambers and the organization of cardiac myocytes [1]. Cardiac myocytes and coronary vessels are embedded in a complex extracellular matrix that consists of collagen, elastin, fibronectin, laminin and proteoglycans [2]. The cardiac extracellular matrix is highly responsive to mechanical loading and plays a central role in the remodeling that occurs in response to altered loading conditions. To understand the functions of the extracellular connective tissue matrix more completely, we need to see them within a broader structural context. In this chapter, we will review the current understanding of myocardial architecture in the hope that it will provide a framework for integrating other more detailed information on the cardiac extracellular matrix found elsewhere in this monograph.

2. The Organization of Myocytes in the Heart

The cardiac muscle cell or myocyte is the main structural component of myocardium, occupying around 70% of ventricular wall volume under normal circumstances. Cardiac myocytes resemble ellipsoid cylinders with a major-axis dimension of 10 to 20 μm and a length of 80 to 100 μm. Myocytes insert end to end and each is connected with several others to form a three-dimensional (3D) network of cells.

The earliest descriptions of cardiac architecture were biased by the presumption that the organization of ventricular myocardium and skeletal muscle are similar and by the use of relatively crude blunt dissection techniques that reinforced that view. Thus, in the first half of the 20th century, the ventricles were seen to consist of discrete muscle bundles that inserted into the atrio-ventricular valve ring and followed a complex helicoidal pathway through the ventricular wall [3, 4]. More recently, the ventricular walls have been incorrectly described as being composed of strap-like muscles [5].

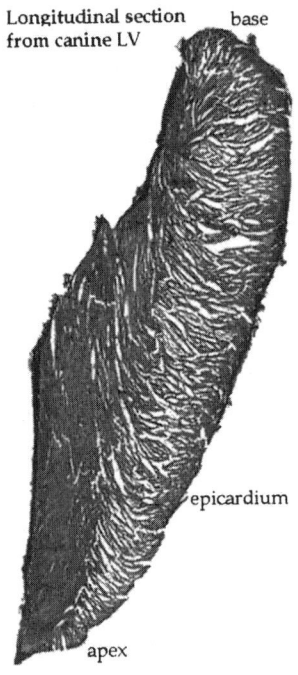

Figure 1. Longitudinal (apex-base) section from the LV of a dog. Section is 100 μm thick. Note the radial spaces or cleavage planes between myocardial laminae.

A very different conception emerged from work reported by Lev [6] and Streeter [7]. Ventricular myocardium was described by these workers as a continuum in which myocyte orientation varied smoothly across the ventricular wall from subepicardium to subendocardium. Detailed measurements of the transmural variation of myocyte orientation have been made at limited numbers of representative ventricular sites in pig [7], dog [8, 9], primate [10] and human [11, 12] hearts. These studies reveal that fiber angle varies by up to 180° transmurally.

It has been widely assumed that ventricular myocardium is transversely isotropic with respect to the myofiber axis, reflecting the view that myocardium is a continuum in which neighbouring myocytes are uniformly coupled. However, macroscopic observations made over a number of years indicate that there is substantial discontinuity in the muscular architecture of the heart. Sections cut from the ventricles of a range of mammalian species have revealed extensive extracellular gaps that are consistent with an ordered arrangement of muscle layers [4, 13-21]. A typical example is given in Figure 1. This thick apex-base transmural section cut from the left ventricular (LV) free wall of a dog heart reveals an array of discrete tissue layers which run from endocardium to epicardium in an approximately radial direction. While there is extensive branching of muscle layers throughout the section, the radial extent of the gaps between adjacent layers is striking - up to 4 or 5 mm in the midwall. There have also been reports of large connective tissue septae in the ventricles of a number of species including man [18, 22-24].

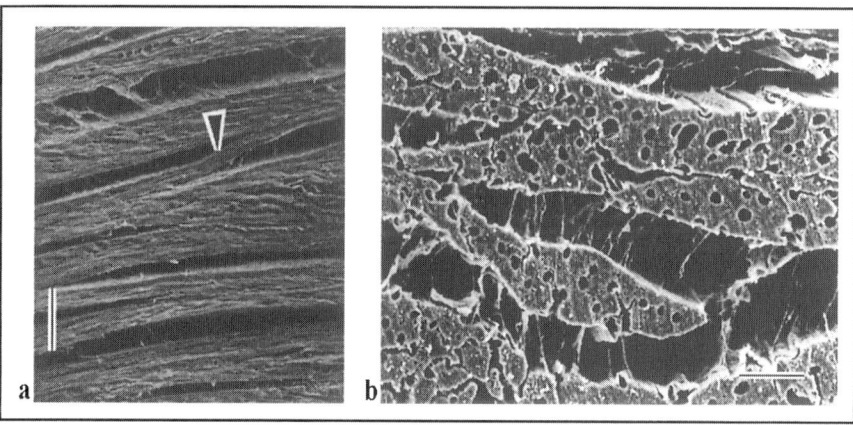

Figure 2. a) SEM of ventricular myocardium sectioned parallel to epicardial surface. Note the laminar arrangement of myocytes. Arrow shows muscle bridge between sheets. Scale bar 100 μm. b) SEM of ventricular myocardium sectioned transverse to myocyte axis. Capillaries and permysial collagen network are evident (Scale bar 25 μm). (From [18], with permission from the American Physiological Society).

Low-power scanning electron microscopy (SEM) has also been used to image ventricular specimens [18]. As shown in Figure 2, myocytes are arranged in discrete layers, separated by cleavage planes and coupled via an extensive extracellular connective tissue network. There is also direct branching between layers with muscle bridges one to two cells thick.

3. The Cardiac Connective Tissue Hierarchy

Studies of the extracellular connective tissue matrix in the heart reveal a complex hierarchical organization that reflects the arrangement of cardiac myocytes and the coronary circulation [23, 25-28]. Caulfield and Borg [25] used SEM to reveal the basic organization of this connective tissue network. They described the following three classes of connective tissue organization: (1) interconnections between myocytes, (2) connections between myocytes and capillaries, and (3) a collagen weave surrounding groups of myocytes. Caulfield and Borg [25] noted that myocytes were connected to adjacent myocytes by numerous bundles of collagen 120 to 150 nm in diameter. These collagen cords were quite straight, with a relatively uniform circumferential distribution and they inserted into the basement membrane of the myocyte close to the Z-line of the sarcomere. Radial collagen cords of similar dimension and appearance also connected myocytes and capillaries. Finally, Caulfield and Borg [25] observed a meshwork of collagen bundles that surrounded groups of three or more myocytes. This external weave was complex, with enclosed myocytes connected to the encompassing network by numerous short, straight, collagen cords. However, Caulfield and Borg [25] observed that adjacent complexes were only loosely coupled by sparse and relatively long collagen fibers.

Robinson and his coworkers [26] went on further to categorize the components of this extracellular connective tissue matrix in an elegant series of studies using light microscopy, together with transmission and scanning electron microscopy. They classified the hierarchy of cardiac connective tissue organization as "endomysium", "perimysium", and "epimysium", using terminology more commonly associated with skeletal muscle.

Endomysium. The cardiac endomysium was seen to incorporate the system of radial collagen cords described by Caulfield and Borg [25], together with a pericellular network of fibers that encompass the myocyte and a lattice of collagen fibrils and microthreads. Robinson and colleagues [26] observed that the radial cords consist of helically wound collagen fibrils that divide close to the cell surface and ramify as part of the pericellular network. In addition to this fine pericellular weave of connective tissue, distinct pericellular cuffs were also seen using light microscopy [26, 27]. It was argued that components of the radial collagen cords also insert into the cell membrane and are tethered to the contractile apparatus at the Z-line [27].

Epimysium. The epimysium was defined as "the sheath of connective tissue that surrounds entire muscles, for example, papillary muscle and trabeculae" [29], and would include components of endocardium and epicardium. It was observed that the epicardium and much of the endocardium consists of a thin layer of endothelial cells overlaying a network of randomly oriented collagen fibrils and elastin fibers.

Perimysium. The term perimysium was used to describe the extensive meshwork of connective tissue that surrounds groups of cells and the connections between contiguous cell bundles. Robinson and colleagues [26] observed large coiled bundles of collagen fibers oriented parallel to the long axis of the myocytes in LV myocardium and papillary muscle. These were associated with groups of cells and were defined as "coiled perimysial fibers". A more detailed study of the organization of the coiled perimysial fibers in rat papillary muscle was later reported [28]. A branched network of coiled perimysial fibers 1 to 10 μm in diameter was seen to diverge from the muscle-tendon interface and a regular array of radial cords interconnected these fibers with adjacent myocytes and with the endocardial membrane surrounding the papillary muscle. In more recent work, MacKenna and colleagues [30] used a confocal laser scanning microscope to investigate the 3D structure of perimysial collagen fibers from rat LV myocardium, describing their conformation as 'planar waviness rotating around the fiber axis'. Similarly, Hanley and co-workers [31] reconstructed 3D images of rat cardiac trabeculae and demonstrated that the perimysial collagen fibers in those structures were also wavy rather than coiled.

4. Collagen Types in the Heart

Collagen is the most abundant component of the extracellular matrix of the heart. There are at least 18 different types of collagen of which five (Types I, III, IV, V and VI) have been identified in myocardium [23, 32]. Whereas types IV and V collagen are components of the basement membrane of cardiac cells, types I and III collagen are the main constituents of the extracellular connective tissue matrix, contributing around 75 to 80% and 15 to 20% of total collagen content, respectively [33, 34].

The mechanical and structural properties of type I and type III collagen are quite different. Type I collagen forms thick parallel rod-like fibers 50-150 nm in diameter and is found in large proportions in tissues requiring high strength such as tendon, bone and skin [32]. Mechanical tests carried out on tendon reveal that type I collagen has a high tensile strength and stiffness. On the other hand, type III collagen is the principal component of the fine collagen fibrils that compose the highly deformable reticular networks characteristic of loose connective tissue and is most abundant in fetal tissues and, in adults, in highly elastic tissues such as skin, blood vessels and lung [32]. Studies with monoclonal antibodies [27, 28, 35-37] indicate that type I and type III collagen are present in both the fine pericellular network of the endomysium and the radial cords that connect myocytes with adjacent myocytes and capillaries. It was demonstrated [27] that type I and type III collagen aggregate in the radial collagen cords to form a copolymer. The intensity of antibody staining for type III collagen is stronger for the pericellular network than for the radial collagen cords. On the basis of these results it could be argued that the fine pericellular networks of the cardiac connective tissue hierarchy are predominantly composed of type III collagen.

It has been proposed that the ratio of type I to type III collagen influences mechanical properties [32]. The proportion of type I collagen to the total of type I and type III collagen is least in the fetal heart and increases with age [33]. In bovine skeletal muscle, Light and Champion [38] found that the proportion of type I collagen to total type I and type III collagen is 84% in epimysium, 72% in perimysium, and 38% in endomysium. In nonhuman primates the proportion of type III collagen to total collagen increases in the evolutionary phase of the response to chronic ventricular pressure overload [34]. This is coincident with the proliferation of fine fibrils in the connective tissue weaves of the endomysium and the perimysium.

Of the minor collagen components in the heart, Type IV collagen is found as an open network in basement membranes where it is involved in regulating molecular transport and cell adhesion, while types V and VI are found in the interstitium where they are associated with other collagen types [32].

The rate of collagen turnover in the normal heart is around 5% per day [32, 39]. Collagen is synthesized by a number of different types of cell in the heart: Fibroblasts and smooth muscle cells produce all cardiac collagen types, while endothelial cells synthesize all cardiac collagen types except type VI and myocytes produce type IV collagen [32]. Collagen is degraded through both intracellular and extracellular pathways, the latter involving neutral metallo-proteinases. Collagen metabolism is highly sensitive to mechanical loading, which can shift the balance between synthesis and degradation, leading to rapid and extensive changes in collagen composition [32]. This is particularly important in the heart remodeling associated with cardiac and cardiovascular diseases such as myocardial infarction, heart failure and hypertension.

5. Three Dimensional Architecture of Ventricular Myocardium

In our laboratory, we have developed morphometric techniques intended to provide comprehensive data on the organization of myocytes, blood vessels, and connective tissue in the normal heart and animal models of cardiac disease. The sections below summarize the results obtained from measurements of myocyte and muscle layer orientation.

Myocyte orientation. We have made extensive measurements of ventricular myocyte orientation in a series of normal dog [8] and pig [40] hearts using a technique which explicitly preserves the registration of fiber orientation with respect to ventricular geometry. The ventricular walls were progressively shaved away in 0.5 mm steps and fiber orientations were mapped out at points on the exposed surface, together with the 3D coordinates of those points. For each heart, this resulted in approximately 10,000 data points describing the fiber orientation throughout right and left ventricles. Our results confirm previous findings and demonstrate surprising consistency at corresponding ventricular sites in different dog hearts. However, there is significant local variation of fiber orientation, particularly in the interventricular septum and toward the boundaries of LV and RV free walls and this information cannot be obtained from the restricted data sets published by previous workers.

Our data is obtained by sectioning the ventricular wall parallel to the epicardial tangent plane to reveal the myocardial layers. Fiber orientation is estimated by measuring the orientation of the myocardial layers in that plane on the assumption that the myocyte axis is parallel to the epicardial tangent plane. However, at some sites such as the ventricular base the myocyte axis has a significant transmural component, defined as the imbrication angle [41]. While our heart model allows for the definition of this angle, there have been no extensive measurements of fiber imbrication in the ventricle to date. However, detailed information on myocyte imbrication can been obtained using the 3D reconstruction techniques such as the extended confocal imaging technique outlined below.

Transmural orientation of muscle layers. Muscle layer orientations have also been recorded systematically in thick longitudinal transverse sections from normal dog [18] and pig [40] hearts. A schematic diagram of the layer organization in these transmural sections is given in Figure 3. Muscle layers are generally normal to the ventricular surfaces in the midwall, but the orientation of layers in subendocardial and subepicardial regions varies at different ventricular sites. For instance, muscle layers approach the endocardial surface of the LV cavity with markedly different orientations in the LV free wall and the interventricular septum. Measurements from sections from right and left ventricles reveal a transmural variation of branching between layers with higher branching densities for sub-epicardial and sub-endocardial sections and a minimum near midwall [18].

A further notable characteristic of the laminar organization seen in the ventricles is the regular appearance of small regions in which layers are perpendicular to the predominant direction. While not particularly extensive in canine myocardium, this intersecting pattern is marked in pig myocardium (unpublished observation). The significance of this variation is not yet known though it is likely to be important for accommodating the significant shear deformation which occurs in myocardium throughout the cardiac cycle, since the perpendicular cleavage planes align closely with the two orthogonal planes of maximum shear [42].

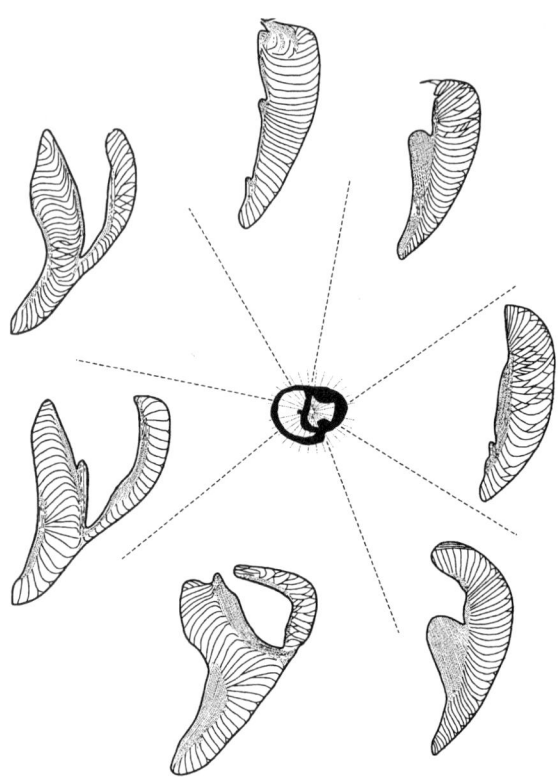

Figure 3. Apex-base transmural sections of the ventricles from a series of circumferential locations in a dog heart. The orientations of myocardial laminae are shown for each section (sketched from original data). (Modified from [18], with permission from the American Physiological Society).

Three dimensional arrangement of muscle layers. Quantitative observations of the 3D organization of muscle layers have been made on segments removed from different LV sites in fixed dog hearts [18]. SEM images for a series of samples from these segments (see Figure 2) reveal consistent patterns in the cellular organization of muscle layers. At this level, the layers appear to be uniform in structure with regular anastomoses between adjacent cells and an ordered interdigitation of cells and capillaries. Layers consist of tightly packed groups of cardiac myocytes aligned so that the cell axis is approximately parallel to the cut edge of the layer and are 4±2 myocytes thick at all transmural sites, with branches between layers generally one to two myocytes thick. In those areas where there are appreciable spaces between adjacent layers, their radial extent appears to be considerable and an extensive network of connective tissue is observed. Long collagen fibers run between muscle layers and insert into the connective tissue network which surrounds them. These perimysial collagen fibers connect adjacent muscle layers and should not be confused with the thicker coiled perimysial fibers [28] that run approximately parallel to the myocyte fiber axis.

Despite the uniformity of muscle layer organization, the architecture of ventricular myocardium is not homogeneous and shows considerable transmural variation in the density of branching between layers. Specifically, there are two- to three-fold fewer branches in the midwall than in subepicardium (or subendocardium), and the extent of cleavage planes is greatest in the midwall. The branching between adjacent layers is relatively sparse in the midwall and the distance between branches along a layer can be of the order of 2 to 4 mm in this region. The length of perimysial fibers connecting adjacent muscle layers was also greatest in the midwall. The mean length of perimysial fibers connecting adjacent layers in the subepicardium varied from 13.2 μm to 19.8 μm, but ranged from 22.3 μm to 49.4 μm in the deep midwall. Moreover the length distribution of these perimysial fibers is considerably wider for the midwall than for the subepicardium.

This work provides more detailed and more systematic information on aspects of myocardial structure and its associated connective tissue hierarchy previously described by Caulfield & Borg [25] and Robinson and colleagues [26]. It is evident that the layers observed in this work may be equated with the "bundles" of myocytes reported by these workers.

More recently, we have developed a new technique to image extended volumes of myocardium [43]. A confocal microscope is used to acquire contiguous sets of z-series images spanning the upper surface of resin embedded myocardial specimens. This surface is then trimmed and the process is repeated to assemble large, high-resolution image volumes. Digital reslicing, segmentation and volume rendering methods can be applied to the resulting volumes to provide quantitative structural information about the 3D organization of myocytes, extracellular collagen matrix and blood vessel network of the heart which has not previously been available.

Figure 4 is a 3D reconstruction of a transmural segment (800 μm x 800 μm x 4.5 mm) cut from the LV free wall of a rat heart perfused with picrosirius red [44]. This image, which consists of 650 million voxels at a spatial resolution of 1.5 μm, provides a striking view of the laminar architecture of the myocardium. The 3D organization of layers varies through the wall to accommodate the transmural variation of fiber orientation. Muscle branches between layers seem to be most dense in subepicardial and subendocardial regions of the specimen, but are relatively sparse in the midwall as reflected by the extent of the cleavage planes between adjacent, midwall muscle layers. The dye, picrosirius red, binds non-sterically to collagen [44] which appears white in Figure 4. The extensive collagen network covering the epicardial surface is evident in this figure, as is the convoluted nature of the endocardial surface. In addition, networks of collagen cords that span cleavage planes to connect adjacent muscle layers are clearly visible in Figure 4. Large collagen cords, which are aligned approximately with the myocyte axis and extend over many cell lengths, are also observed in the cleavage planes. These longitudinal cords have a convoluted rather than a coiled appearance. While these do not fall easily into the classification of Robinson and co-workers, they should be regarded as perimysial since they are associated with more than 2 myocytes along their length [28]. Figure 5 indicates that there is a relatively high density of these cords within muscle layers.

It is possible to segment out and display separate structural components within a myocardial volume (see Figure 6) and this provides a powerful basis for efficient morphometric analysis. For instance, it is a relatively trivial matter to quantify the transmural variation of perimysial collagen once it has been segmented out of the extended volume image as shown in Figure 6b.

Figure 4. Oblique view of extended volume image from LV of rat heart obtained using confocal microscopy. Note the laminar organization and collagen (white) interconnecting layers of myocytes. The epicardial collagen weave is clearly seen along with cleavage planes between myocardial layers. (From [43], with permission from the Royal Microscopical Society).

Using this approach we have shown that perimysial collagen network is densest in the subendocardium and subepicardium and relatively sparse in the midwall [43]. This also reinforces the critical need to account for sample orientation when carrying out quantitative analysis from 2D microscope images. Because the 3D arrangement of the collagen network varies across the ventricular wall, collagen density should be compared between sections only where the principal orientation of collagen bundles are similar [43].

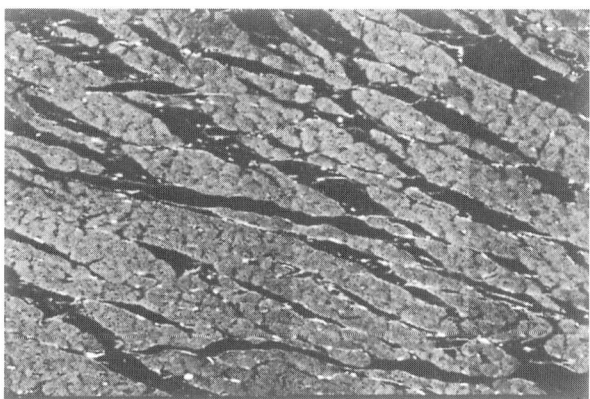

Figure 5. Image slice from LV midwall of rat heart (800x800 μm) illustrating laminar organization of myocytes. Plane of optical section is perpendicular to myocyte axis. White dots are perimysial collagen cords running parallel to myocyte axis.

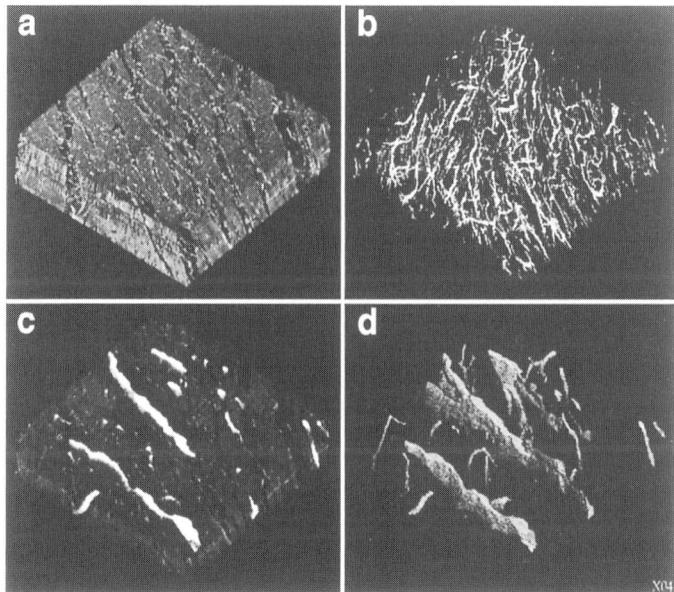

Figure 6. Reconstructed subvolumes (800x800x100 μm). In the upper panel collagen is segmented and rendered (a) with and (b) without background due to myocytes. In the lower panel venous sinuses are segmented and rendered, (c) with and (d) without background. (From [43], with permission from the Royal Microscopical Society).

On the basis of the observations summarized in this chapter, we have developed the conceptual model of myocardial architecture shown in Figure 7. Groups of tightly coupled myocytes are organized into layers with a characteristic transmural direction. These layers twist to accommodate the variation of muscle fiber orientation across the ventricular wall and are

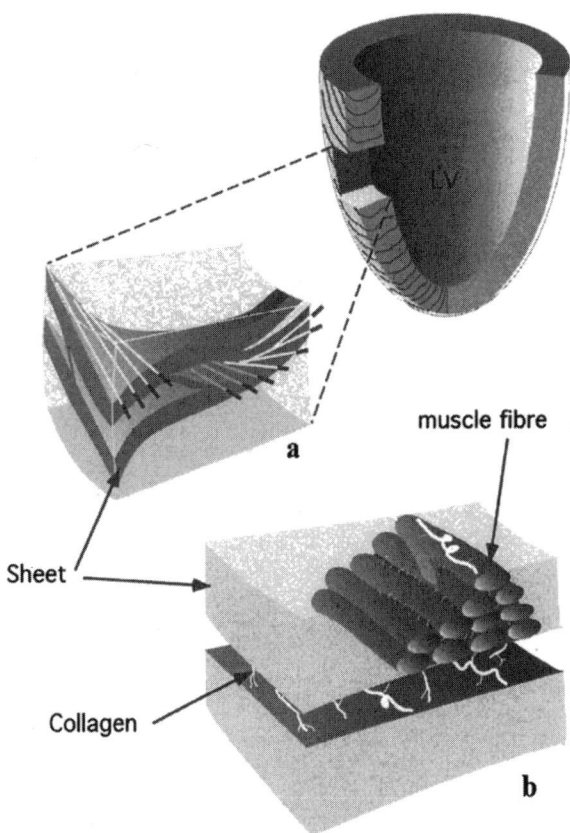

Figure 7. Schematic of cardiac microstructure. (a) A transmural block cut from the ventricular wall shows orientation of fibers. Note the transmural variation in fiber angle. (b) The muscle fibers are shown forming a layer three to four cells thick. Endomysial collagen is shown connecting adjacent cells within a sheet while perimysial collagen links adjacent sheets. (Modified from [18]).

interconnected via relatively sparse muscle bridges. The extracellular connective tissue network of the heart reflects the hierarchy of myocyte architecture and plays a major role in organizing this structure. Endomysial collagen registers the myocytes within a layer. A network of perimysial collagen cords tether adjacent muscle layers and a further population of convoluted perimysial cords is oriented in the direction of the myocyte axis. It should not be imagined that the 3D organization of muscle layers illustrated in this figure is the only possible arrangement: within such a hierarchy of muscle layers a range of various transmural connections is possible.

6. Myocardial Architecture and Cardiac Function

There is a considerable body of indirect experimental evidence that suggests that the characteristics and extent of the extracellular connective tissue matrix are important determinants of diastolic and systolic ventricular function. Robinson and his coworkers [26-29, 45] proposed specific mechanical roles for the cardiac connective tissue matrix based on their anatomical observations. The collagen network was seen as a "strain-locking" system, which resists the extension of myocytes beyond sarcomere lengths of 2.25 µm while allowing

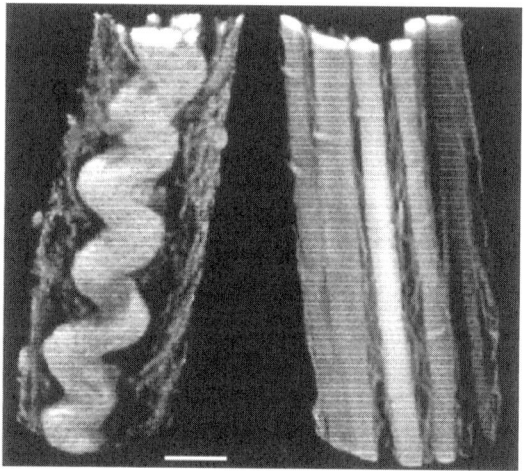

Figure 8. 3D reconstruction of typical permysial collagen fibers from rat ventricular trabeculae fixed at near-resting sarcomere length (left) and at an extended sarcomere length (right). (Scale bar 10 µm.) (From [31], with permission from the Physiological Society).

relatively free extension up to this length. Thus, when myocardium is stretched, the biaxially organized connective tissue networks of the epimysium, perimysium, and endomysium are rearranged to align more closely with the muscle-fiber direction (cargo net hypothesis). In addition, coiled perimysial fibers in ventricular myocardium and papillary muscle, which are obliquely or longitudinally aligned with respect to the long axis of muscle fibers, become less convoluted. Robinson and colleagues [26, 28] noted that the end-points of both these changes in connective tissue configuration are correlated with a striking increase in the axial stiffness of papillary muscles when extension exceeds 15% of slack length. MacKenna and colleagues [30] used 2D light microscopy techniques to demonstrate that, in the rat LV, convoluted perimysial collagen cords aligned with the myocyte axis straighten as sarcomere length approaches 2.25µm. This mechanism is clearly illustrated in Figure 8 [31]. In this case, Hanley and colleagues used confocal microscopy to reconstruct rat ventricular trabeculae in 3D for different axial extensions.

It is most probable that each of the components of the connective tissue hierarchy contributes to the mechanical properties of passive myocardium. Robinson and colleagues argued that the epimysium protects sarcomeres from being overstretched [26]. Caulfield and Borg [25] argued that the short radial collagen cords of the endomysium ensure that adjacent myocytes remain in registration providing a mechanism for uniform distribution of strain at the cellular level. They saw the perimysial network as organizing individual myocytes into a functional unit and providing a framework for distributing the contractile force within the ventricular wall. It is also argued [45] that the extracellular connective tissue matrix stores energy during systole and contributes to the elastic recoil of the ventricles during rapid filling. One of the lines of argument involves a straightforward corollary of the cargo net hypothesis. Since the volume occupied by the contractile lattice is relatively constant, the cross-sectional area of the myocyte increases when shortening occurs. It was argued that this dimensional change will lead to a realignment of the pericellular network of the endomysium with collagen fibrils now preferentially oriented transverse to the muscle-fiber direction. (A similar reorganization was also seen to apply for the perimysial network surrounding groups of muscle cells.) There was further speculation that the systolic deformation of the ventricle could impose tensile load on the collagen cords that tether adjacent myocytes [45] and compress the coiled fibers and tendons of the perimysium [28]. It was postulated that the potential energy storage associated with any of these mechanisms could oppose further shortening and tend to restore the myocardium to its resting state.

It would seem obvious that the major component of passive myocardial stiffness resides within the extracellular connective tissue matrix. However, it is now evident that intracellular structures also contribute significantly to myocardial stiffness. Of particular interest is the huge protein titin which has been viewed as an adjustable spring which actively modulates aspects of myocardial stiffness [46]. The intercellular and extracellular domains should not, of course, be seen as independent: they are coupled both in parallel and in series through transmembrane proteins such as integrins which connect the extracellular matrix to the cytoskeleton. These molecules transmit force between the domains, but also play a role as mechano-receptors, transducing force information and activating cellular signal pathways [2].

Recent bioengineering analyses provide overwhelming evidence that the effectiveness of cardiac mechanical function is intimately associated with the 3D arrangement of myocytes in the heart. The ventricles experience substantial dimensional change throughout the cardiac cycle and because they are thick-walled structures, there are significant transmural gradients of strain. In the LV, for example, end-diastolic circumferential extension can be more than three times greater at the endocardium than the epicardium and, in general, is significantly less than longitudinal extension at all transmural sites. It has been argued that the changing orientation of myoctyes across the ventricular wall and the associated torsion which occurs as the heart contracts act to normalize sarcomere length changes throughout the cardiac cycle and hence normalize myocyte stress and work across the ventricular wall [47, 48]. It is also argued that transmural coupling through the connective tissue matrix is necessary for contraction of the spirally oriented myocytes to be translated into effective pumping action [49]. Experimental work has shown that when strains are calculated with respect to a local structurally-oriented coordinate system that sarcomere strains are indeed uniform across the ventricular wall during diastole [50] and systole [51-53].

The laminar organization of ventricular myocardium also appears to play a critical role in accommodating shear deformations in the heart wall. This hypothesis accompanied a number of early observations of laminar structure in the heart [13, 14, 17]. Caulfield and Borg also stated that the loose collagen connections between groups of myocytes facilitate the large shearing deformations that are necessary for normal diastolic and systolic ventricular function [25]. More specifically, it has been argued the substantial changes in ventricular wall thickness that occur during the cardiac cycle are possible only because of slippage and rearrangement of adjacent muscle layers in the subendocardial regions of the ventricles. Spotnitz and colleagues investigated this idea and demonstrated that the transmural orientation of cleavage planes between muscle layers became progressively more oblique as the wall thinned during diastole [20]. Three-dimensional strain analysis reveals substantial shearing between adjacent subendocardial layers in systole [51, 52, 54] and diastole [52]. The wall slippage mechanism has also been invoked to explain aspects of the regional ventricular dilatation and local wall thinning which follow myocardial infarction [17, 19, 21]. The shear properties of passive LV myocardium have been related to its laminar structure using midwall specimens in which the principal features of myocardial architecture were preserved [55]. This study indicates that passive myocardium has nonlinear, viscoelastic, anisotropic shear properties with least resistance to simple shear displacements imposed in the plane of the myocardial layers. Analyses of the data suggests that simple shear deformation is resisted by elastic elements aligned with the microstructural axes of the tissue, and that the inherent shear stiffness of passive myocardium is relatively small.

The laminar organization of ventricular myocardium is believed also to affect the propagation of electrical activation through the heart and this point has been raised in an associated context by Spach and co-workers [24, 56]. If it is assumed that muscle layers are electrically insulated by the connective tissue surrounding the layers, depolarization can spread to adjacent laminae only via direct muscle branches. Therefore, the path required to traverse a given distance perpendicular to the layers will be convoluted, particularly in the mid-ventricular wall where the frequency of branching is relatively low. This would not influence normal cardiac electrical activation since the endocardial surface of the ventricles are rapidly excited via the Purkinje fiber network and the radial arrangement of muscle layers will ensure an efficient transmural spread of activation. However, the structural anisotropy of ventricular tissue may well contribute to the patterns of electrical activation associated with various arrhythmias. Computer modeling studies in which the effect of the laminar structure was investigated indicate that this is the case [57]. Furthermore, this study also indicated that activation immediately after, and in the region of, an ectopic stimulus proceeds in a non-uniform manner and generates fractionated extracellular electrograms because of the discontinuities introduced by the laminar structure. This study also provided an exciting new insight into the mechanism of action of a defibrillating shock. When such a shock is applied across the heart wall, depolarizing membrane potential steps occur at each boundary between myocardial laminae, and subsequently depolarizing the bulk of the myocardium within a few milliseconds of the shock, a necessary condition for effective defibrillation.

In conclusion, the architecture of the heart has a profound effect on its mechanical and electrical function. The cardiac extracellular matrix plays a central role in the maintenance and adaptation of this structure in the normal heart and is intimately linked with the remodeling and associated impairment of cardiac function that commonly accompanies diseases such as heart failure and hypertension. In order to understand these processes more fully we need conceptual and experimental models that will enable us to integrate the complex interactions between structure and function to which the cardiac extracellular matrix contributes at the whole-heart level. Before this can be done, however, more detailed and more quantitative information about the cellular and molecular processes that underlie the biology of the cardiac extracellular matrix is required.

References

1.   Costa, K.D., J. Holmes, and A.D. McCulloch, Modeling cardiac mechanical properties

in three dimensions. Phil Trans R Soc Lond, 2001. A359: p. 1233-1250.

2.   Bishop, J.E. and G. Lindahl, Regulation of cardiovascular collagen synthesis by mechanical load. Cardiovasc Res, 1999. 42: p. 27-44.

3.   Mall, F.P., On the muscular architecture of the ventricles of the human heart. Am J Anat, 1911. 11: p. 211-266.

4.   Robb, J.S. and R.C. Robb, The normal heart. Anatomy and physiology of the structural units. Am Heart J, 1942. 23: p. 455-467.

5.   Torrent-Guasp, F.F., W.F. Whimster, and K. Redmann, A silicone rubber mould of the heart. Technol Health Care, 1997. 5: p. 13-20.

6.   Lev, M. and C.S. Simkins, Architecture of the human ventricular myocadium. Technique for study using a modification of the Mall-MacCallum method. Lab Invest, 1956. 5: p. 396-409.

7.   Streeter, D.D., Jr. and D.L. Bassett, An engineering analysis of myocardial fiber orientation in pig's left ventricle in systole. Anat Rec, 1966. 155: p. 503-511.

8.   Nielsen, P.M., et al., Mathematical model of geometry and fibrous structure of the heart. Am J Physiol, 1991. 260: p. H1365-78.

9.   Streeter, D.D., Jr., et al., Fiber orientation in the canine left ventricle during diastole and systole. Circ Res, 1969. 24: p. 339-47.

10.   Ross, M.A. and D.D. Streeter, Jr., Nonuniform subendocardial fiber orientation in the normal macaque left ventricle. Eur J Cardiol, 1975. 3: p. 229-47.

11.   Fox, C.C. and G.M. Hutchins, The architecture of the human ventricular myocardium. Johns Hopkins Med J, 1972. 130: p. 289-99.

12.   Greenbaum, R.A., et al., Left ventricular fibre architecture in man. Br Heart J, 1981. 45: p. 248-63.

13.   Feneis, H., Das Gefuge des Herzmuskels bei Systole und Diastole. Morph Jahrb, 1943. 89: p. 371-406.

14.   Hort, W., Untersuchungen uber die Muskelfaserdehnung und das Gefuge des Myokards in der rechten Herzkammerwand des Meerschweinchens. Virchows Arch Pathol Anat Physiol Klin Med, 1957. 329: p. 649-731.

15.   Hort, W., Makroskopische und mikrometrische Untersuchungen am Myokard verschieden stark gefullter linker Kammern. Virchows Arch Path Anat, 1960. 333: p. 523-564.

16.   Weitz, G., Uber das unterschiedliche Verhalten der lage der Herzmuskelfasern in kontrahiertem und dilatierem Zustand. Med Klin Munich, 1951. 46: p. 1031-1032.

17.   Grimm, A.F., K.V. Katele, and H.L. Lin, Fiber bundle direction in the mammalian heart. An extension of the "nested shells" model. Basic Res Cardiol, 1976. 71: p. 381-8.

18.   LeGrice, I.J., et al., Laminar structure of the heart: ventricular myocyte arrangement and connective tissue architecture in the dog. Am J Physiol, 1995. 269: p. H571-82.

19.   Olivetti, G., et al., Side-to-side slippage of myocytes participates in ventricular wall remodeling acutely after myocardial infarction in rats. Circ Res, 1990. 67: p. 23-34.

20.   Spotnitz, H.M., et al., Cellular basis for volume related wall thickness changes in the rat left ventricle. J Mol Cell Cardiol, 1974. 6: p. 317-31.

21.   Weisman, H.F., et al., Cellular mechanisms of myocardial infarct expansion. Circulation, 1988. 78: p. 186-201.

22.   Abrahams, C., J.S. Janicki, and K.T. Weber, Myocardial hypertrophy in Macaca fascicularis. Structural remodeling of the collagen matrix. Lab Invest, 1987. 56: p. 676-83.

23.   Bashey, R.I., A. Martinez-Hernandez, and S.A. Jimenez, Isolation, characterization, and localization of cardiac collagen type VI. Associations with other extracellular matrix components. Circ Res, 1992. 70: p. 1006-17.

24.   Spach, M.S. and P.C. Dolber, Relating extracellular potentials and their derivatives to

anisotropic propagation at a microscopic level in human cardiac muscle. Evidence for electrical uncoupling of side-to-side fiber connections with increasing age. Circ Res, 1986. 58: p. 356-71.

25. Caulfield, J.B. and T.K. Borg, The collagen network of the heart. Lab Invest, 1979. 40: p. 364-72.

26. Robinson, T.F., L. Cohen-Gould, and S.M. Factor, Skeletal framework of mammalian heart muscle. Arrangement of inter- and pericellular connective tissue structures. Lab Invest, 1983. 49: p. 482-98.

27. Robinson, T.F., et al., Morphology, composition, and function of struts between cardiac myocytes of rat and hamster. Cell Tissue Res, 1987. 249: p. 247-55.

28. Robinson, T.F., et al., Coiled perimysial fibers of papillary muscle in rat heart: morphology, distribution, and changes in configuration. Circ Res, 1988. 63: p. 577-92.

29. Robinson, T.F., et al., Structure and function of connective tissue in cardiac muscle: collagen types I and III in endomysial struts and pericellular fibers. Scanning Microsc, 1988. 2: p. 1005-15.

30. MacKenna, D.A., J.H. Omens, and J.W. Covell, Left ventricular perimysial collagen fibers uncoil rather than stretch during diastolic filling. Basic Res Cardiol, 1996. 91: p. 111-22.

31. Hanley, P.J., et al., 3-Dimensional configuration of perimysial collagen fibres in rat cardiac muscle at resting and extended sarcomere lengths. J Physiol, 1999. 517 : p. 831-7.

32. Bishop, J.E. and G.J. Laurent, Collagen turnover and its regulation in the normal and hypertrophying heart. Eur Heart J, 1995. 16: p. 38-44.

33. Medugorac, I. and R. Jacob, Characterisation of left ventricular collagen in the rat. Cardiovasc Res, 1983. 17: p. 15-21.

34. Weber, K.T., et al., Collagen remodeling of the pressure-overloaded, hypertrophied nonhuman primate myocardium. Circ Res, 1988. 62: p. 757-65.

35. Borg, T.K., et al., Morphological and chemical characteristics of the connective tissue network during normal development and hypertrophy. J Mol Cell Cardiol, 1986. 18: p. 247.

36. Borg, T.K., R.E. Gay, and L.D. Johnson, Changes in the distribution of fibronectin and collagen during development of the neonatal rat heart. Coll Relat Res, 1982. 2: p. 211-8.

37. Borg, T.K., et al., Alteration of the connective tissue network of striated muscle in copper deficient rats. J Mol Cell Cardiol, 1985. 17: p. 1173-83.

38. Light, N. and A.E. Champion, Characterization of muscle epimysium, perimysium and endomysium collagens. Biochem J, 1984. 219: p. 1017-26.

39. Weber, K.T. and C.G. Brilla, Pathological hypertrophy and cardiac interstitium. Fibrosis and renin-angiotensin-aldosterone system. Circulation, 1991. 83: p. 1849-65.

40. Stevens, C., et al., Ventricular mechanics in diastole: material parameter sensitivity. J Biomech, 2003. 36: p. 737-48.

41. Streeter, D.D., Jr., Gross morphology and fiber geometry of the heart In: Handbook of Physiology, edited by R.M. Berne. N.S.a.S.R.G. Baltimore: American Physiological Society, Williams and Wilkins Company, 1979: p. 61-112.

42. Arts, T., et al., Relating myocardial laminar architecture to shear strain and muscle fiber orientation. Am J Physiol, 2001. 280: p. H2222-9.

43. Young, A.A., et al., Extended confocal microscopy of myocardial laminae and collagen network. J Microsc, 1998. 192 : p. 139-50.

44. Dolber, P.C. and M.S. Spach, Conventional and confocal fluorescence microscopy of collagen fibers in the heart. J Histochem Cytochem, 1993. 41: p. 465-9.

45. Robinson, T.F., S.M. Factor, and E.H. Sonnenblick, The heart as a suction pump. Sci Am, 1986. 254: p. 84-91.

46. Granzier, H. and S. Labeit, Cardiac titin: an adjustable multi-functional spring. J Physiol, 2002. 541: p. 335-42.

47. Arts, T., R.S. Reneman, and P.C. Veenstra, A model of the mechanics of the left ventricle. Ann Biomed Eng, 1979. 7: p. 299-318.

48. Arts, T., et al., Relation between left ventricular cavity pressure and volume and systolic fiber stress and strain in the wall. Biophys J, 1991. 59: p. 93-102.

49. Ingels, N.B., Jr., Myocardial fiber architecture and left ventricular function. Technol Health Care, 1997. 5: p. 45-52.

50. Omens, J.H., K.D. May, and A.D. McCulloch, Transmural distribution of three-dimensional strain in the isolated arrested canine left ventricle. Am J Physiol, 1991. 261: p. H918-28.

51. Costa, K.D., et al., Laminar fiber architecture and three-dimensional systolic mechanics in canine ventricular myocardium. Am J Physiol, 1999. 276: p. H595-607.

52. Takayama, Y., K.D. Costa, and J.W. Covell, Contribution of laminar myofiber architecture to load-dependent changes in mechanics of LV myocardium. Am J Physiol, 2002. 282: p. H1510-20.

53. Waldman, L.K., et al., Relation between transmural deformation and local myofiber direction in canine left ventricle. Circ Res, 1988. 63: p. 550-62.

54. LeGrice, I.J., Y. Takayama, and J.W. Covell, Transverse shear along myocardial cleavage planes provides a mechanism for normal systolic wall thickening. Circ Res, 1995. 77: p. 182-93.

55. Dokos, S., et al., Shear properties of passive ventricular myocardium. Am J Physiol, 2002. 283: p. H2650-9.

56. Spach, M.S., P.C. Dolber, and J.F. Heidlage, Influence of the passive anisotropic properties on directional differences in propagation following modification of the sodium conductance in human atrial muscle. A model of reentry based on anisotropic discontinuous propagation. Circ Res, 1988. 62: p. 811-32.

57. Hooks, D.A., et al., Cardiac microstructure: implications for electrical propagation and defibrillation in the heart. Circ Res, 2002. 91: p. 331-8.

Chapter 2

Extracellular Matrix and Cardiac Remodeling

Bodh I. Jugdutt
University of Alberta, Edmonton, Alberta

1. Introduction

Cumulative evidence over the last 3 decades suggests that the cardiac interstitium or extracellular matrix (ECM), especially the extracellular collagen matrix (ECCM), plays an important role in cardiac remodeling during various cardiac diseases including myocardial infarction (MI) and heart failure [1-4]. A unique feature of the healthy heart is the ability of the pumping chambers to return to the ideal shape at end-diastole despite repetitive changes during every cardiac cycle throughout life. This physiologic remodeling or property of resuming functional shape depends largely on the integrity of the ECCM, the intricate network of fibrillar collagens found in the ECM. In a broad sense, pathophysiologic cardiac remodeling refers to changes in cardiac structure, shape and function following stress or damage [5-7]. To remodel is defined in the Webster dictionary as " to alter the structure, to remake" and in the Oxford dictionary as "to model again and differently, reconstruct and reorganize." The Oxford dictionary defines "model" as "representation in three dimensions of the proposed structure" and "to model" as "to fashion and shape." This review will focus on the role of the ECCM in pathologic cardiac remodeling and specifically left ventricular (LV) remodeling.

2. Cardiac extracellular matrix

2.1. Cardiac ECM, ECCM and fibroblasts

In the heart, as in other organs [8, 9], the specialized parenchymal cells (cardiomyocytes) are supported by an ECM made up of an intricate macromolecular network of fibers and different cell types of mesenchymal origin including fibroblasts, endothelial cells (cardiac and vascular), smooth muscle cells, blood-borne cells (macrophages and others), pericytes and neurons, bathed in a gel-type ground substance (Table 1). The supporting extracellular network, composed mainly of fibrillar collagen [2], is referred to as the ECCM in this review.

Table 1. Composition of the cardiac extracellular matrix

Component
- Collagen fibrils (types I, III)
- Elastin
- Cells
    Fibroblasts, macrophages,plasma cells, others (vessels; nerves)
- Gel matrix (ground substance),hydrophilic glycosaminoglycans (GAGs), glycoproteins
- Other molecules, growth factors, cytokines, enzymes

In the healthy heart, it is important to note that cardiomyocytes make up 25 to 35% of the cell number [2, 10, 11], ~90% of myocardial mass [12-14] and 67 to 75 % of the myocardial volume [2, 15] while nonmyocytes account for about 67 to 75% of the cell number [11, 15, 16] and ~33% of the myocardial volume [2, 15]. The nonmyocytes make up 75% of the total number of cells [11] or 90 to 95 % of the non-myocyte fraction of cardiac cells [12-14]. Vascular tissue, together with the lumens, was reported to occupy about 60% of the extracellular space in one study [17] and 13% of the volume fraction in another study [18]. Although there are differences among various studies, probably related to the use of different methodologies, it is clear that fibroblasts are one of the dominant cell types in mammalian myocardium (Table 2) and they play a major role in cardiac remodeling.

Most of the matrix macromolecules (of which collagen is the principal structural protein) are produced by fibroblasts [8, 9]. The cardiomyocytes which mediate the heart's pump function throughout life are still the center of attention. However, the cardiac fibroblasts which mediate the formation of the ECCM, essential for maintaining structural and functional integrity [19-21], are gaining increasing attention as potential targets.

2.2. The cardiac ECCM and collagen types

The collagen molecule typically consists of a central core of long, stiff, triple-stranded helices in which 3 α chains (composed of triplet amino acid sequences, usually glycine-proline-hydroxyproline) are wound around each

Table 2. Relative proportion of myocytes and nonmyocytes in myocardium

Group	By cell number	By cell volume	By cell mass
Cardiomyocyte	25% (11)* 30-35% (16) 33% (2)	~75% (11) ~67% (2) 67% (15) 80% (18)	~90% (12,14)
Nonmyocyte	75% (11)* 65-70% (16) 67% (2)	~33% (2) 33% (15) § 20% (18) [13% vascular] †	~10% (12,13,14) [90-95% fibroblasts] ‡

Reference number in parenthesis
* Connective tissue nuclei; † includes lumen (volume fraction)
‡ fibroblasts as % of nonmyocyte fraction; § mostly fibroblasts

other to form a superhelix [9]. Since each collagen alpha chain is encoded by a separate gene, there are several collagen types (Table 3). The major fibrillar collagens in the heart are collagen types I and III and they are synthesized in cardiac fibroblasts [10]. Approximately 85% of the total myocardial collagen is type I and 11% is type III [1]. Collagen type I, the dominant fibrillar collagen, is associated mainly with thick fibers which confer tensile strength and resistance to stretch and deformation [3]. Collagen type III is associated with thin fibers which confer resilience [22].

Types I and III collagen maintain the structural integrity of the myocytes [3]. They maintain alignment of myofibrils within myocytes via collagen-integrin-cytoskeleton-myofibril connections, and thereby allow for the cyclic translation of myocyte shortening into coordinated contraction of the ventricular chambers. It follows that ECCM remodeling may result in significant LV dysfunction. Other components of the ECM (such as the polysaccharide gel, elastin, fibronectin, laminin and proteoglycans) also mediate important functions that have been reviewed [8, 9]. The elastin content in myocardium is modest compared to that in distensible blood vessels [2, 22].

2.3. Collagen synthesis

In general, collagen biosynthesis [8, 9] involves at least 8 enzymatic steps: i) intracellular synthesis of pro-alpha chains; ii) hydroxylation of selected prolines and lysines; iii) glycosylation of selected hydroxyl serines; iv) formation of procollagen triple helixes; v) secretion into the extracellular space; vi) conversion into less soluble molecules by removal of propeptides; vii) self-assembly into collagen fibrils; and viii) aggregation into fibers.

Table 3. Collagen types and potential relevance in cardiovascular remodeling

Type	Polymerized Form	Distribution	Remodeling*
I	fibrillar	Bone, skin, tendon, ligaments, cornea, internal organs (heart), ~ 90% of body collagen	+++
II	fibrillar	Cartilage, intervertebral disc, notochord, vitreous humor	+
III	fibrillar	Skin, blood vessels, internal organs (heart); frequently associated with type I	+++
IV	network forming (sheetlike)	Basal laminae	+
V	fibrillar (with type I)	As type I	+
VI	beaded filaments	Widespread	?
VII	network forming (anchoring)	Beneath stratified squamous epithelia	?
VIII	lattice (hexagonal)	Endothelial cell	+
IX	fibril-associated (lateral)	Cartilage, vitreous body	+
X	short-chain (as type VIII)	Cartilage	+
XI	fibrillar (with type II)	As type II	+
XII	fibril-associated (lateral)	Tendon, ligaments, some other tissues	+
XIII	alternative splicing	Widespread, low quantity	?
XIV	fibril associated (interrupted triple helix)	Skin, tendon	?
XV	interrupted triple helix	Fibroblasts, smooth muscle cells	?
XVI	interrupted triple helix	Fibroblasts, keratinocytes	?
XVII	interrupted triple helix	Dermal-epidermal junction	?
XVIII	interrupted triple helix	Highly vascular tissues	+
XIX	five triple-helical domains	Cultured skin fibroblasts, tumor cells	-

*Role in remodeling: documented definite (+++); less definite (+); probable (?); not known (-).

A key enzyme in collagen synthesis is prolyl 4-hydroxylase (P4H), which catalyzes hydroxylation of proline residues on α monomers to yield thermally stable triple helical protocollagen molecules that are then secreted into the ECCM [8, 9] . Co-factors for the enzyme leading to the formation of 4-hydroxyproline include Fe^{2+}, 2-oxoglutarate, O_2, and ascorbic acid or vitamin C. Vitamin C regulates collagen synthesis and stimulates synthesis of pro-collagen mRNAs [23].

Several growth factors and cytokines also influence collagen gene expression [8, 9] and synthesis (Table 4). Transforming growth factor-β (TGF-β) stimulates collagen synthesis at transcriptional and post-transcriptional levels as well as overall protein synthesis [24]. TGF-β_1 is a mitogen and growth factor in various tissues including the myocardium [25], stimulates cardiac fibroblasts and their conversion to myofibroblasts (MyoFbs) [14], plays a major role in ECCM synthesis [4], and mediates several of its mitogenic and growth effects by inducing connective tissue growth factor (CTGF) expression [26]. Insulin growth factor (IGF-1, IGF-2) increases collagen and overall protein synthesis

[8]. Interleukin-1 (IL-1) can both increase and decrease collagen synthesis [8]. Tumor necrosis factor-α (TNF-α) and interferon-γ inhibit collagen gene transcription [27].

Several hormones and enzymes also influence collagen synthesis (Table 4). Glucocorticoids inhibit procollagen gene transcription [8]. Other steroid hormones and parathyroid hormone also decrease collagen synthesis [28]. Inhibition of P4H prevents collagen synthesis [29]. It is important to note that even a mild reduction of about 20% in 4-hydroxyproline content is sufficient to reduce the 'melting' temperature of the helices below the physiological level of 37°C [30], thereby decreasing the physical stability of collagen, its resistance to proteolysis, its secretion with ECCM, and its ability to interact with other ECCM components.

2.4. Collagen degradation

In general, the orderly degradation of ECCM is critical for growth, development, morphogenesis, remodeling and repair [8, 9]. Two main classes of locally secreted extracellular proteolytic enzymes control ECCM degradation: Ca^{++} or Zn^{++} dependent matrix metalloproteinases (MMPs) and serine proteases. In the heart, collagen degradation is mediated mainly by Zn^{++}-dependent MMPs, most of whom have been cloned [31, 32]. Matrix metalloproteinases include

Table 4. Effect of some cytokines, growth factors, hormones, enzymes and other agents on myocardial collagen matrix

Increase Collagen	Decrease Collagen
Angiotensin II (AngII)	Bradykinin (BK)
Aldosterone (Aldo)	Matrix metalloproteinase (MMP)
Endothelin-1 (ET$_1$)	Tumor necrosis factor-α (TNF-α)
Angiotensin-converting enzyme (ACE)	Catecholamines
Prolyl 4-hydroxylase (P4H)	Parathyroid hormone (PTH)
Tissue inhibitor of metalloproteinase (TIMP)	Thyroid hormone (TH)
Retinoids (Vitamin A)	Glucocorticoids
Transforming growth factor-β$_1$(TGF-β$_1$)	Interleukin-1 (IL-1)
Connective tissue growth factor (CTGF)	
Platelet derived growth factor (PDGF)	
Epidermal growth factor (EGF)	
Transforming growth factor-alpha (TGF-alpha)	
Basic fibroblast growth factor (bFGF)	
Insulin-like growth factor (IGF)	
Interleukin-1 (IL-1)	
Steroids, progesterone	
Growth hormone (GH)	

collagenases, gelatinases, stromelysins, membrane type MMPs (MT-MMPs) and putative MMPs such as PUMP-1 (Table 5).

The collagenases include interstitial collagenase 1 (MMP-1), neutrophil collagenase 2 (MMP-8), and collagenase 3 (MMP-13, a homolog of human MMP-1). These collagenases are highly specific in their action and cleave particular proteins (primarily fibrillar collagen) at specific sites thereby destroying structural integrity with a minimum amount of proteolysis. The major substrates for MMP-1 and MMP-8 are fibrillar collagen types I, II and III. The major substrate for MMP-13 is collagen type I. The gelatinases include MMP-2 and MMP-9 and degrade denatured fibrillar collagens and collagen types IV and V. An important concept is that once collagenase is bound to the fibril and has begun its attack of the ECCM, it would continue to act until all the substrate has been completely degraded unless it is inhibited or controlled [33]. A natural protective mechanism against uncontrolled collagenase degradation

Table 5. Matrix metalloproteinases and potential relevance in myocardial remodeling

Enzyme	MMP	Enzyme (MW=kDa)	Remodeling*
Collagenases	MMP-1	Interstitial collagenase (52/42)	+
	MMP-8	Neutrophil collagenase (85/64)	+
	MMP-13	Collagenase-3 (52/42)	+
Gelatinases	MMP-2	Gelatinase A (72/66), type IV collagenase	+
	MMP-9	Gelatinase B (92/84), type V collagenase	+
Stromelysins	MMP-3	Stromelysin 1 (57/45)	+
	MMP-10	Stromelysin 2 (54/44)	?
	MMP-11	Stromelysin 3 (64/46)	-
Membrane-type	MMP-14	MT1-MMP (66/54)	+
		(MT2-MT3-MT4-MMPs)	?
Others	MMP-7	Matrilysin, PUMP-1 (28/19)	?
	MMP-12	Metalloelastase (54/22)	?

*Role in cardiac remodeling: documented (+), probable (?), or not known (-).

is provided by tissue mediated inhibition of collagenase activity [33].

In the heart, endogenous inhibitors of MMPs termed tissue inhibitors of metalloproteinases (TIMPs) are co-expressed and form tight complexes [34].

Both MMPs and TIMPs are secreted by cardiac fibroblasts and their gene expression is tightly controlled at the transcription level [34]. Several studies [34] have demonstrated: i) transcription of MMP-1 and TIMP mRNAs in fibroblast-like cells; ii) MMP genes (particularly those for collagenase) in fibroblasts, endothelial cells and polymorphonuclear cells; iii) involvement of TIMPs in MMP activation and stimulation of fibroblast growth; and iv) modulation of the synthesis and secretion of pro-MMPs and TIMPs by cytokines, polypeptide growth factors, hormones, steroids, and phorbol esters. Collagenase produced by cardiac fibroblasts has been demonstrated in the perimysium (fibers around cardiomyocytes and strands between myofibrils) and endomysium (consisting of struts that connect myocytes to each other and to capillaries) of the ECCM [35]. Increased activation of latent collagenases and decreased levels of TIMPs can result in increased ECCM degradation and structural remodeling. A fine balance between collagenase and its inhibitors is therefore necessary for normal ECCM remodeling and function. In general, the net proteolytic activity of MMPs depends on 3 main factors: i) transcriptional control; ii) activated control; and iii) inhibitory control. These 3 steps have been recently targeted in animal [36-41] and human [42-45] studies.

3. Cardiac ECCM and remodeling

3.1. Cardiac hypertrophy and ECCM remodeling

In general, growth of the nonmyocytes can lead to structural remodeling of the myocardium and vasculature. Pathological myocardial fibrosis, associated with excessive ECCM deposition (Table 6), is a serious complication of chronic heart disease associated with hypertension and pressure overload. It generally results from an imbalance between collagen biosynthesis and degradation (e.g,. increased collagen synthesis or decreased degradation). However, the rate of myocardial collagen synthesis has been characterized to be fairly slow (0.56% per day in dog ventricles), slower than that of non-collagen protein (7.2% per day) [46]. In addition, the half-life of myocardial collagens is long (between 80 and 120 days), 10 times longer than that of non-collagen proteins [1]. These data suggest that excessive ECCM and increased interstitial fibrosis usually develop gradually, although this might be accelerated in certain conditions. This is an important observation given that any potential therapeutic intervention may need to be pursued over prolonged periods of time. Importantly, the replacement of ECCM after degradation is fairly slow, creating windows of vulnerability for adverse structural remodeling after acute MI or intractable heart failure where enhanced degradation occurs.

3.2. Pressure overload

Chronic pressure overload (Table 6) results in concentric hypertrophic remodeling associated with significant changes in the ECCM and fibrillar collagens [1]. In the spontaneously hypertensive rat (SHR), myocyte hypertrophy is associated with ECCM remodeling involving interstitial fibrosis and increased fibrillar collagen that contributes to diastolic stiffness, and medial wall thickening of intramyocardial vessels that contribute to impaired coronary reserve [47-49]. Some studies in SHR reported increased collagen concentration and synthesis while others found only modest fibrosis [49]. However, even with a modest increase in collagen concentration, a significant increase in collagen type III and a decrease in the type I/III ratio appears to mark the transition to heart failure [50]. In LV pressure overload hypertrophy, ECCM remodeling consists of an increase in the number and size of collagen fibers, increase in the concentration and content of collagen, increase in the proportion of collagen type III versus type I and altered alignment, and increase in tendon thickness and density of the perimysial weave [1].

In human aortic stenosis with heart failure, myocardial collagen concentration increases by as much as 3- to 6-fold [51]. In dogs with chronic pressure overload, non-collagenous protein synthesis increases by 75% after 5 days and normalizes by 2 weeks while collagen protein synthesis increases slowly and steadily by 6- to 8-fold (at the rate of 4% per day) and remains 3-fold greater at 2 and 4 weeks [46]. In the rat model of cardiac hypertrophy, protocollagen P4H activity increases within days and is followed by increased proline incorporation into hydroxyproline [52], while fibroblast proliferation peaks days after the rise in collagen synthesis [53]. Collagen degradation also increases with pressure overload, but it normalizes before the increase in collagen synthesis [1]. Accumulation of collagen has been correlated with LV dysfunction. An increase in collagen volume fraction of 20% or more in hypertrophied myocardium is associated with both interstitial and perimuscular fibrosis, and increased diastolic ventricular stiffness and decreased systolic stiffness [1, 2, 54]. In the dog model of pressure overload, both collagen content and cross-linking of collagen types I and III increase [55]. In hypertensive hypertrophy, increased ECCM or fibrosis correlated with LV diastolic dysfunction [56, 57]. Taken together, these findings supported the concept that fibrosis "encases myocytes in cement" [54] and promotes myocardial dysfunction and failure. Several studies suggest that humoral (such as the renin-angiotensin-aldosterone system or RAAS) rather than hemodynamic factors mediate fibroblast growth and collagen deposition [48, 54]. Angiotensin II (AngII) and aldosterone have been linked to the cardiac fibrosis [4] and are suitable targets for decreasing fibrosis and improving LV function and outcome.

3.3. Chronic volume overload

Chronic volume overload (e.g., post MI; dilated cardiomyopathy; congestive heart failure; aortic or mitral regurgitation; chronic tachycardia) results in hypertrophic cardiac remodeling that is eccentric [2-4]. This is associated with either no change in collagen concentration in some models (e.g., chronic anemia, arterio-venous fistula or atrial septal defect), or clear increases (Table 6) as in post-MI rats [58, 59]. Evidence suggests that in volume overload, remodeling of the myocyte compartment is influenced mainly by the degree of hemodynamic factors while remodeling of the non-myocyte compartment is influenced in major part by humoral factors, such as the RAAS [2-4]. After large MI in rats, early ECCM remodeling is rapidly followed by LV chamber enlargement, non-infarct zone (NIZ) hypertrophy, further ECCM remodeling in the infarct zone (IZ) and NIZ, and adverse global LV structural remodeling [58, 60-62]. In human dilated cardiomyopathy, total collagen and collagen type I/III ratio increase [63]. LV diastolic dysfunction after MI accompanies systolic dysfunction and develops early, so that its relation to the ECCM is quite complex [64].

Table 6. Cardiac diseases associated with increased ECCM

Mainly pressure overload
-Hypertension
    -Essential
    -Genetic
    -Aldosterone
    -Renovascular
-Aortic stenosis
-Aortic coarctation
-Pulmonary artery banding
Mainly volume overload
-Myocardial infarction
-Dilated cardiomyopathy
-Chronic/end-stage heart failure
-Rapid ventricular pacing
-Tachycardia-induced heart failure
-Aortic (or mitral) regurgitation
Other
-Chronic hypothyroidism

4. Cardiac remodeling after injury

4.1. Cardiac remodeling after MI

Structural LV remodeling and healing after MI are complex, inter-related, dynamic and time-dependent processes that progress in parallel over weeks and months [5-7, 59, 60, 65]. They involve changes in i) LV structure, shape and topography at the regional global levels [66], ii) myocytes and nonmyocytes at the cellular level [14, 61, 67, 68], iii) proteins, cytokines, growth factors and autocrine, paracrine and intracrine factors (Table 4) at the subcellular level [69-75], and iv) the ECCM [15, 76-87].

Figure 1. Factors influencing ventricular fibrosis and remodeling after myocardial infarction.

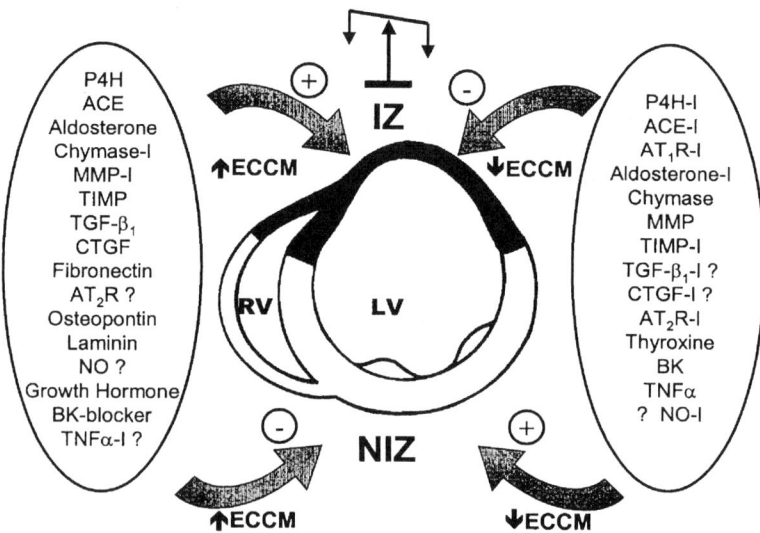

Black = IZ, infarct zone; white = NIZ, non-infarct zone. AT_1R = angiotensin II type 1 receptor; AT_2R = angiotensin II type 2 receptor; BK = bradykinin; I = inhibitor; NO = nitric oxide; LV = left ventricle; RV = right ventricle; other abbreviations as in the text.

4.2. Healing and remodeling after MI

In general, the consequence of remodeling in the heart is considered good or bad depending on its impact on function [5, 7, 61]. In no other cardiac disease process than MI is cardiac remodeling more dramatic and significant in terms of effects on cardiac dysfunction, disability and death [5, 7, 60, 66]. Thus, the structural remodeling after the extreme stress of a MI results in a cycle of cardiac dysfunction, shape deformation, and more dysfunction, and is considered adverse [5, 7]. Although the healing process that follows a MI attempts to repair the damage, preserve structure, and restore shape and function it often goes awry and results in adverse remodeling and further dysfunction (Figure 2). Thus, after large, transmural and anterior MI, healing is associated with marked adverse remodeling of the various cardiac structures (including the ECCM) over time and can lead to a vicious cycle of LV dysfunction, shape deformation, chamber enlargement, congestive heart failure, disability and death [5, 7].

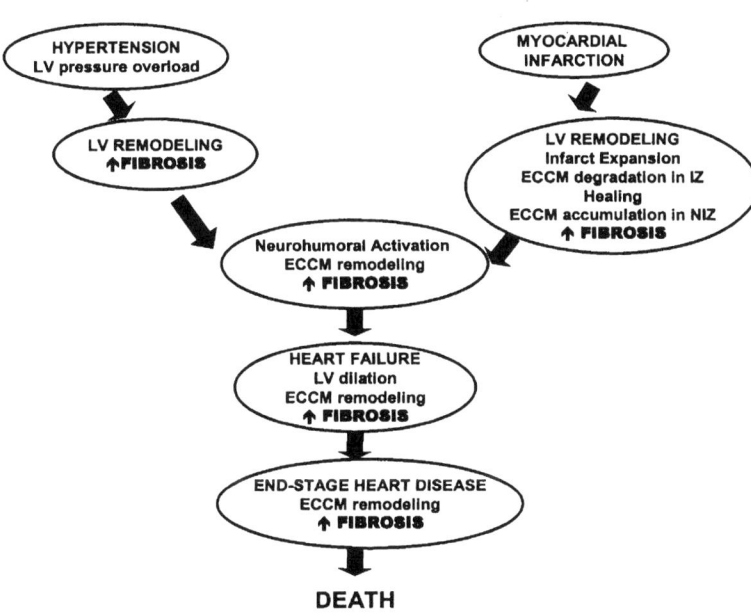

Figure 2. Diagram showing the central position of fibrosis and ECCM remodeling in the progression to heart failure and death in 2 major cardiac diseases. ECCM = extracellular collagen matrix; LV = left ventricular; IZ = infarct zone; NIZ = non-infarct zone.

4.3. ECCM degradation and remodeling after MI

Several studies have demonstrated ECCM disruption after early MI or early ischemia-reperfusion [76-87]. In canine [88-95] or rat [59] MI studies from our laboratory, no statistically significant decreases in IZ collagen content were detected over the first few days compared to sham animals using the hydroxyproline assay [96]; very early collagen was not measured in those studies. However, rapid early activation of latent MMPs and ECCM degradation has been demonstrated after acute MI in other studies [31]. Significant reduction in the amount of collagen was detected in the IZ after 1, 2 and 3 hours of ischemia in rats [82].

Evidence of early decrease or damage in IZ collagen has also been documented using more delicate microscopic methods in rats, dogs, pigs and humans [77, 78, 83, 97-99]. In the rat, MMP-1 appears on day 2 and peaks at day 7 [84]. TIMP-1 mRNA increases by 6 hours and TIMPs peak on day 2 and then decline over 14 days [84], supporting the view that TIMP-1 transcription is regulated by MMP-1 activation. In the dog, ECCM disruption with complete loss of the collagen weave and struts [77] and increased collagenase activity is associated with myocardial stunning [83]. In the pig, ECCM disruption after MI begins within minutes [97], with subendocardial disruption of collagen fibers within 20 minutes in the IZ, followed by disappearance of glycoproteins by 40 minutes, and subsequent loss of the epimysial fibers and endomysial struts, and increase in sarcomere lengths to 3.6 mm. In rats with moderate MI, early collagen degradation is associated with increased collagenase activity and is inhibited in leucopenic rats, suggesting the involvement of inflammatory cells [98]. The finding of increased MMP-1 mRNA activity 1 week after the MI suggests a linkage between the increased collagenase activity and late LV enlargement [98]. In the IZ of the pig after MI [87], cumulative myocardial fluorogenic MMP 2/9 activity was detectable within 10 minutes and increased significantly after 1 hour and markedly by 2, while zymographic MMP activity (consistent with MMP-9) increased markedly by 3 hours. In another pig model of ischemia-reperfusion without MI or inflammatory cell infiltration, active MMP-9 increased, myocardial collagenase increased but remained in latent form, MMP-2 and TIMPs were unaffected, ECCM ultrastructure was unchanged, and MMP inhibition with GM-2487 did not reduce myocardial expansion [100]. The latter findings suggest that interaction of MMPs and inflammatory cells is involved in early infarct expansion.

4.4. ECCM remodeling and inflammation after acute MI

Cumulative evidence suggests that ECCM remodeling after MI is linked to the inflammatory and cellular responses. Inflammation and collagen deposition during postinfarct healing have been studied [5, 59, 88]. After MI in dogs [69], edema and glycosaminoglycans (GAGs) increase over the first 5 days, cell infiltration and fibroblasts peak on day 3, procollagen P4H activity increases between days 5 and 12 and levels off thereafter. Hydroxyproline increases progressively by 5-fold or more in the IZ over 6 weeks and this is associated with significant LV structural remodeling [5, 59, 88, 95].

Early LV remodeling after MI is associated temporally with acute and chronic inflammation [5-7, 59] and rapid early ECCM degradation in the IZ [77, 78, 80, 82, 84] which is blocked in leucopenic animals [98]. The early inflammation after MI, with or without reperfusion, has been shown to involve the release of cytokines such as TNF-α and IL-1 which activate MMPs and contribute to LV remodeling [101]. Increased catecholamine, AngII and endothelin levels in heart failure also lead to increased MMPs [101].

Several MMPs that are expressed in the myocardium after MI contribute to post-MI remodeling [38, 41, 43-45, 102] and heart failure [43, 103]. Thus, heart failure has been shown to be associated with i) increased MMP-3 and MMP-9, decreased MMP-1 and no change in MMP-2 [44], and ii) decreased TIMP-1, TIMP-3 and TIMP-4, and no change in TIMP-2 [43]. In addition, MMP inhibition was shown to attenuate LV dilation after MI-induced heart failure in mice [37] and tachycardia-induced heart failure in pigs [36]. After reperfused MI in pigs, immunoreactive MMP-9 has been localized to neutrophils [104]. After MI in MMP-9 knock out mice, i) ECCM accumulation and macrophage infiltration are decreased [41], and ii) cardiac rupture and angiogenesis are decreased, and heart failure increased [38], implicating MMP-9 and ECCM in cardiac rupture and heart failure.

The early disruption of the ECCM in the IZ leads to myocyte slippage [79] and contributes to IZ expansion with regional stretching, thinning and dilation [65, 105, 106], thus causing deformation of normal ventricular shape and contributing further to regional and global dysfunction [66].

4.5. ECCM remodeling and repair after MI

Early ECCM disruption after MI is followed by gradual reparative fibrosis of the IZ [85, 88] and interstitial fibrosis of the NIZ [85, 107-109] over a variable period of weeks and months (Figures 3, 4). Collagen deposition in the IZ is but one aspect of repair of the damaged ventricular wall with a firm scar [88]. The healing process from the acute MI to scar formation takes several weeks to months, depending on the size of the heart and species (Figure 4) and the infarct characteristics, such as infarct size, transmurality, location, species, and healing conditions [5-7, 88]. The interval to scar formation for small to moderate sized infarcts ranges between 2 to 3 weeks in the rat [59], 4 to 6 weeks in the dog [59, 88], and 6 weeks to 6 months in humans [110, 111]. Large transmural infarcts take longer to heal than small subendocardial infarcts [5-7, 91, 95, 110, 111]. Healing is more rapid in reperfused than non-reperfused infarcts [112] and this in part may be due to smaller and more subendocardial infarcts [113, 114]. In the IZ, collagen content increases 6 to 10 fold during

36

healing and reaches a plateau between 2 and 6 weeks in the dog [88] and 1 to 3 weeks in the rat [59]. In the NIZ, collagen content increases more gradually to between 2 to 3 fold and is associated with interstitial fibrosis [95]. In healed reperfused infarcts, collagen content is slightly lower than in non-reperfused infarcts [114].

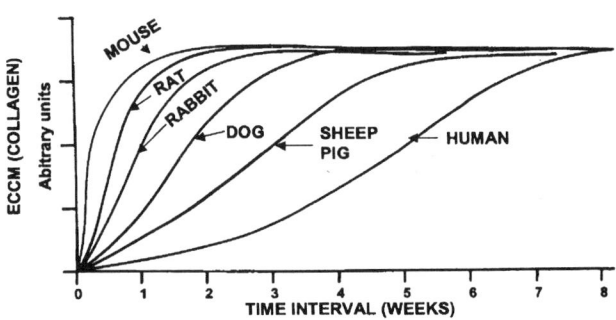

Figure 3. Duration of healing measured by time to collagen plateau for moderate MI size. ECCM = extracellular collagen matrix.

4.6. Myofibroblasts, healing and scar after MI

An important factor during post MI healing is the phenotypic conversion of fibroblasts to specialized MyoFbs that contain sarcomeric α-smooth muscle actin and mediate scar contraction [14, 115-122]. MyoFbs are prominent at sites of repair and are attached to fibrillar collagen [118-122]. Regulatory signals (such as AngII, TGF-β_1, angiotensin-converting-enzyme) [25], and endothelin and AngII receptors found on MyoFbs facilitate MMP activity and ECCM remodeling [2, 116, 117]. TGF-β_1 in activated macrophages may regulate the conversion of fibroblasts to MyoFbs. MyoFbs contain angiotensinogen, cathepsin D and ACE, and can generate AngII and thereby stimulate TGF-β_1 and regulate ECCM turnover and fibrosis. More importantly, the infarct scar containing persistent MyoFbs appears to behave as a living, dynamic structure [118]. In human MI scars, MyoFbs remain active and were shown to persist up to 17 years [119]. This is quite different from dermal scars where MyoFbs are transient. In rat MI, MyoFbs appear early (between 2 and 4 days) and are mainly located in and around the IZ [122] and persist for at least 4 weeks [118]. The MyoFbs were shown to produce types I and III procollagen mRNAs and proteins [85, 120, 121], and procollagen type I mRNA remains elevated 90 days after MI [85]. MyoFbs appear to be the dominant cells involved in excessive ECCM formation associated with cardiac fibrosis [115-122] and are removed by apoptosis [115-122] . It has been suggested that MyoFbs are involved in scar contraction post-MI [105, 118]. One may speculate that MyoFbs in the mature scar show similar organization as collagen fibers whose orientation corresponds to that of normal muscle layers [123]. Such orientation of MyoFbs in the scar might resist infarct scar expansion. MyoFbs are thought to arise from interstitial fibroblasts at repair sites or adventitial fibroblasts of coronary vessels [124, 125]. The development of MyoFbs appears to be influenced by traction and wall tension [122].

5. Cardiac remodeling and dysfunction

5.1. Structural LV remodeling and dysfunction

Local and remote architectural changes during LV structural and ECCM remodeling after MI profoundly alter regional and global shape and function [5-7]. Sequential changes (e.g., ECCM disruption, LV dilation, hypertrophy, and fibrosis) span the phases of injury, healing, and repair [5-7]. Early IZ expansion contributes to regional dilation that precedes global dilation [60, 65, 66, 91, 105, 106]. Mechanical forces and other factors (Table 7) that increase wall stress influence IZ remodeling throughout its course [5-7, 126, 127]. Susceptibility to IZ deformation and distension depends on the characteristics of the myocardial substrate, stage of healing, the status of ECCM and the pharmacologic milieu [5-7]. Transmural MIs are especially prone to expansion and rupture [5-7, 91, 95, 105, 106]. Sequential in vivo changes in LV remodeling and function during healing post MI (Figure 3), before and after the IZ collagen plateau to scar formation [5-7] and beyond, have been tracked using quantitative two-dimensional echocardiography [66, 126, 128-135]. Such longitudinal studies showed that i) appropriate therapy (Table 8) could modify ECCM and exert different effects on IZ collagen [90, 92, 93, 114, 136-141], ii) some therapies could produce unexpected effects [142], and be potentially harmful, as illustrated with nitrates [143], anti-inflammatory agents [90, 131], and ACE-Is [95, 144, 145] during the healing phase. Several studies in patients after MI showed progressive LV enlargement up to 1 year [126, 132], or 3 years [134]. In rats with moderate-sized MIs, LV enlargement progressed over months [62, 146]. However, many studies favored early and prolonged therapy for maximum impact on outcome [5-7, 66, 126, 128-135, 144, 145]. After an MI,

ECCM deposition in the IZ is a potential target for strengthening the infarct scar (Figure 1).

Table 7. Determinants of ventricular remodeling postinfarction

1. Infarct characteristics
 Infarct size, transmurality, location, type (± reperfusion), age
2. Infarct healing
 Inflammatory response (cytokines; growth factors; fibroblasts; myofibroblasts; MMPs; TIMPs)
 ECCM deposition and remodeling (synthesis; breakdown; amount; type; orientation; organization; cross-links)
 Collateral circulation; coronary vascular remodeling; angiogenesis; reperfusion
3. Wall stress (Laplace effect)
 Intracavitary distending forces (push & stretch); preload; afterload
 Intramural distending forces (pull & stretch): contractility; heart rate
 External restraining forces (push & resist):
 Pericardium and fluid; pericardial pressure; extracardiac structures
 Activation of cytokines, growth factors, MMPs; TIMPs
4. Progressive LV dilation
 Mechanical distending and restraining forces
 Nutrient flow to scar and noninfarct zones
 ECCM remodeling
 Hypertrophy of noninfarct zone
 Remodeling of LV shape, LV twist & untwist
 Remodeling of intracavitary flow pattern
 Readjustments in systolic and diastolic function
 Adaptive changes in right ventricle and atria
 Remodeling of vascular structures

ECCM, extracellular collagen matrix; MMP, matrix metalloproteinase; LV, left ventricular; TIMP, tissue inhibitor of metalloproteinase.

5.2. ECCM and mechanical strength of the healing heart

The mechanical strength of myocardial tissue after MI is determined by various characteristics of the ECCM, including its amount, type and organization. The total amount of collagen after MI increases 2.5 to 12-fold in the IZ over several weeks in the dog, yielding a firm scar that resists distension [59, 88, 89]. A 2 to 3-fold increase in NIZ collagen [95, 140] also contributes to stiffness [56, 64]. Although collagen types I and III mRNA levels increase on day 2 in the rabbit [147], the proteins increase later, by day 3 in the rat [99].

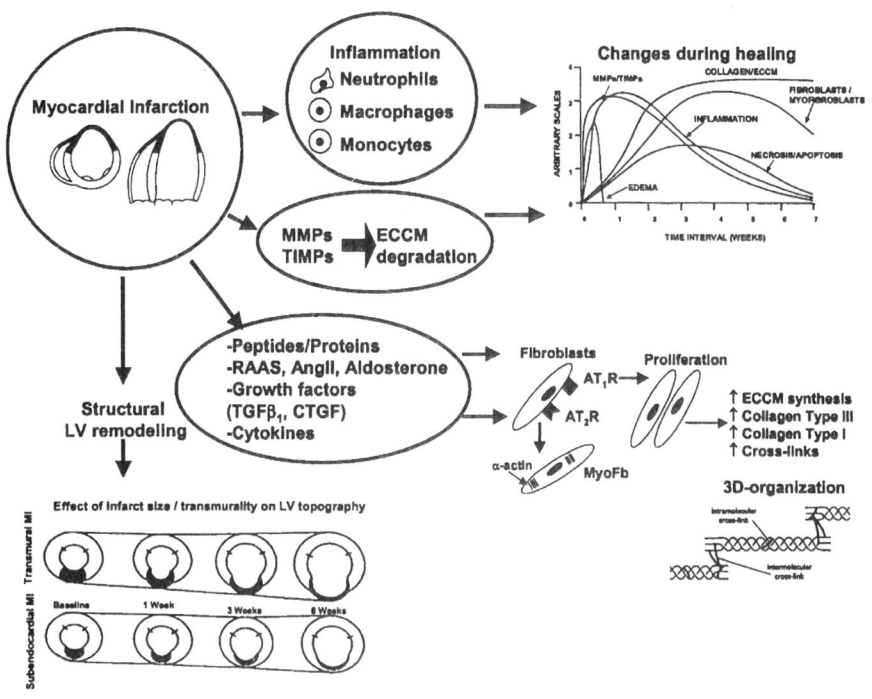

Figure 4. Temporal evolution of molecular, cellular and structural changes during healing and remodeling after myocardial infarction. Top: cascade from infarction to ECCM deposition and organization. Bottom left: topographic changes in short axis LV images (echocardiograms) after subendocardial versus transmural MI. AT_1R = angiotensin II type 1 receptor; AT_2R = angiotensin II type 2 receptor; ECCM = extracellular collagen matrix; AngII = angiotensin II; CTGF = connective tissue growth factor; TGF-β_1 = Transforming growth factor-β_1; MI = myocardial infarction; MMPs = matrix metalloproteinases; MyoFb = myofibroblast, TIMPs - tissue inhibitors of metalloproteinases.

New collagen, mostly thin type III, is immature and susceptible to stretch even by 15 weeks [148]. The infarcted left ventricle is more distensible in the early 2-week window after MI [89]. The findings suggest greater susceptibility to chamber dilation over the early weeks after MI. Subsequent collagen maturation, with development of intermolecular cross-links, loss of water and ground substance, and conversion to type I and increased collagen type I/III ratio result in greater resistance to distension [123]. In the rat, increased collagen type III persists for 21 days and increased type I for more than 90 days after MI, so that the type I/III ratio increases between days 21 and 90 [85]. By 13 weeks,

Table 8. Potential strategies for limiting remodeling post-myocardial infarction

1. Infarction
- Prevent or limit infarction
- Limit infarct expansion
- Limit transmural extension
- Limit reperfusion injury
- LV unloading during infarction
- Avoid paradoxical J-curve effect/hypoperfusion

2. Healing
- Promote normal healing
- Maintain infarct-related artery patency
- Maintain non-infarct-related artery patency
- Increase collateral flow, promote angiogenesis
- Preserve ECCM and integrity
- Preserve architectural framework
- Prevent scar thinning
- Avoid agents that impair infarct healing or decrease infarct collagen (e.g., anti-inflammatory drugs)
- Avoid excessive contractile pull (e.g., inotropes) during early phase
- Avoid excessive exercise and loading during early phase

3. Deformation forces
- Reduce wall stress (e.g., LV unloading)
- Avoid excessive contractile pull with inotropes
- Avoid excessive exercise and loading
- External restraint devices

4. Progressive dilation
- Limit chamber dilation
- Prevent inappropriate hypertrophy
- Protect architectural framework
- Preserve shape; prevent sphericity
- Preserve perfusion to noninfarct and scar zones

ECCM, extracellular collagen matrix; LV, left ventricular.

more cross-links in the IZ than NIZ contribute to its tensile strength [149]. During early post-infarct healing, 3-dimensional disorganization of the ECCM results in myocyte disarray at the infarct border and weakens coupling between the developing scar and the viable NIZ [123]. This finding might explain spontaneous [150] or postmortem [89] ruptures and IZ distension [123] after MI.

5.3. ECCM and heart failure

End-stage heart failure in humans is associated with significant increase in ECCM, myocyte loss across the LV wall, eccentric hypertrophy and decrease in the LV mass to volume ratio [151], quantitative increases in collagen types I, III, IV and VI, fibronectin, laminin and vimentin [152], decrease in collagen type I/III ratio [153], and loss of collagen cross-links [2]. Mice with a genetic defect in collagen cross-linking have impaired contractility [154]. Loss of collagen cross-links after injury is associated with progressive LV dilation [155]. Recently, LV mechanical unloading using the LV assist device (LVAD) in heart failure patients resulted in a downregulation of MMPs, increased TIMPs, decreased ECCM damage and increased cross-links [103], supporting the concept that LV unloading limits remodeling. It is important to note that small decreases in levels of collagen can lead to LV dilation [156] or even rupture [78].

5.4. ECCM organization and resistance to distension

In 1907, Holmgren demonstrated the extensive cardiac network of extracellular fibrils [157]. In 1935, Nagle used the terms epi-, peri- and endomysium to characterize the structural components of the ECCM in muscle [158]. In 1979, Caulfield and Borg [19] described the 3 levels of 3-dimensional organization in the ECCM network: the epimysium (fibers covering muscle), the perimysium providing structural support and the endomysium. They proposed that this organization mediates several functions including mechanical support and coupling of myocytes [19].

Subsequently, the IZ scar after MI was recognized as being alive rather than dead [118], and consisting of anisotropic and trilayered tissue [123], with different 3-dimensional orientation of each layer [159]. In addition, the stiffness of the IZ scar increases over the first 2 weeks and decreases at 6 weeks in sheep [160]. Passive LV stiffness during early filling increases over 6 weeks after MI and the post-mortem "rupture threshold" of the fresh infarcted left ventricle decreases after 14 days [89] in the dog. As IZ collagen increases between day 7 and 42 post-MI, the IZ scar (mostly subendocardial) was shown to undergo shrinkage and thinning with little further stretching [89]. In the pig, the IZ scar after MI (mostly transmural) contracts over the first 3 weeks [161], likely mediated by MyoFbs in the IZ [119]. These findings suggest that ECCM remodeling attempts to restore mechanical strength and resistance to distension of the scar during healing post-MI.

5.5. The ECCM and cardiac dilation

In 1960, Hort [156] postulated that the ECCM prevents excessive cardiac dilation, and implied that ECCM dissolution promotes dilation. Subsequent studies provided proof of that concept. Thus, inhibition of the ECCM by β-amino-propionitrile (BAPN) in newborn rats resulted in ventricular aneurysms [19]. Cardiac dilation was shown to be associated with a rearrangement of muscle bundles and ECCM rather than overstretch of sarcomeres [162]. In vivo or in vitro perfusion of rat hearts with oxidized glutathione for 3 hours produced marked decrease and damage of ECCM and extensive dilation that persisted for 6 months [163]. A single infusion of adriamycin in rats induced a loss of ECCM by 3-6 weeks that persisted for 6 months and preceded scar formation [81]. Perfusion of rat hearts with collagenase for 60 minutes decreased collagen content, damaged endomysial struts, and increased ventricular volume and sarcomere lengths [164]. Treatment of rats with the lathyrogen BAPN inhibited collagen cross-linking and prevented increase in tension [165]. Loss of ECCM 1-2 hours after MI in dogs was associated with regional IZ dilation [98]. Strut rupture and myocyte slippage were linked to early IZ dilation after MI [80, 99] and progressive and global LV dilation in heart failure [81]. Finally, collagen loss was linked to cardiac rupture [38, 78].

Recent studies have monitored cardiac ECCM turnover during repair using serological markers [76, 166-168]. Elevated levels of procollagen type III amino-terminal peptide (PIIINP) and other serum markers were correlated with poor outcome in patients with congestive heart failure receiving conventional therapy including ACE-Is, and the decreased levels after spironolactone therapy were associated with improved outcome [169].

5.6. ACE, AngII, ECCM and chymase

Several clinical trials indicate that coronary reperfusion and ACE-Is are effective in reducing LV dilation and LV dysfunction after MI [170, 171]. The aldosterone antagonist, spironolactone was shown to limit collagen turnover in congestive heart failure and improve overall outcome [169]. Weber has reviewed the importance of AngII in the fibrogenic component of tissue repair, the local production of tissue AngII after injury, the presence of AngII receptors on cardiac myocytes and fibroblasts, and their roles in ECCM turnover [4]. Local ACE is markedly elevated in high turnover sites (such as normal heart valves, coronary artery adventitia) and the IZ after MI [4, 120, 121]. AngII receptors, predominantly AngII type 1 (AT_1), are upregulated between day 3 and 8 weeks after MI [4], suggesting that they are involved in fibrosis. AngII type 2 (AT_2) receptors [172], which are re-expressed after MI [173] and upregulated in heart failure, are more abundant in human than rat hearts and are expressed in fibroblast-like cells [172], suggesting that they too are involved in cardiac fibrosis. In cultured cardiac rat fibroblasts, AngII stimulates collagen synthesis by both AT_1 and AT_2 receptors, and inhibits collagenase activity which can be attenuated by AT_2 receptor blockade while aldosterone stimulates collagen synthesis via AT_1 receptors but has no effect on collagenase activity [174]. The finding that significant amounts of AngII are formed via alternate pathways such as chymase in dogs and humans [175] has triggered several clinical trials using AT_1 receptor blockade to achieve more complete AngII blockade in patients with hypertension, heart failure, and MI [176, 177]. However, during chronic ACE inhibition, chymase activity increases and plasma AngII levels are elevated [178, 179], and this may explain the observed continued LV enlargement in some studies [134].

Given the importance of AngII and ACE in ECCM turnover, healing, and repair after injury, it is not surprising that ACE-Is and AT_1 receptor antagonists inhibit fibrosis in the IZ and NIZ after MI [5-7, 95, 176]. ACE-I induced decrease in fibrosis is clearly beneficial for the NIZ and in non-infarcted hearts with fibrosis. It reverses myocyte hypertrophy and fibrosis in SHRs [47-50], decreases collagen content [50, 180], reverses the type I/III ratio [50], and increases collagenase activity by inducing myocardial MMP-1 and MMP-2 [181]. It also retards collagen formation in young rats [182], inhibits aortic ACE and collagen deposition in SHR [116], reduces IZ collagen in dogs [92-95], reduces NIZ collagen in rats treated 1 week after MI [183], prevents fibroblast proliferation and fibrosis when given early after MI in rats [107], and prevents pericardial fibrosis [4]. In humans, it increases LV distensibility and reduces pericardial constraint [184]. In a large study of canine MI, deaths over 7 weeks correlated with greater infarct size, LV volume and lower IZ collagen [95]. Moreover, in the 6 week IZ scars after dog MI, ACE-Is increased the collagen type I/III ratio [95], which may be beneficial [95, 180]. AT_1 receptor blockade also prevents fibrosis after MI in rats [185]. These findings support the concept that ultimate outcome represents a balance of effects [186].

The antifibrotic effects of ACE-Is, aldosterone antagonists and AT_1 receptor blockers are expected to improve LV diastolic function due to excessive ECCM [176, 177]. However, diastolic dysfunction after MI begins early and its evolution is fairly complex [64]. Cumulative experimental evidence indicates that the IZ scar is a live structure [118, 119]. This suggests that long-term exposure to some vasoactive and other agents used after MI [5-7] might potentially decrease [95], alter [50, 95], or even damage [77, 181] the ECCM in the IZ and NIZ, and impair healing [187].

5.7. Other therapies and the ECCM

Beta-adrenergic blockers do not appear to alter myocardial collagen [188]. Since little or no collagen is found at sites of rupture [78], several approaches were proposed for enhancing healing and collagen synthesis in the IZ scar post-MI and prevent rupture. Growth hormone has been used to stimulate post-MI healing [189] and has been shown to increase IZ scar collagen and reduce LV aneurysm formation [190]. Interestingly, nitrates preserved IZ collagen and decreased LV remodeling after MI in the dog but did not prevent the decrease in IZ collagen content associated with an ACE-I [93]. However, nitrate therapy after reperfusion in dog MI prevented the decrease in collagen content of the IZ scars seen after reperfusion [114, 138]. Although digitoxin increases P4H activity in rat MI [52], digoxin therapy after dog MI did not alter collagen content of the IZ scars [139]. Endothelin (ET) peptides, which are mitogenic to fibroblasts and increase collagen types I and III synthesis in cultured rat cardiac fibroblasts, can be blocked by the antagonist PED-3512-PI [116]. Endothelin-1 has been shown to decrease collagenase activity [116] while ET blockade after MI impaired healing [187]. Bradykinin (BK), which is increased by ACE-Is, was shown to increase collagenase activity and decrease collagen type I synthesis [31] in cultured rat cardiac fibroblasts. Bradykinin can also exert a further antifibroblast effect via nitric oxide (NO) [191] and thereby decrease IZ collagen.

5.8. ECCM and growth factors

Growth factors such as TGF-β and CTGF have been considered as potential targets for fibrosis. Early after MI, TGF-β_1 [70] and CTGF [72] expression are increased and TGF-β induces CTGF expression in both cardiac fibroblasts and cardiomyocytes [72, 73]. CTGF increases the expression of collagen type I, fibronectin and integrin in fibroblasts [191]. In the IZ after MI in rats, increased CTGF expression appears to be related to MyoFbs in the infarct margin [72]. CTGF is also markedly upregulated in human hearts with ischemic cardiomyopathy and to a lesser extent in dilated cardiomyopathy [73]. These overall findings suggest that CTGF is an important mediator of TGF-β signaling, and abnormal expression of the CTGF gene can be used as a marker of fibrosis [73].

5.9. ECCM and P4H inhibition

Prolyl 4-hydroxylase (P4H) inhibition is another potential approach against fibrosis. In rat MI, a decrease in NIZ fibrosis after AT_1 receptor blockade was associated with a decrease in immunoreactive P4H, suggesting that the expression of cardiac P4H may be regulated by AngII via AT_1 receptor activation [192]. The P4H inhibitor FG041, given for 4 weeks after MI in rats, was recently shown to improve LV systolic function and reduce systolic area after isoproterenol administration in vivo, and decrease collagen types I and III in the NIZ ex vivo [36]. The hydroxyproline/proline ratio in the NIZ was shown to increase by 64% of the MI control group and was close to the normal value in sham hearts in the MI-FG041 hearts [29]. However, the ratio was reduced by 24% in the IZ of the MI-FG041 hearts but diastolic LV dimensions were not reported in that study [29] and longer term consequences of P4H inhibition have not been studied. Two other major questions still need to be addressed. First, some P4H inhibitors are designed to block the interaction of the X-Pro-Gly regions with the active site of the enzyme, and their effects on the 20 or so collagen proteins and many more non-collagen proteins that possess triple helical regions containing 4-hydroxyproline [30] are not known. Second, other P4H inhibitors are designed to competitively inhibit 2-oxoglutarate needed for hydroxylation but 2-oxoglutarate is also needed for intermediary metabolism and other reactions. Selective inhibitors therefore need to be developed.

5.10. ECCM and reperfusion

Reperfusion therapy after MI also modifies the ECCM [77, 78]. However, the interactions are complex. The early inflammation and cellular response after MI and reperfusion [193] are associated with increased collagenases and collagen degradation [31, 35]. In contrast to transmural MI with extensive early ECCM damage and rapid aneurysm formation [91, 95], non-transmural MIs are associated with spared epicardial rims of viable myocardium, less IZ expansion and better prognosis [5-7, 114]. Reperfusion therapy, even late, produces overall benefits including non-transmural infarcts that are associated with less IZ remodeling and global dilation [114], and accelerated healing [112]. However, reperfusion is also associated with decreased IZ collagen [114], persistent LV dysfunction [113], earlier ruptures [149], and a tendency for increased ruptures when reperfusion is done after 17 hours of onset of MI [194]. In reperfused rat MI, the density of the cross-links is reduced [195] and ECCM disruption is greater than in non-reperfused MI [196]. Streptokinase therapy also decreases myocardial collagen [197]. The combination of reperfusion with antifibrotic agents could therefore create a double jeopardy under certain conditions, suggesting the need for caution.

6. Conclusions

There is considerable evidence indicating that excessive ECCM can impair myocardial function in several cardiac diseases, so that antifibrotic agents might be beneficial in those conditions. Long-term inhibition of P4H or suppression of TGF-β_1 and CTGF represent two potential targets for decreasing ECCM and fibrosis in diseased myocardium. The ECCM also plays a major role in healing and remodeling after MI [4-7, 88, 118], which remains the leading cause of death and disability. After MI, however, the situation is complicated by the presence of an IZ and a NIZ (Figure 1), and the fact that ECCM disruption and damage during IZ healing, NIZ hypertrophy and interstitial fibrosis, and dilation involving both the IZ and NIZ all contribute to adverse LV structural remodeling. The sequence of MI, LV dysfunction, and progressive LV dilation underscores the fact that bad remodeling outweighs the good, and ECCM damage plays a pivotal role in this cycle. A major goal of therapy after MI is to prevent, control, and reverse all bad remodeling. One approach is to find adjunctive therapies that might protect the ECCM. MMPs can now be targeted with genetic approaches [198, 199]. Local gene therapy, applied to specific areas of the heart and IZ or NIZ regions might provide a safe and effective method for regional control of cardiac fibrosis.

Acknowledgements

Supported in part by a grant from the Canadian Institutes of Health Research, Ottawa, Ontario. I am grateful to Catherine Jugdutt for assistance with manuscript preparation.

References

1. Weber, K.T., et al., Collagen remodeling of the pressure-overloaded, hypertrophied nonhuman primate myocardium. Circ Res, 1988. 62: p. 757-65.
2. Weber, K.T., Cardiac interstitium in health and disease: the fibrillar collagen network. J Am Coll Cardiol, 1989. 13: p. 1637-52.
3. Weber, K.T., et al., Fibrillar collagen and remodeling of dilated canine left ventricle. Circulation, 1990. 82: p. 1387-401.
4. Weber, K.T., Extracellular matrix remodeling in heart failure: a role for de novo angiotensin II generation. Circulation, 1997. 96: p. 4065-82.
5. Jugdutt, B.I., Prevention of ventricular remodelling post myocardial infarction: timing and duration of therapy. Can J Cardiol, 1993. 9: p. 103-14.
6. Jugdutt, B., Modification of left ventricular remodeling after myocardial infarction. In: The Failing Heart., Dhalla, N.S., et al., eds. Philadelphia, PA: The Lippincott-Raven Publishers, 1995: p. 231-245.
7. Jugdutt, B., Prevention of ventricular remodeling after myocardial infarction and in congestive heart failure. Heart Failure Reviews, 1996. 1: p. 115-129.
8. Miller, E., Gay,S., Collagen structure and function. In: Wound Healing. Biochemical and clinical aspects, Cohen, I.K., et al., eds. Philadelphia, PA: WB Saunders Co., 1992: p. 130-151.
9. Alberts, B., et al., Molecular biology of the cell. Garland Publishing Co., 1994.
10. Olivetti, G., P. Anversa, and A.V. Loud, Morphometric study of early postnatal development in the left and right ventricular myocardium of the rat. II. Tissue composition, capillary growth, and sarcoplasmic alterations. Circ Res, 1980. 46: p. 503-12.
11. Zak, R., Development and proliferative capacity of cardiac muscle cells. Circ Res, 1974. 35: p. S17-26.
12. Eghbali, M., et al., Collagen chain mRNAs in isolated heart cells from young and adult rats. J Mol Cell Cardiol, 1988. 20: p. 267-76.
13. Eghbali, M., et al., Localization of types I, III and IV collagen mRNAs in rat heart cells by in situ hybridization. J Mol Cell Cardiol, 1989. 21: p. 103-13.

14. Eghbali, M., et al., Cardiac fibroblasts are predisposed to convert into myocyte phenotype: specific effect of transforming growth factor beta. Proc Natl Acad Sci U S A, 1991. 88: p. 795-9.

15. Cleutjens, J.P., The role of matrix metalloproteinases in heart disease. Cardiovasc Res, 1996. 32: p. 816-21.

16. Nag, A.C., Study of non-muscle cells of the adult mammalian heart: a fine structural analysis and distribution. Cytobios, 1980. 28: p. 41-61.

17. Frank, J.S. and G.A. Langer, The myocardial interstitium: its structure and its role in ionic exchange. J Cell Biol, 1974. 60: p. 586-601.

18. Weinberg, E.O., et al., Angiotensin-converting enzyme inhibition prolongs survival and modifies the transition to heart failure in rats with pressure overload hypertrophy due to ascending aortic stenosis. Circulation, 1994. 90: p. 1410-22.

19. Caulfield, J.B. and T.K. Borg, The collagen network of the heart. Lab Invest, 1979. 40: p. 364-72.

20. Robinson, T.F., L. Cohen-Gould, and S.M. Factor, Skeletal framework of mammalian heart muscle. Arrangement of inter- and pericellular connective tissue structures. Lab Invest, 1983. 49: p. 482-98.

21. Robinson, T.F., et al., Morphology, composition, and function of struts between cardiac myocytes of rat and hamster. Cell Tissue Res, 1987. 249: p. 247-55.

22. Burton, A., Relation of structure to function of the tissues of the wall of vessels. Physiol Rev, 1954. 34: p. 619-642.

23. Murad, S., et al., Regulation of collagen synthesis by ascorbic acid. Proc Natl Acad Sci U S A, 1981. 78: p. 2879-82.

24. Penttinen, R.P., S. Kobayashi, and P. Bornstein, Transforming growth factor beta increases mRNA for matrix proteins both in the presence and in the absence of changes in mRNA stability. Proc Natl Acad Sci U S A, 1988. 85: p. 1105-8.

25. Eghbali, M., Cellular origin and distribution of transforming growth factor-beta in the normal rat myocardium. Cell Tissue Res, 1989. 256: p. 553-8.

26. Grotendorst, G.R., H. Okochi, and N. Hayashi, A novel transforming growth factor beta response element controls the expression of the connective tissue growth factor gene. Cell Growth Differ, 1996. 7: p. 469-80.

27. Meldrum, D.R., Tumor necrosis factor in the heart. Am J Physiol, 1998. 274: p. R577-95.

28. Oikarinen, A.I., et al., Modulation of collagen metabolism by glucocorticoids. Receptor-mediated effects of dexamethasone on collagen biosynthesis in chick embryo fibroblasts and chondrocytes. Biochem Pharmacol, 1988. 37: p. 1451-62.

29. Nwogu, J.I., et al., Inhibition of collagen synthesis with prolyl 4-hydroxylase inhibitor improves left ventricular function and alters the pattern of left ventricular dilatation after myocardial infarction. Circulation, 2001. 104: p. 2216-21.

30. Brodsky, B. and N.K. Shah, Protein motifs. 8. The triple-helix motif in proteins. FASEB J, 1995. 9: p. 1537-46.

31. Tyagi, S.C., Proteinases and myocardial extracellular matrix turnover. Mol Cell Biochem, 1997. 168: p. 1-12.

32. Woessner, JF.Jr., The matrix metalloproteinase family. In Matrix Metalloproteinases. Parks, W. and Mecham, R., eds. San Diego,CA: Academic Press, 1998: p. 1-14.

33. Jeffrey, J., Collagen Degradation. In Wound Healing. Biochemical and clinical aspects, 1992: p. 177-194.

34. Tyagi, S.C., et al., Co-expression of tissue inhibitor and matrix metalloproteinase in myocardium. J Mol Cell Cardiol, 1995. 27: p. 2177-89.

35. Montfort, I. and R. Perez-Tamayo, The distribution of collagenase in normal rat tissues. J Histochem Cytochem, 1975. 23: p. 910-20.

36. Spinale, F.G., et al., Matrix metalloproteinase inhibition during the development of congestive heart failure : effects on left ventricular dimensions and function. Circ Res, 1999. 85: p. 364-76.

37. Rohde, L.E., et al., Matrix metalloproteinase inhibition attenuates early left ventricular enlargement after experimental myocardial infarction in mice. Circulation, 1999. 99: p. 3063-70.

38. Heymans, S., et al., Inhibition of plasminogen activators or matrix metalloproteinases prevents cardiac rupture but impairs therapeutic angiogenesis and causes cardiac failure. Nat Med, 1999. 5: p. 1135-42.

39. Roten, L., et al., Effects of gene deletion of the tissue inhibitor of the matrix metalloproteinase-type 1 (TIMP-1) on left ventricular geometry and function in mice. J Mol Cell Cardiol, 2000. 32: p. 109-20.

40. Nagatomo, Y., et al., Differential effects of pressure or volume overload on myocardial MMP levels and inhibitory control. Am J Physiol, 2000. 278: p. H151-61.

41. Ducharme, A., et al., Targeted deletion of matrix metalloproteinase-9 attenuates left ventricular enlargement and collagen accumulation after experimental myocardial infarction. J Clin Invest, 2000. 106: p. 55-62.

42. Gunja-Smith, Z., et al., Remodeling of human myocardial collagen in idiopathic dilated cardiomyopathy. Role of metalloproteinases and pyridinoline cross-links. Am J Pathol, 1996. 148: p. 1639-48.

43. Li, Y.Y., et al., Differential expression of tissue inhibitors of metalloproteinases in the failing human heart. Circulation, 1998. 98: p. 1728-34.

44. Thomas, C.V., et al., Increased matrix metalloproteinase activity and selective upregulation in LV myocardium from patients with end-stage dilated cardiomyopathy. Circulation, 1998. 97: p. 1708-15.

45. Spinale, F.G., et al., A matrix metalloproteinase induction/activation system exists in the human left ventricular myocardium and is upregulated in heart failure. Circulation, 2000. 102: p. 1944-9.

46. Bonnin, C.M., M.P. Sparrow, and R.R. Taylor, Collagen synthesis and content in right ventricular hypertrophy in the dog. Am J Physiol, 1981. 241: p. H708-13.

47. Brilla, C.G., J.S. Janicki, and K.T. Weber, Cardioreparative effects of lisinopril in rats with genetic hypertension and left ventricular hypertrophy. Circulation, 1991. 83: p. 1771-9.

48. Brilla, C.G., J.S. Janicki, and K.T. Weber, Impaired diastolic function and coronary reserve in genetic hypertension. Role of interstitial fibrosis and medial thickening of intramyocardial coronary arteries. Circ Res, 1991. 69: p. 107-15.

49. Brilla, C.G., L. Matsubara, and K.T. Weber, Advanced hypertensive heart disease in spontaneously hypertensive rats. Lisinopril-mediated regression of myocardial fibrosis. Hypertension, 1996. 28: p. 269-75.

50. Mukherjee, D. and S. Sen, Collagen phenotypes during development and regression of myocardial hypertrophy in spontaneously hypertensive rats. Circ Res, 1990. 67: p. 1474-80.

51. Oldershaw, P.J., et al., Correlations of fibrosis in endomyocardial biopsies from patients with aortic valve disease. Br Heart J, 1980. 44: p. 609-11.

52. Turto, H., Collagen metabolism in experimental cardiac hypertrophy in the rat and the effect of digitoxin treatment. Cardiovasc Res, 1977. 11: p. 358-66.

53. Skosey, J.L., et al., Biochemical correlates of cardiac hypertrophy. V. Labeling of collagen, myosin, and nuclear DNA during experimental myocardial hypertrophy in the rat. Circ Res, 1972. 31: p. 145-57.

54. Jalil, J.E., et al., Fibrillar collagen and myocardial stiffness in the intact hypertrophied rat left ventricle. Circ Res, 1989. 64: p. 1041-50.

55. Iimoto, D.S., J.W. Covell, and E. Harper, Increase in cross-linking of type I and type III collagens associated with volume-overload hypertrophy. Circ Res, 1988. 63: p. 399-408.

56. Covell, J.W., Factors influencing diastolic function. Possible role of the extracellular matrix. Circulation, 1990. 81: p. III155-8.

57. Douglas, P.S. and B. Tallant, Hypertrophy, fibrosis and diastolic dysfunction in early canine experimental hypertension. J Am Coll Cardiol, 1991. 17: p. 530-6.

58. van Krimpen, C., et al., Angiotensin I converting enzyme inhibitors and cardiac remodeling. Basic Res Cardiol, 1991. 86: p. 149-55.

59. Jugdutt, B.I., M.J. Joljart, and M.I. Khan, Rate of collagen deposition during healing and ventricular remodeling after myocardial infarction in rat and dog models. Circulation, 1996. 94: p. 94-101.

60. Pfeffer, M.A. and E. Braunwald, Ventricular remodeling after myocardial infarction. Experimental observations and clinical implications. Circulation, 1990. 81: p. 1161-72.

61. Anversa, P., G. Olivetti, and J.M. Capasso, Cellular basis of ventricular remodeling after myocardial infarction. Am J Cardiol, 1991. 68: p. 7D-16D.

62. Pfeffer, J.M., et al., Progressive ventricular remodeling in rat with myocardial infarction. Am J Physiol, 1991. 260: p. H1406-14.

63. Marijianowski, M.M., et al., Dilated cardiomyopathy is associated with an increase in the type I/type III collagen ratio: a quantitative assessment. J Am Coll Cardiol, 1995. 25: p. 1263-72.

64. Balghith, M. and B.I. Jugdutt, Assessment of diastolic dysfunction after acute myocardial infarction using Doppler echocardiography. Can J Cardiol, 2002. 18: p. 69-77.

65. Michorowski, B., P. Senaratne, and B. Jugdutt, Myocardial infarct expansion. Cardiovasc Rev & Rep, 1987. 8: p. 42-47.

66. Jugdutt, B.I., Identification of patients prone to infarct expansion by the degree of regional shape distortion on an early two-dimensional echocardiogram after myocardial infarction. Clin Cardiol, 1990. 13: p. 28-40.

67. Gerdes, A.M., et al., Structural remodeling of cardiac myocytes in patients with ischemic cardiomyopathy. Circulation, 1992. 86: p. 426-30.

68. Villarreal, F. and N. Kim, Modulation of cardiac fibroblast function by growth factors and mechanical stimuli. Cardiovasc Pathol, 1998. 7: p. 145-151.

69. Judd, J.T. and B.C. Wexler, Prolyl hydroxylase and collagen metabolism after experimental mycardial infarction. Am J Physiol, 1975. 228: p. 212-6.

70. Casscells, W., et al., Transforming growth factor-beta 1 in normal heart and in myocardial infarction. Ann N Y Acad Sci, 1990. 593: p. 148-60.

71. Murry, C.E., et al., Macrophages express osteopontin during repair of myocardial necrosis. Am J Pathol, 1994. 145: p. 1450-62.

72. Ohnishi, H., et al., Increased expression of connective tissue growth factor in the infarct zone of experimentally induced myocardial infarction in rats. J Mol Cell Cardiol, 1998. 30: p. 2411-22.

73. Chen, M.M., et al., CTGF expression is induced by TGF- beta in cardiac fibroblasts and cardiac myocytes: a potential role in heart fibrosis. J Mol Cell Cardiol, 2000. 32: p. 1805-19.

74. Deswal, A., et al., Cytokines and cytokine receptors in advanced heart failure: an analysis of the cytokine database from the Vesnarinone trial (VEST). Circulation, 2001. 103: p. 2055-9.

75. Deten, A., et al., Changes in extracellular matrix and in transforming growth factor beta isoforms after coronary artery ligation in rats. J Mol Cell Cardiol, 2001. 33: p. 1191-207.

76. Sekita, S., et al., Studies on collagen in the experimental myocardial infarction. Jpn Circ J, 1985. 49: p. 171-8.

77.   Zhao, M.J., et al., Profound structural alterations of the extracellular collagen matrix in postischemic dysfunctional ("stunned") but viable myocardium. J Am Coll Cardiol, 1987. 10: p. 1322-34.

78.   Factor, S.M., et al., Alterations of the myocardial skeletal framework in acute myocardial infarction with and without ventricular rupture. A preliminary report. Am J Cardiovasc Pathol, 1987. 1: p. 91-7.

79.   Weisman, H.F., et al., Cellular mechanisms of myocardial infarct expansion. Circulation, 1988. 78: p. 186-201.

80.   Caulfield, J. and P. Wolkowicz, A mechanism for cardiac dilatation. Heart Failure, 1990. 6: p. 138-150.

81.   Olivetti, G., et al., Side-to-side slippage of myocytes participates in ventricular wall remodeling acutely after myocardial infarction in rats. Circ Res, 1990. 67: p. 23-34.

82.   Takahashi, S., A.C. Barry, and S.M. Factor, Collagen degradation in ischaemic rat hearts. Biochem J, 1990. 265: p. 233-41.

83.   Charney, R.H., et al., Collagen loss in the stunned myocardium. Circulation, 1992. 85 p. 1483-90.

84.   Cleutjens, J.P., et al., Regulation of collagen degradation in the rat myocardium after infarction. J Mol Cell Cardiol, 1995. 27: p. 1281-92.

85.   Cleutjens, J.P., et al., Collagen remodeling after myocardial infarction in the rat heart. Am J Pathol, 1995. 147: p. 325-38.

86.   Uusimaa, P., et al., Collagen scar formation after acute myocardial infarction: relationships to infarct size, left ventricular function, and coronary artery patency. Circulation, 1997. 96: p. 2565-72.

87.   Etoh, T., et al., Myocardial and interstitial matrix metalloproteinase activity after acute myocardial infarction in pigs. Am J Physiol, 2001. 281: p. H987-94.

88.   Jugdutt, B.I. and R.W. Amy, Healing after myocardial infarction in the dog: changes in infarct hydroxyproline and topography. J Am Coll Cardiol, 1986. 7: p. 91-102.

89.   Jugdutt, B.I., Left ventricular rupture threshold during the healing phase after myocardial infarction in the dog. Can J Physiol Pharmacol, 1987. 65: p. 307-16.

90.   Jugdutt, B.I., Effect of nitroglycerin and ibuprofen on left ventricular topography and rupture threshold during healing after myocardial infarction in the dog. Can J Physiol Pharmacol, 1988. 66: p. 385-95.

91.   Jugdutt, B.I., et al., Functional impact of remodeling during healing after non-Q wave versus Q wave anterior myocardial infarction in the dog. J Am Coll Cardiol, 1992. 20: p. 722-31.

92.   Jugdutt, B.I., et al., Effect of enalapril on ventricular remodeling and function during healing after anterior myocardial infarction in the dog. Circulation, 1995. 91: p. 802-12.

93.   Jugdutt, B.I., et al., Combined captopril and isosorbide dinitrate during healing after myocardial infarction. Effect on ventricular remodeling, function, mass and collagen. J Am Coll Cardiol, 1995. 25: p. 1089-96.

94.   Jugdutt, B.I., Effect of captopril and enalapril on left ventricular geometry, function and collagen during healing after anterior and inferior myocardial infarction in a dog model. J Am Coll Cardiol, 1995. 25: p. 1718-25.

95.   Jugdutt, B.I., A. Lucas, and M.I. Khan, Effect of angiotensin-converting enzyme inhibition on infarct collagen deposition and remodelling during healing after transmural canine myocardial infarction. Can J Cardiol, 1997. 13: p. 657-68.

96.   Bergman, I., Two improved ad simplified methods for the spectrophotometric determination of hydroxyproline. Anal Chem, 1963. 35: p. 1961-1965.

97.   Sato, S., et al., Connective tissue changes in early ischemia of porcine myocardium: an ultrastructural study. J Mol Cell Cardiol, 1983. 15: p. 261-75.

98.   Cannon, R.O., 3rd, et al., Early degradation of collagen after acute myocardial infarction

in the rat. Am J Cardiol, 1983. 52: p. 390-5.

99. Whittaker, P., D.R. Boughner, and R.A. Kloner, Role of collagen in acute myocardial infarct expansion. Circulation, 1991. 84: p. 2123-34.

100. Lu, L., et al., Matrix metalloproteinases and collagen ultrastructure in moderate myocardial ischemia and reperfusion in vivo. Am J Physiol, 2000. 279: p. H601-9.

101. Mann, D.L. and F.G. Spinale, Activation of matrix metalloproteinases in the failing human heart: breaking the tie that binds. Circulation, 1998. 98: p. 1699-702.

102. Cheung, P.Y., et al., Matrix metalloproteinase-2 contributes to ischemia-reperfusion injury in the heart. Circulation, 2000. 101: p. 1833-9.

103. Li, Y.Y., et al., Downregulation of matrix metalloproteinases and reduction in collagen damage in the failing human heart after support with left ventricular assist devices. Circulation, 2001. 104: p. 1147-52.

104. Danielsen, C.C., H. Wiggers, and H.R. Andersen, Increased amounts of collagenase and gelatinase in porcine myocardium following ischemia and reperfusion. J Mol Cell Cardiol, 1998. 30: p. 1431-42.

105. Eaton, L.W., et al., Regional cardiac dilatation after acute myocardial infarction: recognition by two-dimensional echocardiography. N Engl J Med, 1979. 300: p. 57-62.

106. Erlebacher, J.A., et al., Early dilation of the infarcted segment in acute transmural myocardial infarction: role of infarct expansion in acute left ventricular enlargement. J Am Coll Cardiol, 1984. 4: p. 201-8.

107. van Krimpen, C., et al., DNA synthesis in the non-infarcted cardiac interstitium after left coronary artery ligation in the rat: effects of captopril. J Mol Cell Cardiol, 1991. 23: p. 1245-53.

108. Litwin, S.E., et al., Contractility and stiffness of noninfarcted myocardium after coronary ligation in rats. Effects of chronic angiotensin converting enzyme inhibition. Circulation, 1991. 83: p. 1028-37.

109. Volders, P.G., et al., Interstitial collagen is increased in the non-infarcted human myocardium after myocardial infarction. J Mol Cell Cardiol, 1993. 25: p. 1317-23.

110. Mallory, G., P. White, and S.-S. J, The speed of healing of myocardial infarction: a study of the pathologic anatomy in 72 cases. Am Heart J, 1939. 18: p. 647-671.

111. Fishbein, M.C., D. Maclean, and P.R. Maroko, The histopathologic evolution of myocardial infarction. Chest, 1978. 73: p. 843-9.

112. Boyle, M.P. and H.F. Weisman, Limitation of infarct expansion and ventricular remodeling by late reperfusion. Study of time course and mechanism in a rat model. Circulation, 1993. 88: p. 2872-83.

113. Kim, C.B. and E. Braunwald, Potential benefits of late reperfusion of infarcted myocardium. The open artery hypothesis. Circulation, 1993. 88: p. 2426-36.

114. Jugdutt, B.I., Effect of reperfusion on ventricular mass, topography, and function during healing of anterior infarction. Am J Physiol, 1997. 272: p. H1205-11.

115. Vracko, R. and D. Thorning, Contractile cells in rat myocardial scar tissue. Lab Invest, 1991. 65: p. 214-27.

116. Guarda, E., et al., Effects of endothelins on collagen turnover in cardiac fibroblasts. Cardiovasc Res, 1993. 27: p. 2130-4.

117. Villarreal, F.J., et al., Identification of functional angiotensin II receptors on rat cardiac fibroblasts. Circulation, 1993. 88: p. 2849-61.

118. Sun, Y. and K.T. Weber, Infarct scar: a dynamic tissue. Cardiovasc Res, 2000. 46: p. 250-6.

119. Willems, I.E., et al., The alpha-smooth muscle actin-positive cells in healing human myocardial scars. Am J Pathol, 1994. 145: p. 868-75.

120. Sun, Y., et al., Cardiac angiotensin converting enzyme and myocardial fibrosis in the rat. Cardiovasc Res, 1994. 28: p. 1423-32.

121. Campbell, S.E., J.S. Janicki, and K.T. Weber, Temporal differences in fibroblast proliferation and phenotype expression in response to chronic administration of angiotensin II or aldosterone. J Mol Cell Cardiol, 1995. 27: p. 1545-60.
122. Sun, Y. and K.T. Weber, Angiotensin converting enzyme and myofibroblasts during tissue repair in the rat heart. J Mol Cell Cardiol, 1996. 28: p. 851-8.
123. Whittaker, P., D.R. Boughner, and R.A. Kloner, Analysis of healing after myocardial infarction using polarized light microscopy. Am J Pathol, 1989. 134: p. 879-93.
124. Sappino, A.P., W. Schurch, and G. Gabbiani, Differentiation repertoire of fibroblastic cells: expression of cytoskeletal proteins as marker of phenotypic modulations. Lab Invest, 1990. 63: p. 144-61.
125. Skalli, O., et al., Myofibroblasts from diverse pathologic settings are heterogeneous in their content of actin isoforms and intermediate filament proteins. Lab Invest, 1989. 60: p. 275-85.
126. Jugdutt, B.I. and J.W. Warnica, Intravenous nitroglycerin therapy to limit myocardial infarct size, expansion, and complications. Effect of timing, dosage, and infarct location. Circulation, 1988. 78: p. 906-19.
127. Jugdutt, B.I., Intravenous nitroglycerin unloading in acute myocardial infarction. Am J Cardiol, 1991. 68: p. 52D-63D.
128. Pfeffer, M.A., et al., Effect of captopril on progressive ventricular dilatation after anterior myocardial infarction. N Engl J Med, 1988. 319: p. 80-6.
129. Sharpe, N., et al., Treatment of patients with symptomless left ventricular dysfunction after myocardial infarction. Lancet, 1988. 1: p. 255-9.
130. Jugdutt, B.I., B.L. Michorowski, and C.T. Kappagoda, Exercise training after anterior Q wave myocardial infarction: importance of regional left ventricular function and topography. J Am Coll Cardiol, 1988. 12: p. 362-72.
131. Jugdutt, B.I. and C.A. Basualdo, Myocardial infarct expansion during indomethacin or ibuprofen therapy for symptomatic post infarction pericarditis. Influence of other pharmacologic agents during early remodelling. Can J Cardiol, 1989. 5: p. 211-21.
132. St John Sutton, M., et al., Quantitative two-dimensional echocardiographic measurements are major predictors of adverse cardiovascular events after acute myocardial infarction. The protective effects of captopril. Circulation, 1994. 89: p. 68-75.
133. Pfeffer, M.A., et al., Effect of captopril on mortality and morbidity in patients with left ventricular dysfunction after myocardial infarction. Results of the survival and ventricular enlargement trial. The SAVE Investigators. N Engl J Med, 1992. 327: p. 669-77.
134. Gaudron, P., et al., Progressive left ventricular dysfunction and remodeling after myocardial infarction. Potential mechanisms and early predictors. Circulation, 1993. 87: p. 755-63.
135. Pfeffer, M.A., et al., Early versus delayed angiotensin-converting enzyme inhibition therapy in acute myocardial infarction. The healing and early afterload reducing therapy trial. Circulation, 1997. 95: p. 2643-51.
136. Jugdutt, B.I., B.L. Schwarz-Michorowski, and M.I. Khan, Effect of long-term captopril therapy on left ventricular remodeling and function during healing of canine myocardial infarction. J Am Coll Cardiol, 1992. 19: p. 713-21.
137. Jugdutt, B.I. and M.I. Khan, Effect of prolonged nitrate therapy on left ventricular remodeling after canine acute myocardial infarction. Circulation, 1994. 89: p. 2297-307.
138. Jugdutt, B.I., et al., Impact of left ventricular unloading after late reperfusion of canine anterior myocardial infarction on remodeling and function using isosorbide-5-mononitrate. Circulation, 1995. 92: p. 926-34.
139. Jugdutt, B.I., et al., Effect of prolonged inotropic stimulation on ventricular remodeling

during healing after myocardial infarction in the dog: mechanistic insights. J Am Coll Cardiol, 1996. 27: p. 1787-95.

140. Jugdutt, B.I. and S. Musat-Marcu, Opposite effects of amlodipine and enalapril on infarct collagen and remodelling during healing after reperfused myocardial infarction. Can J Cardiol, 2000. 16: p. 617-25.

141. Jugdutt, B.I., et al., Vascular remodeling during healing after myocardial infarction in the dog model: effects of reperfusion, amlodipine and enalapril. J Am Coll Cardiol, 2002. 39: p. 1538-45.

142. Jugdutt, B.I., Delayed effects of early infarct-limiting therapies on healing after myocardial infarction. Circulation, 1985. 72: p. 907-14.

143. Jugdutt, B.I., Myocardial salvage by intravenous nitroglycerin in conscious dogs: loss of beneficial effect with marked nitroglycerin-induced hypotension. Circulation, 1983. 68: p. 673-84.

144. Group, C.T.S., Effects of enalapril on mortality in severe congestive heart failure: results of the Cooperative North Scandinavian Enalapril Survival study. N Engl J Med, 1987. 316: p. 1429-1435.

145. Pouleur, H., et al., Effects of long-term enalapril therapy on left ventricular diastolic properties in patients with depressed ejection fraction. SOLVD Investigators. Circulation, 1993. 88: p. 481-91.

146. Pfeffer, J.M., M.A. Pfeffer, and E. Braunwald, Influence of chronic captopril therapy on the infarcted left ventricle of the rat. Circ Res, 1985. 57: p. 84-95.

147. Knowlton, A.A., et al., Rapid expression of fibronectin in the rabbit heart after myocardial infarction with and without reperfusion. J Clin Invest, 1992. 89: p. 1060-8.

148. Connelly, C.M., et al., Reversible and irreversible elongation of ischemic, infarcted, and healed myocardium in response to increases in preload and afterload. Circulation, 1991. 84: p. 387-99.

149. McCormick, R.J., et al., Regional differences in LV collagen accumulation and mature cross-linking after myocardial infarction in rats. Am J Physiol, 1994. 266: p. H354-9.

150. Honan, M.B., et al., Cardiac rupture, mortality and the timing of thrombolytic therapy: a meta-analysis. J Am Coll Cardiol, 1990. 16: p. 359-67.

151. Beltrami, C.A., et al., Structural basis of end-stage failure in ischemic cardiomyopathy in humans. Circulation, 1994. 89: p. 151-63.

152. Schaper, J. and B. Speiser, The extracellular matrix in the failing human heart. Basic Res Cardiol, 1992. 87: p. 303-9.

153. Mukherjee, D. and S. Sen, Alteration of collagen phenotypes in ischemic cardiomyopathy. J Clin Invest, 1991. 88: p. 1141-6.

154. Capasso, J.M., T.F. Robinson, and P. Anversa, Alterations in collagen cross-linking impair myocardial contractility in the mouse heart. Circ Res, 1989. 65: p. 1657-64.

155. Woodiwiss, A.J., et al., Reduction in myocardial collagen cross-linking parallels left ventricular dilatation in rat models of systolic chamber dysfunction. Circulation, 2001. 103: p. 155-60.

156. Hort, W., Untersuchungen zur funktionellen Morphologie des Bindegewebsgerustes und der Blutgefasse der linken Herzkammerwand. Virchows Arch Path Anat, 1960. 333: p. 565-581.

157. Holmgren, E., Ueber die trophospongien der quergestreiften muskelfasern, nebst bemerkungen uber den allgemeinen bau dieser fasern. Arch Mikrosk Anat, 1907. 71: p. 165-174.

158. Nagle, E., Die mechanischen Eigenschaften von Perimysium internum und Sarkolemm bei den quergestreiften Muskelfasern. Z Zellforsch Mikrosk Anat, 1935. 22: p. 694-706.

159. Wickline, S.A., et al., Structural remodeling of human myocardial tissue after infarction. Quantification with ultrasonic backscatter. Circulation, 1992. 85: p. 259-68.

54

160.   Gupta, K.B., et al., Changes in passive mechanical stiffness of myocardial tissue with aneurysm formation. Circulation, 1994. 89(5): p. 2315-26.

161.   Holmes, J.W., et al., Scar remodeling and transmural deformation after infarction in the pig. Circulation, 1994. 90: p. 411-20.

162.   Linzbach, A., Heart failure from the point of view of quantitative anatomy. Am J Cardiol, 1960. 5: p. 370-382.

163.   Caulfield, J.B. and P. Wolkowicz, Inducible collagenolytic activity in isolated perfused rat hearts. Am J Pathol, 1988. 131: p. 199-205.

164.   MacKenna, D.A., et al., Contribution of collagen matrix to passive left ventricular mechanics in isolated rat hearts. Am J Physiol, 1994. 266: p. H1007-18.

165.   Bing, O.H., et al., The effect of lathyrogen beta-amino proprionitrile (BAPN) on the mechanical properties of experimentally hypertrophied rat cardiac muscle. Circ Res, 1978. 43: p. 632-7.

166.   Sato, Y., et al., Measuring serum aminoterminal type III procollagen peptide, 7S domain of type IV collagen, and cardiac troponin T in patients with idiopathic dilated cardiomyopathy and secondary cardiomyopathy. Heart, 1997. 78: p. 505-8.

167.   Host, N.B., et al., The aminoterminal propeptide of type III procollagen provides new information on prognosis after acute myocardial infarction. Am J Cardiol, 1995. 76: p. 869-73.

168.   Klappacher, G., et al., Measuring extracellular matrix turnover in the serum of patients with idiopathic or ischemic dilated cardiomyopathy and impact on diagnosis and prognosis. Am J Cardiol, 1995. 75: p. 913-8.

169.   Zannad, F., et al., Limitation of excessive extracellular matrix turnover may contribute to survival benefit of spironolactone therapy in patients with congestive heart failure: insights from the randomized aldactone evaluation study (RALES). Rales Investigators. Circulation, 2000. 102: p. 2700-6.

170.   Latini, R., et al., ACE inhibitor use in patients with myocardial infarction. Summary of evidence from clinical trials. Circulation, 1995. 92: p. 3132-7.

171.   Cohn, J.N., R. Ferrari, and N. Sharpe, Cardiac remodeling--concepts and clinical implications: a consensus paper from an international forum on cardiac remodeling. Behalf of an International Forum on Cardiac Remodeling. J Am Coll Cardiol, 2000. 35: p. 569-82.

172.   Matsubara, H., Pathophysiological role of angiotensin II type 2 receptor in cardiovascular and renal diseases. Circ Res, 1998. 83: p. 1182-91.

173.   Nio, Y., et al., Regulation of gene transcription of angiotensin II receptor subtypes in myocardial infarction. J Clin Invest, 1995. 95: p. 46-54.

174.   Brilla, C.G., et al., Collagen metabolism in cultured adult rat cardiac fibroblasts: response to angiotensin II and aldosterone. J Mol Cell Cardiol, 1994. 26: p. 809-20.

175.   Urata, H., et al., Angiotensin II-forming pathways in normal and failing human hearts. Circ Res, 1990. 66: p. 883-90.

176.   Jugdutt, B., M. Crawford, and W.C. Saunders, Angiotensin II receptor blockers. Cardiology Clinics Annual of Drug Therapy, 1998. 2: p. 1-17.

177.   Jugdutt, B. and A. Kimchi, New advances in the use of AT1 receptor blockers (ARBs). In, Proceedings of the 2nd International Congress on Heart Disease-New trends in Research, Diagnosis and Treatment. Medimond Medical Publications, 2001: p. 531-538.

178.   Rousseau, M.F., et al., Progression of left ventricular dysfunction secondary to coronary artery disease, sustained neurohormonal activation and effects of ibopamine therapy during long-term therapy with angiotensin-converting enzyme inhibitor. Am J Cardiol, 1994. 73: p. 488-93.

179.   Baruch, L., et al., Augmented short- and long-term hemodynamic and hormonal effects of an angiotensin receptor blocker added to angiotensin converting enzyme inhibitor

therapy in patients with heart failure. Vasodilator Heart Failure Trial (V-HeFT) Study Group. Circulation, 1999. 99: p. 2658-64.

180.   Sen, S. and F.M. Bumpus, Collagen synthesis in development and reversal of cardiac hypertrophy in spontaneously hypertensive rats. Am J Cardiol, 1979. 44(5): p. 954-8.

181.   Tyagi, S.C., A. Ratajska, and K.T. Weber, Myocardial matrix metalloproteinase(s): localization and activation. Mol Cell Biochem, 1993. 126: p. 49-59.

182.   Keeley, F.W., A. Elmoselhi, and F.H. Leenen, Enalapril suppresses normal accumulation of elastin and collagen in cardiovascular tissues of growing rats. Am J Physiol, 1992. 262: p. H1013-21.

183.   Michel, J.B., et al., Hormonal and cardiac effects of converting enzyme inhibition in rat myocardial infarction. Circ Res, 1988. 62: p. 641-50.

184.   Rousseau, M.F., et al., Effects of benazeprilat on left ventricular systolic and diastolic function and neurohumoral status in patients with ischemic heart disease. Circulation, 1990. 81: p. III123-9.

185.   Smits, J.F., et al., Angiotensin II receptor blockade after myocardial infarction in rats: effects on hemodynamics, myocardial DNA synthesis, and interstitial collagen content. J Cardiovasc Pharmacol, 1992. 20: p. 772-8.

186.   Ertl, G. and B. Jugdutt, ACE inhibition after myocardial infarction: can megatrials provide answers? Lancet, 1994. 344: p. 1068-9.

187.   Nguyen, Q.T., et al., Endothelin A receptor blockade causes adverse left ventricular remodeling but improves pulmonary artery pressure after infarction in the rat. Circulation, 1998. 98: p. 2323-30.

188.   Caspari, P.G., K. Gibson, and P. Harris, Changes in myocardial collagen in normal development and after beta blockade. Recent Adv Stud Cardiac Struct Metab, 1975. 7: p. 99-104.

189.   Gudbjarnason, S., et al., Stimulation of reparative processes following experimental myocardial infarction. Arch Intern Med, 1966. 118: p. 33-40.

190.   Castagnino, H.E., et al., Preservation of the myocardial collagen framework by human growth hormone in experimental infarctions and reduction in the incidence of ventricular aneurysms. Int J Cardiol, 1992. 35: p. 101-14.

191.   Scott-Burden, T., et al., Platelet-derived growth factor suppresses and fibroblast growth factor enhances cytokine-induced production of nitric oxide by cultured smooth muscle cells. Effects on cell proliferation. Circ Res, 1992. 71: p. 1088-100.

192.   Ju, H., et al., Effect of AT1 receptor blockade on cardiac collagen remodeling after myocardial infarction. Cardiovasc Res, 1997. 35: p. 223-32.

193.   Jordan, J.E., Z.Q. Zhao, and J. Vinten-Johansen, The role of neutrophils in myocardial ischemia-reperfusion injury. Cardiovasc Res, 1999. 43: p. 860-78.

194.   Group, L.S., Late Assessment of Thrombolytic Efficacy (Late) study with alteplase 6024 hours after onset of acute myocardial infarction. Lancet, 1993. 25; 342: p. 759-766.

195.   Morita, M., et al., Effects of late reperfusion on infarct expansion and infarct healing in conscious rats. Am J Pathol, 1993. 143: p. 419-30.

196.   Przyklenk, K. and R.A. Kloner, Superoxide dismutase plus catalase improve contractile function in the canine model of the "stunned myocardium". Circ Res, 1986. 58: p. 148-56.

197.   Peuhkurinen, K.J., et al., Thrombolytic therapy with streptokinase stimulates collagen breakdown. Circulation, 1991. 83: p. 1969-75.

198.   Pyo, R., et al., Targeted gene disruption of matrix metalloproteinase-9 (gelatinase B) suppresses development of experimental abdominal aortic aneurysms. J Clin Invest, 2000. 105: p. 1641-9.

199.   Bigatel, D.A., et al., The matrix metalloproteinase inhibitor BB-94 limits expansion of experimental abdominal aortic aneurysms. J Vasc Surg, 1999. 29: p. 130-8.

Chapter 3

Pro-Inflammatory Cytokines and Cardiac Extracellular Matrix: Regulation of Fibroblast Phenotype

R. Dale Brown, M. Darren Mitchell, Carlin S. Long
University of Colorado Health Sciences Center and Denver Health Medical Center, Denver, Colorado, U.S.A.

1. Introduction

1.1 Inflammation and wound healing in myocardial injury and failure

Heart failure remains the leading cause of death in industrial societies, with 400,000-700,000 new cases in the US alone each year [1]. Annual direct expenditures for heart failure in the US have been estimated at $20-$40 billion, making it one of the costliest health problems in this country. Although coronary artery disease is the predominant cause, heart failure may also result from long-standing pressure-volume overload, infectious myocarditis, alcohol abuse, or inborn genetic abnormalities. Regardless of etiology, however, all causes of heart failure converge on a final common pathway of diminished contractile performance and pathophysiological remodeling. Classically described as chamber dilation and wall thinning, cardiac remodeling also manifests as fibrosis: an increased mass of the interstitial compartment. Fibrosis results in increased myocardial stiffness, and contributes significantly to the loss of myocardial contractile function [2].

The macroscopic functional alterations observed in the failing heart arise from fundamental changes in the constituent cell types within the heart and their genetic programs. The reduction in myocardial contractility reflects cellular hypertrophy and an altered program of contractile and Ca^{2+} handling gene expression by the cardiac myocyte [3]. Cardiac fibrosis, on the other hand, reflects hyperplasia and altered production of connective tissue proteins by the cardiac fibroblast, the principal cell type of the non-myocyte cell population of the heart [4, 5]. Although the cell and molecular biology of the cardiac myocyte has been a focus of intensive research since the early 1980s, knowledge of cardiac non-myocytes is much less advanced, despite constituting a numerical majority of cells in the heart.

The responses of the heart to injury share many features in common with wound healing and fibrosis observed in other tissues, including lung, liver, kidney, and skin [6-10]. The sequence of events which occur in response to injury can be summarized briefly as follows: hemostasis; recruitment of circulating immune-inflammatory cells; macrophage activation; activation of fibroblasts and formation of a provisional matrix; and remodeling of the granulomatous scar [11, 12]. These events reflect transitions of the participating cell type(s) to activated phenotypes having fundamentally different biological functions from the corresponding quiescent cells in normal tissue. These cellular phenotypes, arising from coordinated programs of gene expression, are regulated by specific cytokines and growth factors. The respective receptor signal transduction cascades converge on limited sets of nuclear transcription factors, which act on characteristic sets of genes. Synergistic or antagonistic interactions between cytokines are mediated by overlapping sets of transcription factors [13, 14]. Signal transduction mechanisms for individual cytokines have been authoritatively reviewed [15-22].

The initial steps in the progression of wound healing are dominated by cells derived from the blood (platelets, lymphocytes, neutrophils, macrophages), which act acutely to limit the extent of injury and clear the wound of pathogens and damaged cells [11, 12]. The subsequent phases of wound healing are dominated by resident parenchymal cells, particularly fibroblasts, and are directed to the repair and rebuilding of functionally contiguous intact tissue. These processes are coordinated by the principal inflammatory cytokines, interferon-γ, interleukin-1 (IL-1), tumor necrosis factor-α (TNF-α), and interleukin-6 (IL-6), and by fibrogenic growth factors including angiotensin II, aldosterone, and transforming growth factor-β (TGF-β). TGF-β is widely recognized to play a central role in tissue restoration [23]. Ultimately this process extends beyond the original site of injury, resulting in diffuse tissue involvement, which further compromises organ performance. Many of the same cytokines and growth factors seen during normal wound healing also appear in the fibrotic state [24, 25]. Cardiac fibrosis thus can be viewed as inappropriate wound remodeling leading to pathophysiological accumulation of connective tissue [8].

1.2 Pro-inflammatory cytokines in heart failure

These general considerations lead to a dualistic view of pro-inflammatory cytokines in the heart. Acute activation of pro-inflammatory cytokines is seen as a physiologically adaptive response to injury, whereas chronic cytokine activation has been hypothesized to contribute to the progression of heart failure [26]. Several lines of evidence have accumulated to support this hypothesis [27-30]. First, elevated circulating concentrations of TNF-α [31, 32] and its soluble receptors [33-35], IL-6 [32, 36], and IL-1 [36, 37] have been found in heart failure patients. This is true for ischemic injury, infectious myocarditis, and idiopathic dilated cardiomyopathy. Prospective studies on a large patient population have shown that elevated concentrations of TNF-α ligand and receptor molecules, and to a lesser extent IL-6, predict mortality [34, 35]. Although the primary sources of cytokines have not been established with certainty, it appears likely that cytokine production within the failing myocardium contributes significantly [33, 38, 39]. Conversely, conventional therapeutic approaches used for clinical management of heart failure have been shown to reverse the expression of inflammatory cytokines. These agents include the Ca^{2+} channel antagonist amlodipine [40], β-adrenergic receptor antagonists [41], inhibitors of the renin-angiotensin system [42, 43], and phosphodiesterase inhibitors [42, 44, 45]. Mechanical unloading of failing myocardium with left ventricular assist devices was also shown to produce a similar reversal of the pro-inflammatory cytokine expression [46].

Second, experimental administration of pro-inflammatory cytokines recapitulates many of the sequelae occurring in myocardial injury or heart failure [47]. Thus, acute infusion of TNF-α in healthy human volunteers causes a rapid and reversible depression of myocardial contractile performance [48]. Experiments with isolated rodent and feline cardiomyocytes demonstrate that these effects reflect a disturbance of Ca^{2+} handling [49]. Similar results have been obtained with IL-1 [37]. More prolonged exposure to TNF-α or IL-1 leads to cardiomyocyte hypertrophy, and reversion to a fetal program of expression of Ca^{2+} handling and contractile protein isoforms, similar to changes observed in human heart failure [24, 31, 50-52]. These alterations in myocyte structure and function, in concert with profound cytokine-induced changes in the interstitial compartment, result in the progressive and cumulative structural remodeling of the heart which accompanies decompensated heart failure.

2. Pro-Inflammatory Cytokines and Cardiac Extracellular Matrix Metabolism

Structural remodeling of the myocardium is a key event in the progression of heart failure regardless of etiology, and an important determinant of declining myocardial performance [53]. This remodeling initially includes hypertrophy, chamber dilation, and wall thinning, and subsequently fibrosis and increased mechanical stiffness. Studies in rodents have suggested an important causal role for TNF-α in this process. Bozkurt et al. [54] showed that a 15-day TNF-α infusion by osmotic minipump in rats elicited an initial rapid decline in myocardial contractile performance which was reversed upon TNF-α withdrawal, consistent with the reversible depression of E-C coupling *in vitro*. Structural changes in the heart mimicking the response to infarction included myocyte hypertrophy, left ventricular dilation, and wall thinning. These effects followed more slowly and were not reversed during the recovery period. Importantly, an early structural change was the rapid decrease in collagen content and loss of organization of the interstitial collagen network, as noted by electron microscopy. The authors proposed that the degradation of extracellular matrix (ECM) elicited by TNF-α plays a causal role in the ensuing reorganization of cardiomyocytes within the chamber wall. This proposal was supported in subsequent studies by Sivasubramanian et al. [55] using transgenic mice for cardiac specific over-expression of TNF-α. These mice develop dilated cardiac hypertrophy, followed by overt heart failure. Examination of young (4 wk) mice revealed an initial elevated expression of matrix metalloproteinase (MMP) activity and depressed expression of tissue inhibitors of metalloproteinases (TIMPs). The net enhancement of ECM degradative activity accompanied histological disorganization of myofibrils and loss of fibrillar collagen, and preceded the onset of statistically significant chamber dilation beginning at 8 weeks of age. By 12 weeks of age, these mice showed increased collagen deposition, together with elevated expression of TIMP and decreased MMP activity. No further increase in chamber dilation was detected following the normalization of MMP/TIMP expression and increased deposition of ECM.

These findings were extended by Feldman and colleagues in a separate mouse strain transgenic for cardiac-specific TNF-α over-expression [56, 57]. Comparison of young (12 weeks) and older (48 weeks) animals revealed elevated expression of collagens I and III and MMPs, together with evidence for denatured collagen fibrils indicative of excess degradative activity. The perturbation of ECM metabolism was progressive with age, and paralleled the sequence of myocardial hypertrophy, dilation, and eventual fibrosis, resulting in contractile dysfunction. Neutralization of circulating TNF-α by adenoviral mediated gene transfer of a TNF-α binding protein-immunoglobulin chimera (sTNFRI:Fc) reversed the systemic inflammatory response and improved the myocardial pathology in young animals, but did not improve myocardial performance in older animals. In a separate study, treatment of TNF-α over expressing mice with a pharmacologic inhibitor of MMPs (Batimistat, BB-94) gave similar results of improved myocardial hypertrophy and function in younger mice but not older mice [58]. To date, these transgenic approaches related to cardiac extracellular matrix have not been reported for the other proinflammatory cytokines. This represents an important area for future research.

Cytokine regulation of ECM metabolism thus represents an important contributor to structural remodeling of the injured or failing myocardium. MMP activation and net ECM degradation may be proximal responses to the acute elevation of pro-inflammatory cytokines in response to injury. This phase is followed by compensatory collagen resynthesis, ultimately leading to cardiac fibrosis in the failing heart. These results are consistent with surgical myocardial infarction in rats, where MMP induction and collagen breakdown predominate initially [59-61], followed by elevations of both pro-inflammatory and pro-fibrotic (e.g. TGF-β) cytokines, as well as increased collagen production [62].

3. Regulation of Fibroblast Phenotype by Pro-Inflammatory Cytokines

The foregoing discussion emphasizes the importance of ECM metabolism and homeostasis in cardiac structure and function, and of the regulatory role of the pro-inflammatory cytokines. In the heart as in other tissues, quiescent fibroblasts in normal tissue primarily are responsible for steady state turnover of extracellular matrix [63, 64]. Many cytokines and growth factors exert direct actions on fibroblastic cells. These cells respond to both inflammatory cytokines and fibrogenic growth factors with distinct functional phenotypes, and produce additional autocrine-paracrine mediators which amplify the signals from the primary stimuli. In response to inflammatory stimuli at sites of injury, the fibroblast undergoes a profound phenotypic transition, resulting in chemotactic migration from the wound margin into the zone of injury, accelerated degradation and provisional replacement of damaged extracellular matrix, and induction of additional autocrine and paracrine mediators including IL-6, IL-8, TGF-β, prostaglandins, and nitric oxide (NO) [64]. As the inflammatory response subsides, fibroblast proliferation and sustained remodeling of the granulation tissue leads to formation of a mature collagen-rich scar. Myofibroblasts, which are specially differentiated fibroblasts exhibiting contractile properties and expressing smooth muscle α-actin, play an important role in contracture of the granulomatous wound [61, 65, 66].

Despite these conserved features of the host-defense and wound healing paradigms, fibroblasts from different tissues exhibit a striking diversity in their specific responses to pro-inflammatory cytokines. It is our thesis that these differences reflect the specific physiological functions of fibroblasts in individual organ systems and differential utilization of potential signaling pathways by the pro-inflammatory cytokines. To substantiate this hypothesis, we evaluated the primary research literature on the actions of IL-1, TNF-α, interferon-γ, and IL-6 to regulate fibroblast phenotype in skin, joint synovium, lung, and liver. Insufficient data were available from kidney for a complete characterization. These responses are compared to the heart, as described later. We chose the principal phenotypic endpoints related to wound healing and fibrosis, including proliferation, chemotaxis, synthesis of extracellular matrix, and expression of MMPs and TIMPs. Results are summarized in Table 1. In some cases, conflicting observations are reported. The reasons for this variability may reflect the phenotypic state of the fibroblasts in the source tissue, or the specific culture conditions. For example, many studies utilize fibroblasts from chronic inflamed or fibrotic tissues. Further, the denuded environment of conventional tissue culture may itself recreate aspects of the wound site.

Skin fibroblast responses to IL-1 are characterized by increased proliferation, potentiated migration to platelet derived growth factor (PDGF), increased MMP synthesis, and altered expression and remodeling of ECM [67-70]. Synovial fibroblasts respond to IL-1 with increased proliferation and migration [71, 72]. IL-1 has complex effects on collagen synthesis, mediated by IL-6 or prostaglandin E2, consistent with hyperplastic remodeling of rheumatoid synovial pannus [73-76]. In lung fibroblasts, IL-1 exerts highly variable effects on proliferation, with increases, decreases, or no effect noted. IL-1 augments migration of these cells to PDGF, and consistently increases net collagen synthesis with concomitant increase in MMP expression [77-79]. Responses of hepatic stellate fibroblastic cells to IL-1 are incompletely characterized. IL-1 is reported to be a mitogenic growth factor ([80], but see [81]). Both increases [82] and decreases [83] in collagen synthesis have been reported, with increased MMP expression [84]. Effects of IL-1 on TIMP expression in these tissues are variable and may be mediated indirectly by IL-6 ([69, 85, 86]; see below).

Like IL-1, TNF-α exerts complex effects on fibroblast behavior in many tissues. TNF-α increases proliferation of skin fibroblasts and potently increases migration [87-89]. Decreased collagen synthesis is consistently observed, accompanied by robust elevation in MMP expression and attenuated TIMP-1 expression [89-91]. Thus, TNF-α promotes ECM degradation in skin, whereas IL-1 tends to increase ECM deposition. In rheumatoid synovial fibroblasts, TNF-α is well established to stimulate cell proliferation and to induce synoviocyte production of immune cell chemoattractants [92, 93]. TNF-α attenuates collagen synthesis, and strongly upregulates MMP expression, contributing to erosive joint damage in rheumatoid arthritis [76, 94, 95]. Thus, actions of IL-1 are thought to predominate in rheumatoid synovial fibroblasts, whereas TNF-α contributes more importantly to recruitment and activation of inflammatory cell infiltrates [96, 97]. Responses of lung to TNF-α are complex, as seen with IL-1. Variable and context-dependent effects on lung fibroblast proliferation have been reported [98, 99]. TNF-α augments PDGF induced fibroblast migration, and upregulates cell adhesion molecules (CAMs) important for neutrophil invasion [79, 100]. Increased ECM synthesis is most commonly

Table 1. Tissue specific fibroblast responses to pro-inflammatory cytokines

Table 1. Tissue specific fibroblast responses to pro-inflammatory cytokines

Cytokine	Response	Skin	Synovium	Lung	Liver	Heart
			Organ			
IL-1	Proliferation	↑	↑↑	↔, ↑, ↓	↑	↓↓
	Chemotaxis	↑	↑↑	↑	n.a.	↑↑
	ECM Synthesis	↑, ↓	↑, ↓	↑	↑, ↓	↓
	MMP Expression	↑↑	↑↑	↑	↑	↑↑
	TIMP Expression	↔	↑	↔, ↑	n.a.	↓
TNF	Proliferation	↑	↑↑	↑, ↓	↑, ↓	↓
	Chemotaxis	↑↑	↑	↑	↑	↑
	ECM Synthesis	↓↓	↓	↑↑, ↓	↓↓	↓
	MMP Expression	↑↑	↑↑	↑	↑	↑↑
	TIMP Expression	↓	n.a.	n.a.	↑	↓
IFN	Proliferation	↓	↑, ↓	↓↓	↓	↓↓
	Chemotaxis	n.a.	n.a.	n.a.	n.a.	↓
	ECM Synthesis	↓↓	↓↓	↓↓	↓↓	↓
	MMP Expression	↓	↓	n.a.	n.a.	↔
	TIMP Expression	n.a.	↔	n.a.	n.a.	n.a.
IL-6	Proliferation	↓	↔	↑	↑	↔
	Chemotaxis	n.a.	↑	↔	n.a.	↔
	ECM Synthesis	↑↑	↔	↑	↑	↔
	MMP Expression	↑, ↓	↔	↑	n.a.	↔
	TIMP Expression	↑, ↓	↑↑	↑	n.a.	n.a.

↑, increase; ↑↑, increase that is substantial in magnitude or highly reproducible; ↓, decrease; ↓↓, decrease that is substantial in magnitude or highly reproducible; ↔, no effect; n.a., data not available. See text for details.

observed ([101, 102]; reviewed in [103]). MMP expression also is increased as part of ECM remodeling [79, 102]. Consistent with *in vitro* data, excessive TNF-α activation in lung leads to increased ECM deposition and fulminant fibrosis [101]. This action may reflect synergism between TNF-α and TGF-β. Such interactions also are important in liver, where TNF-α has modest effects on regulation of fibroblast proliferation by itself, but synergizes strongly with TGF-β and other factors to increase proliferation *in vitro* and *in vivo* [104-107]. Distinct from the lung, however, hepatic fibroblasts show less exuberant ECM accumulation, with consistently depressed collagen synthesis and concerted upregulation of MMPs and TIMPs [108-110]. TNF-α is important to stimulate hepatic fibroblast production of chemoattractants for leukocyte recruitment [80, 111].

In contrast to the other pro-inflammatory cytokines, interferon-γ largely inhibits proliferation [109, 112-115], and profoundly suppresses collagen synthesis [82, 116-118] in fibroblasts from skin, synovium, lung, and liver. Interferon-γ appears to act on multiple targets in the cell cycle to inhibit proliferation [119]. Inhibition of collagen synthesis occurs by direct action at an interferon-γ response element in the collagen I promoter, and opposes the actions of TGF-β to increase collagen transcription [117]. Interferon-γ is reported to decrease MMP expression in skin [120] and synovium [121], whereas effects on MMPs and TIMPS in other tissues are minimal or not reported. Importantly, interferon-γ stands as the major physiological antagonist to the pro-fibrotic actions of TGF-β, and inhibitor of TGF-β induced transformation of fibroblasts to the myofibroblast phenotype [64, 66].

IL-6 is conventionally described as a pro-inflammatory cytokine because of its actions to promote immune cell differentiation and induction of acute phase proteins at the liver. In peripheral cell types including fibroblasts, however, IL-6 exerts anti-inflammatory and pro-fibrogenic actions, and is thought to be protective against the pro-inflammatory actions of IL-1 and TNF-α [22]. IL-6 is anti-proliferative in dermal fibroblasts and inhibits proliferation to IL-1 and TNF-α [122]. There are a few reports of modest IL-6 elicited increases in proliferation in synovium, lung, and liver fibroblasts [81, 123, 124]. The most prominent action of IL-6 in fibroblasts is to promote net ECM deposition; however, the mechanistic balance between collagen synthesis and expression of MMPs and TIMPs varies among tissues (reviewed in [125, 126]). IL-6 increases ECM deposition in dermal fibroblasts primarily through enhanced collagen synthesis, with weaker and variable effects on MMP and TIMP expression [68, 127-129]. In synovial fibroblasts, by contrast, the dominant action of IL-6 is upregulation of TIMP-1 whereas biosynthesis of ECM and MMPs are unaffected [130, 131]. IL-6 promotes matrix deposition in lung fibroblasts by composite effects on collagen synthesis and expression of MMPs and TIMPs [126, 130, 132]. IL-6 is noted as a fibrogenic growth factor in liver, but the mechanism is not fully characterized [80]. Enhanced chemotaxis by IL-6 has been noted in synovial fibroblasts, but not in lung [72, 133].

To summarize, IL-1 and TNF-α orchestrate many aspects of fibroblast phenotype. Paradoxically, these cytokines may be associated with either inflammatory restoration of damaged tissue, or with excessive fibroblast proliferation and fibrosis. Interferon-γ exerts more uniform and selective actions targeted on suppression of fibroblast proliferation and collagen synthesis. IL-6 acts in a support role to modulate ECM metabolism. These observations illustrate the complexities and tissue specificity of fibroblast responses to inflammatory cytokines.

4. Pro-Inflammatory Cytokine Regulation of Cardiac Fibroblasts

Whereas the importance of fibroblasts in controlling tissue remodeling is well established in other fibrotic states, detailed knowledge of the cardiac fibroblast is much less advanced, despite unique features of function and regulation. Moreover, these cells represent an important source of intracardiac growth promoting agonists with substantial effects on myocyte phenotype and function. For these reasons, the cardiac fibroblast represents an important therapeutic target distinct from current approaches directed toward the cardiac myocyte. These arguments provide a compelling rationale for investigating the effects of both circulating and locally produced cytokines and growth factors in this cell population. To address this question, our laboratory has focused on the influence of pro-inflammatory cytokines on the primary phenotypic endpoints of proliferation, chemotaxis, ECM metabolism, and regulation of TGF-β expression, in cultured neonatal rat cardiac fibroblasts [134]. This knowledge forms a foundation for ongoing studies on cytokine signal transduction mechanisms in cardiac fibroblasts.

4.1. Proliferation

Inappropriate fibroblast hyperplasia is a hallmark of fibrosis. Studies from this lab showed that IL-1 inhibits cardiac fibroblast proliferation [135]. Cell cycle analysis demonstrated that IL-1 inhibits progression through the G1/S transition [136]. IL-1 inhibition was mediated by altered expression and activities of several key proteins involved in G1/S progression; namely, decreased phosphorylation of retinoblastoma protein, diminished expression and activities of cyclins D2, D3, E, A, and the affiliated cyclin dependent kinases-2 and -4, and increased expression of p27KIP1 cyclin kinase inhibitor. These actions were a direct consequence of interleukin-1 receptor signaling, rather than occurring indirectly through production of prostaglandins or NO.

A. Autonomous Fibroblast Proliferation

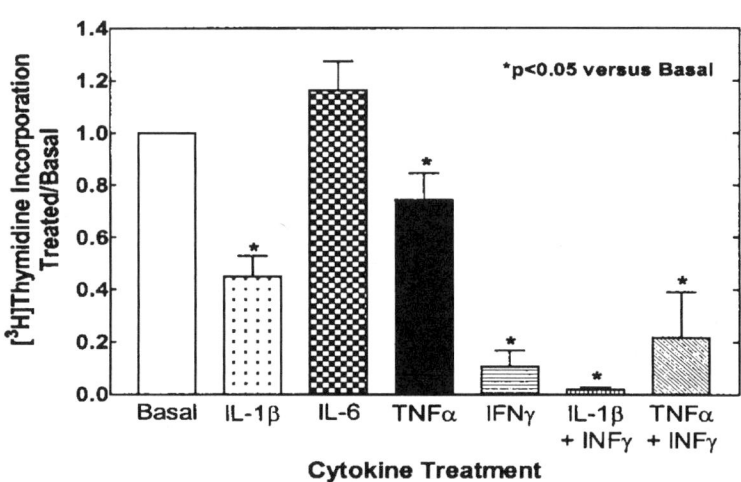

B. Serum-Stimulated Proliferation

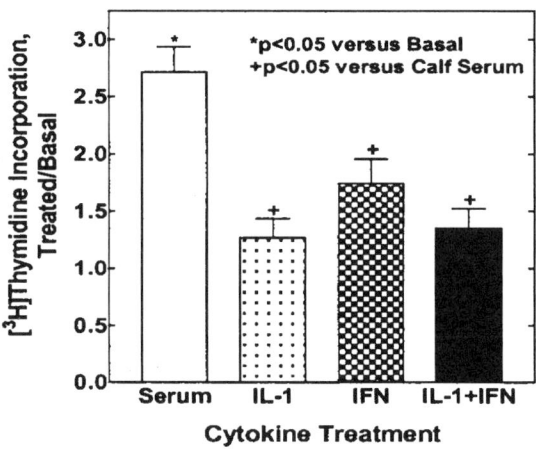

Figure 1. Pro-inflammatory cytokine regulation of cardiac fibroblast proliferation. Confluent cultures of cardiac fibroblasts were labeled with [^{3}H]-thymidine as a marker of cell proliferation. Data were normalized to [^{3}H]-thymidine incorporation in the absence of any stimulation (basal) and represent the results from separate experiments (mean ± SD). Panel A: Effect of cytokine treatment on fibroblast proliferation in the absence of serum stimulation. Panel B: Cytokine inhibition of serum-stimulated fibroblast proliferation.

Effects of additional inflammatory cytokines on cardiac fibroblast proliferation, measured as [³H]-thymidine incorporation, are shown in Figure 1. Cardiac fibroblasts were initially screened for proliferation in the presence of target cytokines. As shown in Figure 1A, these cells exhibit the ability to proliferate autonomously in serum free growth medium, presumably due to autocrine release of mitogenic growth factors [134, 137]. The pro-inflammatory cytokines inhibit autonomous proliferation with relative potencies interferon-γ > IL-1 > TNF-α. The combination of interferon-γ plus either IL-1 or TNF-α inhibits proliferation more than either agent alone. IL-6 had no effect on proliferation. Figure 1B shows cytokine inhibition of serum-stimulated fibroblast proliferation. Interestingly, we note that IL-1 is more effective than interferon-γ to inhibit proliferation under these conditions. Again, additive inhibition is observed with IL-1 plus interferon-γ. Concentration-response studies demonstrate that the inhibitory responses to combinations of cytokines are additive, but not synergistic nor potentiated (data not shown).

4.2. Chemotaxis

Migration of immune-inflammatory cells, closely followed by fibroblasts, from undamaged tissue into the zone of injury, dominates the early phase of the healing process. Cytokines and growth factors have been shown to exert profound effects on immune cell migration, but much less is known about fibroblast migration. In particular we are unaware of previous studies characterizing the chemotactic responses of cardiac fibroblasts to pro-inflammatory cytokines. In order to quantitate cardiac fibroblast migration, we have utilized a modified Boyden chamber assay of chemotaxis in combination with fluorescence staining of migrated cell nuclei with ethidium bromide. Figure 2A shows unstimulated migration of control cultures. Figure 2B demonstrates greatly increased migration over the same interval in cells that were incubated with IL-1. Comparative data for pro-inflammatory cytokines are tabulated in Figure 2C. It can be seen that IL-1 is by far the strongest pro-migratory agonist. TNF-α stimulates migration, but much less robustly. Interferon-γ tends to inhibit migration. IL-6 is ineffective. The regulation of migration is consistent with previous work from our group showing that IL-1 and TNF-α caused upregulation of the cell adhesion molecules ICAM-1 and VCAM in cardiac fibroblasts [138].

4.3. ECM Metabolism

We next turned to define the influence of inflammatory cytokines on ECM metabolism. Initial experiments investigated the synthesis of collagen.

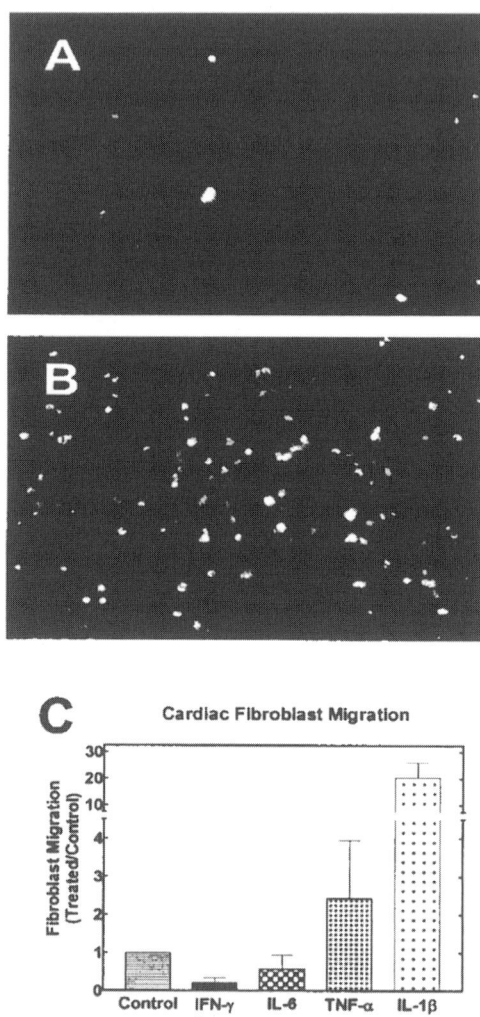

Figure 2. Pro-inflammatory cytokine regulation of cardiac fibroblast migration. Cardiac fibroblasts were treated with cytokines for 24 hr, then re-plated into modified Boyden chambers with indicated cytokine for 18 hr. Migrated cells were visualized by nuclear staining with ethidium bromide, then counted using fluorescence microscopy. Panel A: Untreated cells (Control). Panel B: Treatment with 10 ng/ml IL-1β. Panel C: Results obtained from three separate experiments. Data were normalized to the number of cells migrated in the absence of any stimulation and presented as the means ± S.D.

Biosynthesis of total collagen is measured as incorporation of [³H]-proline into proteins in confluent quiescent cardiac fibroblasts in serum free growth medium and treated with the indicated agonists. Fractional [³H]-proline incorporation into collagen subsequently is determined by digestion with bacterial collagenase [139]. Figure 3 shows that the fibrogenic cytokine TGF-β1 stimulates collagen synthesis. IL-1 or interferon-γ alone partially inhibit collagen synthesis. However, the two cytokines in combination act synergistically to inhibit collagen synthesis.

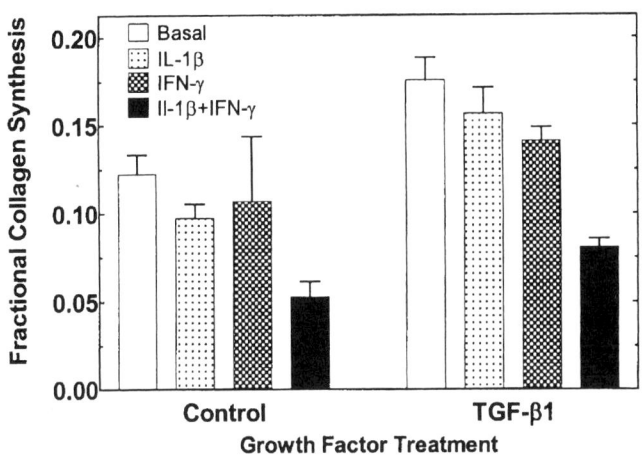

Figure 3. Cytokine regulation of collagen synthesis in cultured cardiac fibroblasts. Biosynthesis of total collagen was measured as incorporation of [³H]-proline, followed by digestion with bacterial collagenase to quantitate the amount of collagen-specific radiolabel. All treatments were 10 ng/ml of the indicated cytokine for 48 hr, with [³H]-proline present during the last 24 hr. Data are shown as mean ± S.D. of triplicate determinations.

These results are in agreement with a report by Siwik et al. [140], confirmed by Sano et al. [141], who showed that IL-1 and TNF-α decrease total collagen synthesis. IL-1 was shown to decrease mRNAs for the major fibrillar collagens, pro-α1(I), pro-α2(I), and pro-α1(III), but to increase expression of non-fibrillar collagens α1(IV), α2(IV), and fibronectin [140]. These authors also found that IL-1 and TNF-α increased the expression and activity of MMPs 2, 3, 9, and 13. Furthermore, Li et al showed that IL-1 and TNF-α decreased expression of the tissue inhibitors of metalloproteinases, TIMP-1 and TIMP-3, and increased expression of ADAM-10, a disintegrin metalloproteinase thought to be involved in TNF-α processing. Decreased collagen synthesis in response to interferon-γ in neonatal rat cardiac fibroblasts was previously noted by Grimm et al. [142]. Taken together, these results suggest that pro-inflammatory cytokines elicit a shift in ECM metabolism in cardiac fibroblasts toward decreased collagen synthesis and increased degradation. In the setting of acute injury, the degradation of ECM will facilitate the chemotactic invasion of fibroblasts into damaged tissue. These responses are consistent with the consequences of inflammatory cytokine activation observed in heart failure, and underscore the importance of the cardiac fibroblast as a key player in ECM remodeling.

4.4. Transforming Growth Factor-β Expression

In contrast to the pro-inflammatory cytokines, TGF-β is well established as a primary fibrogenic growth factor in many organ systems. As noted above, compensatory ECM resynthesis, ultimately resulting in fibrosis, is observed in human heart failure and in experimental animal models. Importantly, many of the actions of the renin-angiotensin system within the myocardium appear to be mediated by increased tissue production of TGF-β [8, 143]. In this regard, our laboratory has previously shown that cardiac fibroblasts predominate over cardiac myocytes as a source of TGF-β in neonatal rat cardiac cell cultures [144]. All three isoforms are produced, with TGF-β3>TGF-β1>>TGF-β2. Stimulation of cardiac fibroblast AT_1 receptors was shown to upregulate TGF-β1 expression. TGF-β1 in turn exerts paracrine effects to increase myocyte hypertrophy, as well as autocrine stimulation of cardiac fibroblast proliferation, perhaps mediated through platelet derived growth factor [145]. Cardiac fibroblast production of autocrine-paracrine signaling molecules also constitutes an important response to pro-inflammatory cytokines. It was therefore of interest to examine the regulation of TGF-β expression by pro-inflammatory cytokines in cardiac fibroblasts.

TGF-β isoform mRNAs were detected by ribonuclease protection assay (Figure 4). Control cultures express predominantly TGF-β3 with lower amounts of TGF-β1. Stimulation with IL-1β markedly enhances TGF-β1mRNA and almost completely abolishes TGF-β3 mRNA. In contrast, TNF-α depresses both TGF-β1and TGF-β3 mRNA levels. Co-treatment with TNF-α plus IL-1 is essentially additive. The IL-1-mediated increase in TGF-β1mRNA is maintained, whereas the suppression of TGF-β3 mRNA is enhanced. Finally, interferon-γ has no effect on TGF-β mRNA compared to control cells, but blocks the actions of IL-1 by a post-transcriptional mechanism (data not shown).

Treatments

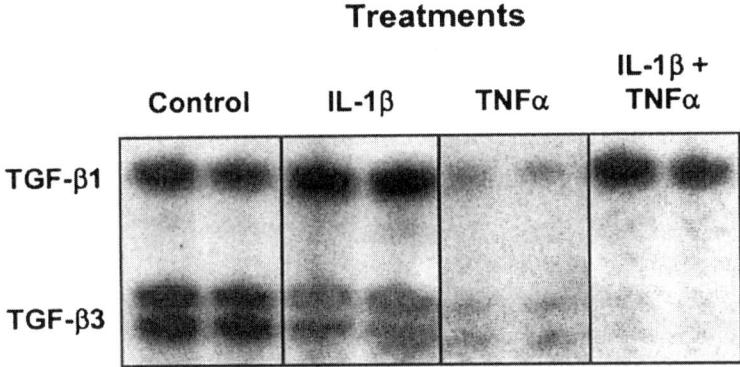

Figure 4. Cytokine regulation of TGF-β expression in cardiac fibroblasts. Confluent fibroblasts were treated for 48 hr with 10 ng/ml of the indicated cytokines prior to RNA isolation. TGF-β mRNAs were detected by ribonuclease protection assay (Pharmingen rck-3 kit) followed by gel electrophoresis and autoradiography. The probe signals for TGF-β1 and TGF-β3 are indicated. Similar results were obtained in two additional experiments.

Thus, IL-1, TNF-α, and interferon-γ exert specific and contrasting effects on TGF-β isoforms.

These results suggest that IL-1 may regulate TGF-β expression as a feedback mechanism to limit ECM degradation in response to injury. The functional consequences of the striking TGF-β isoform switch in response to IL-1 are not known, since both isoforms are presumed to signal through the same TGF-β receptor complex. Interestingly, differential expression of TGF-β isoforms has been shown in healing dermal wounds [146]. Increased TGF-β1 expression is associated with scar formation, whereas TGF-β3 expression diminished scar formation [147]. It will be of some interest to examine the regulation of ECM metabolism by individual TGF-β isoforms in cardiac fibroblasts.

In summary, our findings in neonatal rat cultures demonstrate that pro-inflammatory cytokines convert the cardiac fibroblast phenotype toward inhibition of proliferation, enhanced migration, diminished synthesis and increased degradation of ECM, and altered expression of fibrogenic growth factors. IL-1 appears to be the most potent and pleiotrophic regulator in this system, whereas TNF-α is less effective. Interferon-γ has important but selective effects to inhibit cell proliferation and reduce collagen synthesis. IL-6 is largely ineffective in the functional assays we employed. However, we note that IL-6 actions may be greatly potentiated in the presence of soluble IL-6 receptor (sIL-6R) [131]. This constellation of behaviors is consistent with the mobilization of cardiac fibroblasts that occurs during the acute inflammatory phase of the response to injury, as opposed to the hyperplasia and excessive ECM deposition which accompany the progression to heart failure. These initial responses represent a necessary and adaptive component of wound healing.

5. Prospects for Cytokine Directed Therapies in Heart Failure

The growing recognition of the importance of inflammatory processes in heart disease has provided motivation to develop cytokine-directed therapies. Despite the strong biologically based rationale, this approach has yielded disappointing results to date. An early attempt to suppress inflammatory processes with methylprednisone therapy in acute myocardial infarction increased the incidence of ventricular arrhythmias and increased infarct size [148]. Prednisolone therapy in acute myocarditis showed no improvement of left ventricular ejection fraction or survival [149]. More recently, accumulating evidence supporting a role for TNF-α in heart failure led to evaluation of the TNF-α sequestering receptor-antibody chimera (sTNFRI:Fc, Etanercept) in patients with advanced heart failure (NYHA Class III and IV). This agent effectively reduced plasma concentrations of active TNF-α and preliminary results suggested an improvement in cardiac output and quality of life. However, a multicenter-randomized trial was terminated due to failure to demonstrate significant improvement in morbidity or mortality (reviewed in [53]).

These initial experiences raise several issues, which need to be addressed in order to achieve the goal of anti-cytokine therapy for heart failure. First, it will be important to define the temporal window for effective therapy. It seems likely that cardiac fibroblasts undergo a state transition from physiologically appropriate compensatory responses to proinflammatory cytokines with acute injury, to maladaptive and exacerbating responses in decompensated heart failure. An important observation in this context is that transgenic mice doubly deficient in TNF-α Type I and Type II receptors exhibit increased extents of myocardial infarcts following coronary ligation [150]. On the other hand, the cumulative and progressive nature of heart failure may limit successful outcome to therapy in the advanced stages of the disease [58]. Second, intracardiac delivery of anti-cytokine agents may be necessary rather than the vascular depot approach used with the recombinant TNF-α receptor: antibody chimera. This problem may be surmounted by development of conventional small molecule pharmaceutical agents. Third, additional cytokine targets may be considered. Our data with neonatal rat fibroblasts show that IL-1 exerts consistently greater biological activity than TNF-α. We speculate that while systemic elevation of TNF-α is the overt sign of inflammatory activation, interleukin-1 may contribute importantly to inflammatory damage within the myocardium. Moreover, we show here that interferon-γ potently inhibits fibroblast proliferation and collagen synthesis. Similar observations in the lung and kidney have led to promising initial results using interferon-γ therapy in fibroproliferative lung disease [151, 152]. This avenue should be pursued in the heart. Also related to the question of therapeutic target is the strategic choice between antagonizing pluripotent initiator cytokines such as IL-1 or TNF-α, versus selective inhibition of key signaling intermediates or endpoint responses. For example, we have discussed the importance of cytokine activation of MMPs as a contributor to myocardial remodeling. Studies with transgenic mice deficient in MMPs have shown decreased occurrence of left ventricular dilation [153] and myocardial rupture [154] following surgical infarction. In animal models, pharmacological MMP inhibitors attenuate left ventricular remodeling and preserve contractile function following surgical infarction [155-157]. In closing, it is our view that development of successful therapeutic approaches will depend critically upon understanding the mechanisms of cytokine regulation of fibroblast phenotype in the individual cell, and in the context of the intact heart.

Acknowledgments

Work in the authors' laboratory is supported by NIH (HL59428). The authors are grateful to Mary Atz and Gail Morris for assistance with experimental procedures, and to Dr. Kelly Ambler for assistance with computer graphics and numerous insightful discussions. RDB dedicates this effort to John L. Skosey, M.D., Ph.D., for providing an initial opportunity to learn about the biology of pro-inflammatory cytokines in fibroblasts.

References

1. Braunwald, E., Pathophysiology of Heart Failure. In Heart Disease, Braunwald, E., eds. Philadelphia, PA: WB Saunders Co.,1988: p. 426-448.
2. Eghbali, M. and K.T. Weber, Collagen and the myocardium: fibrillar structure, biosynthesis and degradation in relation to hypertrophy and its regression. Mol Cell Biochem, 1990. 96: p. 1-14.
3. Swynghedauw, B., Molecular mechanisms of myocardial remodeling. Physiol Rev, 1999. 79: p. 215-62.
4. Booz, G.W. and K.M. Baker, Molecular signalling mechanisms controlling growth and function of cardiac fibroblasts. Cardiovasc Res, 1995. 30: p. 537-43.
5. Eghbali, M., Cardiac fibroblasts: function, regulation of gene expression, and phenotypic modulation. Basic Res Cardiol, 1992. 87: p. S83-9
6. Friedman, S.L., Stellate cell activation in alcoholic fibrosis--an overview. Alcohol Clin Exp Res, 1999. 23: p. 904-10.
7. Kupper, T.S. and R.W. Groves, The IL-1 axis and cutaneous inflammation. J Invest Dermatol, 1995. 105: p. S62S-66.

8.   Weber, K.T., Fibrosis, a common pathway to organ failure: angiotensin II and tissue repair. Semin Nephrol, 1997. 17: p. 467-91.

9.   Sime, P.J., et al., Transfer of tumor necrosis factor-alpha to rat lung induces severe pulmonary inflammation and patchy interstitial fibrogenesis with induction of transforming growth factor-beta1 and myofibroblasts. Am J Pathol, 1998. 153: p. 825-32.

10. Zalewski, A. and Y. Shi, Vascular myofibroblasts. Lessons from coronary repair and remodeling. Arterioscler Thromb Vasc Biol, 1997. 17: p. 417-22.

11. Davidson, JM., Wound Repair. In: Inflammation: Basic Principles and Clinical Correlates. Gallin JI, Goldstein RH, Snyderman R,eds. New York,NY: Raven Press, Ltd., 1992: p. 809-819.

12. Weber, K.T., Y. Sun, and L.C. Katwa, Wound healing following myocardial infarction. Clin Cardiol, 1996. 19: p. 447-55.

13. Hill, C.S. and R. Treisman, Transcriptional regulation by extracellular signals: mechanisms and specificity. Cell, 1995. 80: p. 199-211.

14. Manning, A. and Rao A. Agents Targeting Transcription. In: Inflammation Basic Principles and Clinical Correlates. Gallin JI, Snyderman R, eds. Philadelphia, PA: Lippincott Williams & Wilkins, 1999: p. 1159-1176.

15. Boehm, U., et al., Cellular responses to interferon-gamma. Annu Rev Immunol, 1997. 15: p. 749-95.

16. Dinarello, C.A., Biologic basis for IL-1 in disease. Blood, 1996. 87: p. 2095-147.

17. Heinrich, P.C., et al., Interleukin-6-type cytokine signalling through the gp130/Jak/STAT pathway. Biochem J, 1998. 334: p. 297-314.

18. Ledgerwood, E.C., J.S. Pober, and J.R. Bradley, Recent advances in the molecular basis of TNF signal transduction. Lab Invest, 1999. 79: p. 1041-50.

19. Leong, K.G. and A. Karsan, Signaling pathways mediated by tumor necrosis factor alpha. Histol Histopathol, 2000. 15: p. 1303-25.

20. O'Neill, L.A. and C. Greene, Signal transduction pathways activated by the IL-1 receptor family: ancient signaling machinery in mammals, insects, and plants. J Leukoc Biol, 1998. 63: p. 650-7.

21. Roberts, A.B., TGF-beta signaling from receptors to the nucleus. Microbes Infect, 1999. 1: p. 1265-73.

22. Tilg, H., C.A. Dinarello, and J.W. Mier, IL-6 and APPs: anti-inflammatory and immunosuppressive mediators. Immunol Today, 1997. 18(9): p. 428-32.

23. Branton, M.H. and J.B. Kopp, TGF-beta and fibrosis. Microbes Infect, 1999. 1: p. 1349-65.

24. Lange, L. and G F. Schreiner, Immune cytokines and cardiac disease. Trends Cardiovasc Med, 1992. 2: p. 145-151.

25. Li, D. and S.L. Friedman, Liver fibrogenesis and the role of hepatic stellate cells: new insights and prospects for therapy. J Gastroenterol Hepatol, 1999. 14: p. 618-33.

26. Seta, Y., et al., Basic mechanisms in heart failure: the cytokine hypothesis. J Card Fail, 1996. 2: p. 243-9.

27. Lopez, F. and S. Casado, Heart failure, redox alterations, and endothelial dysfunction. Hypertension, 2001. 38: p. 1400-1405.

28. Frangogiannis, N.G., C.W. Smith, and M.L. Entman, The inflammatory response in myocardial infarction. Cardiovasc Res, 2002. 53: p. 31-47.

29. Niebauer, J., Inflammatory mediators in heart failure. Int J Cardiol, 2000. 72: p. 209-13.

30. Sharma, R., A.J. Coats, and S.D. Anker, The role of inflammatory mediators in chronic heart failure: cytokines, nitric oxide, and endothelin-1. Int J Cardiol, 2000. 72: p. 175-86.

31. Levine, B., et al., Elevated circulating levels of tumor necrosis factor in severe chronic heart failure. N Engl J Med, 1990. 323: p. 236-41.

32.   Torre-Amione, G., et al., Proinflammatory cytokine levels in patients with depressed left ventricular ejection fraction: a report from the Studies of Left Ventricular Dysfunction (SOLVD). J Am Coll Cardiol, 1996. 27: p. 1201-6.

33.   Conraads, V.M., J.M. Bosmans, and C.J. Vrints, Chronic heart failure: an example of a systemic chronic inflammatory disease resulting in cachexia. Int J Cardiol, 2002. 85: p. 33-49.

34.   Deswal, A., et al., Cytokines and cytokine receptors in advanced heart failure: an analysis of the cytokine database from the Vesnarinone trial (VEST). Circulation, 2001. 103: p. 2055-9.

35.   Rauchhaus, M., et al., Plasma cytokine parameters and mortality in patients with chronic heart failure. Circulation, 2000. 102: p. 3060-7.

36.   Testa, M., et al., Circulating levels of cytokines and their endogenous modulators in patients with mild to severe congestive heart failure due to coronary artery disease or hypertension. J Am Coll Cardiol, 1996. 28: p. 964-971.

37.   Long, C.S., The role of IL-1 in the failing heart. Heart Fail Rev, 2001. 6: p. 81-94.

38.   Paulus, W.J., How are cytokines activated in heart failure? Eur J Heart Fail, 1999. 1: p. 309-12.

39.   Torre-Amione, G., et al., Tumor necrosis factor-alpha and tumor necrosis factor receptors in the failing human heart. Circulation, 1996. 93: p. 704-11.

40.   Mohler, E.R., 3rd, et al., Role of cytokines in the mechanism of action of amlodipine: the PRAISE Heart Failure Trial. Prospective Randomized Amlodipine Survival Evaluation. J Am Coll Cardiol, 1997. 30: p. 35-41.

41.   Ohtsuka, T., et al., Effect of beta-blockers on circulating levels of inflammatory and anti-inflammatory cytokines in patients with dilated cardiomyopathy. J Am Coll Cardiol, 2001. 37: p. 412-7.

42.   Gullestad, L., et al., Effect of high- versus low-dose angiotensin converting enzyme inhibition on cytokine levels in chronic heart failure. J Am Coll Cardiol, 1999. 34: p. 2061-7.

43.   Tsutamoto, T., et al., Angiotensin II type 1 receptor antagonist decreases plasma levels of tumor necrosis factor alpha, interleukin-6 and soluble adhesion molecules in patients with chronic heart failure. J Am Coll Cardiol, 2000. 35: p. 714-21.

44.   Matsumori, A., et al., Vesnarinone, a new inotropic agent, inhibits cytokine production by stimulated human blood from patients with heart failure. Circulation, 1994. 89: p. 955-8.

45.   Prabhu, S.D., et al., beta-Adrenergic blockade in developing heart failure: effects on myocardial inflammatory cytokines, nitric oxide, and remodeling. Circulation, 2000. 101: p. 2103-9.

46.   Torre-Amione, G., et al., Decreased expression of tumor necrosis factor-alpha in failing human myocardium after mechanical circulatory support : A potential mechanism for cardiac recovery. Circulation, 1999. 100: p. 1189-93.

47.   Feldman, A.M., et al., The role of tumor necrosis factor in the pathophysiology of heart failure. J Am Coll Cardiol, 2000. 35: p. 537-44.

48.   Suffredini, A.F., et al., The cardiovascular response of normal humans to the administration of endotoxin. N Engl J Med, 1989. 321: p. 280-7.

49.   Yokoyama, T., et al., Cellular basis for the negative inotropic effects of tumor necrosis factor-alpha in the adult mammalian heart. J Clin Invest, 1993. 92: p. 2303-12.

50.   Guillen, I., et al., Cytokine signaling during myocardial infarction: sequential appearance of IL-1 beta and IL-6. Am J Physiol, 1995. 269: p. R229-35.

51.   Herskowitz, A., et al., Cytokine mRNA expression in postischemic/reperfused myocardium. Am J Pathol, 1995. 146: p. 419-28.

52.   Yokoyama, T., et al., Tumor necrosis factor-alpha provokes a hypertrophic growth response in adult cardiac myocytes. Circulation, 1997. 95: p. 1247-52.

53. Bradham, W.S., et al., Tumor necrosis factor-alpha and myocardial remodeling in progression of heart failure: a current perspective. Cardiovasc Res, 2002. 53: p. 822-30.

54. Bozkurt, B., et al., Pathophysiologically relevant concentrations of tumor necrosis factor-alpha promote progressive left ventricular dysfunction and remodeling in rats. Circulation, 1998. 97: p. 1382-91.

55. Sivasubramanian, N., et al., Left ventricular remodeling in transgenic mice with cardiac restricted overexpression of tumor necrosis factor. Circulation, 2001. 104: p. 826-31.

56. Kubota, T., et al., Soluble tumor necrosis factor receptor abrogates myocardial inflammation but not hypertrophy in cytokine-induced cardiomyopathy. Circulation, 2000. 101: p. 2518-25.

57. Li, Y.Y., et al., Myocardial extracellular matrix remodeling in transgenic mice overexpressing tumor necrosis factor alpha can be modulated by anti-tumor necrosis factor alpha therapy. Proc Natl Acad Sci U S A, 2000. 97: p. 12746-51.

58. Li, Y.Y., et al., MMP inhibition modulates TNF-alpha transgenic mouse phenotype early in the development of heart failure. Am J Physiol, 2002. 282: p. H983-9.

59. Cleutjens, J.P., et al., Regulation of collagen degradation in the rat myocardium after infarction. J Mol Cell Cardiol, 1995. 27: p. 1281-92.

60. Takahashi, S., A.C. Barry, and S.M. Factor, Collagen degradation in ischaemic rat hearts. Biochem J, 1990. 265: p. 233-41.

61. Sun, Y. and K.T. Weber, Infarct scar: a dynamic tissue. Cardiovasc Res, 2000. 46: p. 250-6.

62. Yue, P., et al., Cytokine expression increases in nonmyocytes from rats with postinfarction heart failure. Am J Physiol, 1998. 275: p. H250-8.

63. Eghbali, M., et al., Collagen chain mRNAs in isolated heart cells from young and adult rats. J Mol Cell Cardiol, 1988. 20: p. 267-76.

64. Postlethwaite, A.E. and Kang, A.H., Fibroblasts and Matrix Proteins. In: Basic Principles and Clinical Correlates., Gallin, J.I., Snyderman R., eds. Philadelphia, PA: Lippincott Williams & Wilkins, 1999: p.227-257.

65. Powell, D.W., et al., Myofibroblasts. II. Intestinal subepithelial myofibroblasts. Am J Physiol, 1999. 277: p. C183-201.

66. Sappino, A.P., W. Schurch, and G. Gabbiani, Differentiation repertoire of fibroblastic cells: expression of cytoskeletal proteins as marker of phenotypic modulations. Lab Invest, 1990. 63: p. 144-61.

67. Heckmann, M., et al., Biphasic effects of interleukin-1 alpha on dermal fibroblasts: enhancement of chemotactic responsiveness at low concentrations and of mRNA expression for collagenase at high concentrations. J Invest Dermatol, 1993. 100: p. 780-4.

68. Kawaguchi, Y., M. Hara, and T.M. Wright, Endogenous IL-1alpha from systemic sclerosis fibroblasts induces IL-6 and PDGF-A. J Clin Invest, 1999. 103: p. 1253-60.

69. Mauviel, A., et al., Uncoordinate regulation of collagenase, stromelysin, and tissue inhibitor of metalloproteinases genes by prostaglandin E2: selective enhancement of collagenase gene expression in human dermal fibroblasts in culture. J Cell Biochem, 1994. 54: p. 465-72.

70. Moon, S.E., et al., Induction of matrix metalloproteinase-1 (MMP-1) during epidermal invasion of the stroma in human skin organ culture: keratinocyte stimulation of fibroblast MMP-1 production. Br J Cancer, 2001. 85: p. 1600-5.

71. Kumkumian, G.K., et al., Platelet-derived growth factor and IL-1 interactions in rheumatoid arthritis. Regulation of synoviocyte proliferation, prostaglandin production, and collagenase transcription. J Immunol, 1989. 143: p. 833-7.

72. Wang, A.Z., et al., Improved in vitro models for assay of rheumatoid synoviocyte chemotaxis. Clin Exp Rheumatol, 1994. 12: p. 293-9.

73. Barchowsky, A., D. Frleta, and M.P. Vincenti, Integration of the NF-kappaB and

mitogen-activated protein kinase/AP-1 pathways at the collagenase-1 promoter: divergence of IL-1 and TNF-dependent signal transduction in rabbit primary synovial fibroblasts. Cytokine, 2000. 12: p. 1469-79.

74. Goldring, M.B. and S.M. Krane, Modulation by recombinant interleukin 1 of synthesis of types I and III collagens and associated procollagen mRNA levels in cultured human cells. J Biol Chem, 1987. 262: p. 16724-9.

75. Ito, A., et al., Effects of interleukin-6 on the metabolism of connective tissue components in rheumatoid synovial fibroblasts. Arthritis Rheum, 1992. 35: p. 1197-201.

76. Rinaldi, N., et al., Loss of collagen type IV in rheumatoid synovia and cytokine effect on the collagen type-IV gene expression in fibroblast-like synoviocytes from rheumatoid arthritis. Virchows Arch, 2001. 439: p. 675-82.

77. Elias, J.A., et al., Cytokine networks in the regulation of inflammation and fibrosis in the lung. Chest, 1990. 97: p. 1439-45.

78. MacFarlane, D.J., C.M. O'Connor, and M.X. Fitzgerald, Collagen production in human lung fibroblasts in response to cytokines. Biochem Soc Trans, 1994. 22: p. 49S.

79. Sasaki, M., et al., Differential regulation of metalloproteinase production, proliferation and chemotaxis of human lung fibroblasts by PDGF, interleukin-1beta and TNF-alpha. Mediators Inflamm, 2000. 9: p. 155-60.

80. Tsukamoto, H., Cytokine regulation of hepatic stellate cells in liver fibrosis. Alcohol Clin Exp Res, 1999. 23: p. 911-6.

81. Toda, K., et al., Induction of hepatic stellate cell proliferation by LPS-stimulated peripheral blood mononuclear cells from patients with liver cirrhosis. J Gastroenterol, 2000. 35: p. 214-20.

82. Tiggelman, A.M., et al., Collagen synthesis by human liver (myo)fibroblasts in culture: evidence for a regulatory role of IL-1 beta, IL-4, TGF beta and IFN gamma. J Hepatol, 1995. 23: p. 307-17.

83. Matsuoka, M., N.T. Pham, and H. Tsukamoto, Differential effects of interleukin-1 alpha, tumor necrosis factor alpha, and transforming growth factor beta 1 on cell proliferation and collagen formation by cultured fat-storing cells. Liver, 1989. 9: p. 71-8.

84. Quinones, S., G. Buttice, and M. Kurkinen, Promoter elements in the transcriptional activation of the human stromelysin-1 gene by the inflammatory cytokine, interleukin 1. Biochem J, 1994. 302 : p. 471-7.

85. Lin, N., T. Sato, and A. Ito, Triptolide, a novel diterpenoid triepoxide from Tripterygium wilfordii Hook. f., suppresses the production and gene expression of pro-matrix metalloproteinases 1 and 3 and augments those of tissue inhibitors of metalloproteinases 1 and 2 in human synovial fibroblasts. Arthritis Rheum, 2001. 44: p. 2193-200.

86. Medina, L., et al., Leukotriene C4 upregulates collagenase expression and synthesis in human lung fibroblasts. Biochim Biophys Acta, 1994. 1224: p. 168-74.

87. Lilli, C., et al., Effects of transforming growth factor-beta1 and tumour necrosis factor-alpha on cultured fibroblasts from skin fibroma as modulated by toremifene. Int J Cancer, 2002. 98: p. 824-32.

88. Postlethwaite, A.E. and J.M. Seyer, Stimulation of fibroblast chemotaxis by human recombinant tumor necrosis factor alpha (TNF-alpha) and a synthetic TNF-alpha 31-68 peptide. J Exp Med, 1990. 172: p. 1749-56.

89. Taniguchi, S., et al., Butylated hydroxyanisole blocks the inhibitory effects of tumor necrosis factor-alpha on collagen production in human dermal fibroblasts. J Dermatol Sci, 1996. 12: p. 44-9.

90. Han, Y.P., Y.D. Nien, and W.L. Garner, Tumor necrosis factor-alpha-induced proteolytic activation of pro-matrix metalloproteinase-9 by human skin is controlled by down-regulating tissue inhibitor of metalloproteinase-1 and mediated by tissue-associated chymotrypsin-like proteinase. J Biol Chem, 2002. 277: p. 27319-27.

91. Reunanen, N., et al., Activation of p38 alpha MAPK enhances collagenase-1 (matrix

metalloproteinase (MMP)-1) and stromelysin-1 (MMP-3) expression by mRNA stabilization. J Biol Chem, 2002. 277: p. 32360-8.

92. Volin, M.V., et al., RANTES expression and contribution to monocyte chemotaxis in arthritis. Clin Immunol Immunopathol, 1998. 89: p. 44-53.

93. Youn, J., et al., Regulation of TNF-alpha-mediated hyperplasia through TNF receptors, TRAFs, and NF-kappaB in synoviocytes obtained from patients with rheumatoid arthritis. Immunol Lett, 2002. 83: p. 85-93.

94. Konttinen, Y.T., et al., Collagenase-3 (MMP-13) and its activators in rheumatoid arthritis: localization in the pannus-hard tissue junction and inhibition by alendronate. Matrix Biol, 1999. 18: p. 401-12.

95. Sun, H.B. and H. Yokota, Reduction of cytokine-induced expression and activity of MMP-1 and MMP-13 by mechanical strain in MH7A rheumatoid synovial cells. Matrix Biol, 2002. 21: p. 263-70.

96. Jenkins, J.K., K.J. Hardy, and R.W. McMurray, The pathogenesis of rheumatoid arthritis: a guide to therapy. Am J Med Sci, 2002. 323: p. 171-80.

97. Joosten, L.A., et al., Protection against cartilage and bone destruction by systemic interleukin-4 treatment in established murine type II collagen-induced arthritis. Arthritis Res, 1999. 1: p. 81-91.

98. Elias, J.A., Tumor necrosis factor interacts with interleukin-1 and interferons to inhibit fibroblast proliferation via fibroblast prostaglandin-dependent and -independent mechanisms. Am Rev Respir Dis, 1988. 138: p. 652-8.

99. Tufvesson, E. and G. Westergren-Thorsson, Alteration of proteoglycan synthesis in human lung fibroblasts induced by interleukin-1beta and tumor necrosis factor-alpha. J Cell Biochem, 2000. 77: p. 298-309.

100. Spoelstra, F.M., et al., Interferon-gamma and interleukin-4 differentially regulate ICAM-1 and VCAM-1 expression on human lung fibroblasts. Eur Respir J, 1999. 14: p. 759-66.

101. Kolb, M., et al., Transient expression of IL-1beta induces acute lung injury and chronic repair leading to pulmonary fibrosis. J Clin Invest, 2001. 107: p. 1529-36.

102. Yang, M. and M. Kurkinen, Different mechanisms of regulation of the human stromelysin and collagenase genes. Analysis by a reverse-transcription-coupled-PCR assay. Eur J Biochem, 1994. 222: p. 651-8.

103. Bienkowski, R.S. and M.G. Gotkin, Control of collagen deposition in mammalian lung. Proc Soc Exp Biol Med, 1995. 209: p. 118-40.

104. Bachem, M.G., et al., Tumor necrosis factor alpha (TNF alpha) and transforming growth factor beta 1 (TGF beta 1) stimulate fibronectin synthesis and the transdifferentiation of fat-storing cells in the rat liver into myofibroblasts. Virchows Arch B Cell Pathol Incl Mol Pathol, 1993. 63: p. 123-30.

105. Diehl, A.M. and R. Rai, Review: regulation of liver regeneration by pro-inflammatory cytokines. J Gastroenterol Hepatol, 1996. 11: p. 466-70.

106. Gallois, C., et al., Role of NF-kappaB in the antiproliferative effect of endothelin-1 and tumor necrosis factor-alpha in human hepatic stellate cells. Involvement of cyclooxygenase-2. J Biol Chem, 1998. 273: p. 23183-90.

107. Knittel, T., et al., Effect of tumour necrosis factor-alpha on proliferation, activation and protein synthesis of rat hepatic stellate cells. J Hepatol, 1997. 27: p. 1067-80.

108. Armendariz-Borunda, J., K. Katayama, and J.M. Seyer, Transcriptional mechanisms of type I collagen gene expression are differentially regulated by interleukin-1 beta, tumor necrosis factor alpha, and transforming growth factor beta in Ito cells. J Biol Chem, 1992. 267: p. 14316-21.

109. Knittel, T., et al., Expression patterns of matrix metalloproteinases and their inhibitors in parenchymal and non-parenchymal cells of rat liver: regulation by TNF-alpha and TGF-beta1. J Hepatol, 1999. 30: p. 48-60.

110. Poulos, J.E., et al., Fibronectin and cytokines increase JNK, ERK, AP-1 activity, and transin gene expression in rat hepatic stellate cells. Am J Physiol, 1997. 273: p. G804-11.

111. Knittel, T., et al., Expression and regulation of cell adhesion molecules by hepatic stellate cells (HSC) of rat liver: involvement of HSC in recruitment of inflammatory cells during hepatic tissue repair. Am J Pathol, 1999. 154: p. 153-67.

112. Alvaro-Gracia, J.M., et al., Mutual antagonism between interferon-gamma and tumor necrosis factor-alpha on fibroblast-like synoviocytes: paradoxical induction of IFN-gamma and TNF-alpha receptor expression. J Clin Immunol, 1993. 13: p. 212-8.

113. Brinckerhoff, C.E. and P.M. Guyre, Increased proliferation of human synovial fibroblasts treated with recombinant immune interferon. J Immunol, 1985. 134: p. 3142-6.

114. Hein, R., et al., Treatment of systemic sclerosis with gamma-interferon. Br J Dermatol, 1992. 126: p. 496-501.

115. Lukacs, N.W., et al., Type 1/type 2 cytokine paradigm and the progression of pulmonary fibrosis. Chest, 2001. 120: p. 5S-8S.

116. Amento, E.P., et al., Influences of gamma interferon on synovial fibroblast-like cells. Induction and inhibition of collagen synthesis. J Clin Invest, 1985. 76: p. 837-48.

117. Ghosh, A.K., et al., Antagonistic regulation of type I collagen gene expression by interferon-gamma and transforming growth factor-beta. Integration at the level of p300/CBP transcriptional coactivators. J Biol Chem, 2001. 276: p. 11041-8.

118. Maguire, M.C., C.M. O'Connor, and M.X. Fitzgerald, Type I and type III collagen mRNA expression in human lung fibroblasts. Biochem Soc Trans, 1994. 22: p. 51S.

119. Harvat, B.L. and A.M. Jetten, Decreased growth inhibitory responses of squamous carcinoma cells to interferon-gamma involve failure to recruit cki proteins into cdk2 complexes. J Invest Dermatol, 2001. 117: p. 1274-81.

120. Varga, J., et al., Control of extracellular matrix degradation by interferon-gamma. The tryptophan connection. Adv Exp Med Biol, 1996. 398: p. 143-8.

121. Unemori, E.N., et al., Stromelysin expression regulates collagenase activation in human fibroblasts. Dissociable control of two metalloproteinases by interferon-gamma. J Biol Chem, 1991. 266: p. 23477-82.

122. Mihara, M., Y. Moriya, and Y. Ohsugi, IL-6-soluble IL-6 receptor complex inhibits the proliferation of dermal fibroblasts. Int J Immunopharmacol, 1996. 18: p. 89-94.

123. Mihara, M., et al., Interleukin-6 (IL-6) induces the proliferation of synovial fibroblastic cells in the presence of soluble IL-6 receptor. Br J Rheumatol, 1995. 34: p. 321-5.

124. Scaffidi, A.K., et al., Oncostatin M stimulates proliferation, induces collagen production and inhibits apoptosis of human lung fibroblasts. Br J Pharmacol, 2002. 136: p. 793-801.

125. Kossakowska, A.E., et al., Interleukin-6 regulation of matrix metalloproteinase (MMP-2 and MMP-9) and tissue inhibitor of metalloproteinase (TIMP-1) expression in malignant non-Hodgkin's lymphomas. Blood, 1999. 94: p. 2080-9.

126. Solis-Herruzo, J.A., et al., Interleukin-6 increases rat metalloproteinase-13 gene expression through stimulation of activator protein 1 transcription factor in cultured fibroblasts. J Biol Chem, 1999. 274: p. 30919-26.

127. Brenneisen, P., et al., Ultraviolet-B induction of interstitial collagenase and stromelyin-1 occurs in human dermal fibroblasts via an autocrine interleukin-6-dependent loop. FEBS Lett, 1999. 449: p. 36-40.

128. Duncan, M.R. and B. Berman, Stimulation of collagen and glycosaminoglycan production in cultured human adult dermal fibroblasts by recombinant human interleukin 6. J Invest Dermatol, 1991. 97: p. 686-92.

129. Sato, T., A. Ito, and Y. Mori, Interleukin 6 enhances the production of tissue inhibitor of metalloproteinases (TIMP) but not that of matrix metalloproteinases by human

fibroblasts. Biochem Biophys Res Commun, 1990. 170: p. 824-9.

130.   Richards, C.D., et al., Selective regulation of metalloproteinase inhibitor (TIMP-1) by oncostatin M in fibroblasts in culture. J Immunol, 1993. 150: p. 5596-603.

131.   Silacci, P., et al., Interleukin (IL)-6 and its soluble receptor induce TIMP-1 expression in synoviocytes and chondrocytes, and block IL-1-induced collagenolytic activity. J Biol Chem, 1998. 273: p. 13625-9.

132.   Lang, D.S., H. Schocker, and S. Hockertz, Effects of crocidolite asbestos on human bronchoepithelial-dependent fibroblast stimulation in coculture: the role of IL-6 and GM-CSF. Toxicology, 2001. 159: p. 81-98.

133.   Kahler, C.M., et al., Influence of neuropeptides on neutrophil adhesion and transmigration through a lung fibroblast barrier in vitro. Exp Lung Res, 2001. 27: p. 25-46.

134.   Long, C.S., C.J. Henrich, and P.C. Simpson, A growth factor for cardiac myocytes is produced by cardiac nonmyocytes. Cell Regul, 1991. 2: p. 1081-95.

135.   Palmer, J.N., et al., Interleukin-1 beta induces cardiac myocyte growth but inhibits cardiac fibroblast proliferation in culture. J Clin Invest, 1995. 95: p. 2555-64.

136.   Koudssi, F., et al., Cardiac fibroblasts arrest at the G1/S restriction point in response to interleukin (IL)-1beta. Evidence for IL-1beta-induced hypophosphorylation of the retinoblastoma protein. J Biol Chem, 1998. 273: p. 25796-803.

137.   Long, C.S., W.E. Hartogensis, and P.C. Simpson, Beta-adrenergic stimulation of cardiac non-myocytes augments the growth-promoting activity of non-myocyte conditioned medium. J Mol Cell Cardiol, 1993. 25: p. 915-25.

138.   Kacimi, R., et al., Expression and regulation of adhesion molecules in cardiac cells by cytokines: response to acute hypoxia. Circ Res, 1998. 82: p. 576-86.

139.   Peterkofsky, B. and R. Diegelmann, Use of a mixture of proteinase-free collagenases for the specific assay of radioactive collagen in the presence of other proteins. Biochemistry, 1971. 10: p. 988-94.

140.   Siwik, D.A., D.L. Chang, and W.S. Colucci, Interleukin-1beta and tumor necrosis factor-alpha decrease collagen synthesis and increase matrix metalloproteinase activity in cardiac fibroblasts in vitro. Circ Res, 2000. 86: p. 1259-65.

141.   Sano, I., et al., OPC-8212, a quinoline derivative, counteracts the reduction in type III collagen mRNA due to lipopolysaccharides in cultured rat cardiac fibroblasts. Jpn Heart J, 2001. 42: p. 125-34.

142.   Grimm, D., et al., Extracellular matrix proteins in cardiac fibroblasts derived from rat hearts with chronic pressure overload: effects of beta-receptor blockade. J Mol Cell Cardiol, 2001. 33: p. 487-501.

143.   Lijnen, P. and V. Petrov, Antagonism of the renin-angiotensin-aldosterone system and collagen metabolism in cardiac fibroblasts. Methods Find Exp Clin Pharmacol, 1999. 21: p. 363-74.

144.   Gray, M.O., et al., Angiotensin II stimulates cardiac myocyte hypertrophy via paracrine release of TGF-beta 1 and endothelin-1 from fibroblasts. Cardiovasc Res, 1998. 40: p. 352-63.

145.   Long, C.S., Autocrine and paracrine regulation of myocardial cell growth in vitro. The TGF-beta paradigm. Trends Cardiovasc Med, 1996. 6: p. 217-226.

146.   Frank, S., M. Madlener, and S. Werner, Transforming growth factors beta1, beta2, and beta3 and their receptors are differentially regulated during normal and impaired wound healing. J Biol Chem, 1996. 27: p. 10188-93.

147.   Shah, M., D.M. Foreman, and M.W. Ferguson, Neutralisation of TGF-beta 1 and TGF-beta 2 or exogenous addition of TGF-beta 3 to cutaneous rat wounds reduces scarring. J Cell Sci, 1995. 108: p. 985-1002.

148.   Roberts, R., V. DeMello, and B.E. Sobel, Deleterious effects of methylprednisolone in patients with myocardial infarction. Circulation, 1976. 53: p. S204-6.

149.   Mason, J.W., et al., A clinical trial of immunosuppressive therapy for myocarditis. The Myocarditis Treatment Trial Investigators. N Engl J Med, 1995. 333: p. 269-75.

150.   Kurrelmeyer, K.M., et al., Endogenous tumor necrosis factor protects the adult cardiac myocyte against ischemic-induced apoptosis in a murine model of acute myocardial infarction. Proc Natl Acad Sci U S A, 2000. 97: p. 5456-61.

151.   Lasky, J.A. and L.A. Ortiz, Antifibrotic therapy for the treatment of pulmonary fibrosis. Am J Med Sci, 2001. 322: p. 213-21.

152.   Oldroyd, S.D., et al., Interferon-gamma inhibits experimental renal fibrosis. Kidney Int, 1999. 56: p. 2116-27.

153.   Ducharme, A., et al., Targeted deletion of matrix metalloproteinase-9 attenuates left ventricular enlargement and collagen accumulation after experimental myocardial infarction. J Clin Invest, 2000. 106: p. 55-62.

154.   Heymans, S., et al., Inhibition of plasminogen activators or matrix metalloproteinases prevents cardiac rupture but impairs therapeutic angiogenesis and causes cardiac failure. Nat Med, 1999. 5: p. 1135-42.

155.   Peterson, J.T., et al., Matrix metalloproteinase inhibition attenuates left ventricular remodeling and dysfunction in a rat model of progressive heart failure. Circulation, 2001. 103: p. 2303-9.

156.   Rohde, L.E., et al., Matrix metalloproteinase inhibition attenuates early left ventricular enlargement after experimental myocardial infarction in mice. Circulation, 1999. 99: p. 3063-70.

157.   Spinale, F.G., et al., Matrix metalloproteinase inhibition during the development of congestive heart failure : effects on left ventricular dimensions and function. Circ Res, 1999. 85: p. 364-76.

Chapter 4

Pharmacology of G-Protein-linked Signaling in Cardiac Fibroblasts

Sara A. Epperson, Åsa B. Gustafsson, Annette M. Gonzalez, Sonia Villegas,
J. Gary Meszaros and Laurence L. Brunton.
University of California San Diego, San Diego, California, U.S.A.

1. Introduction

Cardiac fibroblasts (CFs) constitute about two-thirds of the heart by cell number and about one-sixth by cell mass [1]. CFs have a number of properties with the potential to be pharmacologically regulated to beneficially alter the progress of cardiac disease including cell growth and division, production and remodeling of extracellular matrix, generation of secreted or diffusable local signaling molecules, and mobility [2]. Our lab, amongst many others, has evaluated the capacity of adult rat ventricular fibroblasts to respond to hormones acting via G-protein coupled receptors and to produce cellular messengers such as cyclic AMP, inositol phosphates, NO, and cyclic GMP and to mobilize Ca^{++} [3-10]. One aim of this work has been to characterize the pharmacology of the cardiac fibroblast, suggesting what agents, acting via what mechanisms, could be useful in altering some of these fundamental properties of the cardiac fibroblast. In this article, we review the major receptor-second messenger systems that are operative in the isolated adult rat CF, supported by some recent data from our lab.

2. Characterization of Cardiac Fibroblast Cultures

Most biochemical work on cardiac fibroblasts has been carried out on cells isolated from the hearts of adult male rats. Research papers are not uniformly careful to report the purity and age of these cultures or the precise conditions of their use in experiments. For instance, it is difficult to know how much the purity of CFs from various laboratories differs and to what extent cell purity affects reported data. Since fibroblasts are small and generally isolated by enzymic digestion of ventricular strips in the presence of normal Ca^{++}, contamination with myocytes is usually not a problem; myocytes do not survive a digestion in the presence of millimolar Ca^{++} and, in any event, CFs can be moderately well separated by sedimentation from the less bouyant myocytes and considerably enriched by a period of brief attachment and subsequent removal of less-adherent cells. Smooth muscle cells and endothelial cells could pose a problem. Endothelial cells, in particular, exhibit large accumulations of second messengers [4], so that a modest contamination of a fibroblast population with endothelial cells could give a mistaken interpretation of fibroblast responses. We have assessed the purity of our cardiac fibroblast cultures by immunofluorescent staining with anti-vimentin, anti-Von Willebrand factor, and anti-α sarcomeric actin [7]. These primary cultures are >95% fibroblasts as judged by positive staining for vimentin and lack of staining for von Willebrand factor and sarcomeric actin; only minimal contamination is observed at passage 1 and it is subsequently eliminated by passaging of the cells. It is worth reminding ourselves that CFs are not a clonal cell line but are, rather, primary isolates that may contain multiple sub-types with differing characteristics. There are some data to support this view, such as that of Klett et al. [11] who obtained different populations of CFs by employing different collagenases in the isolation procedure.

Hormonal responses of CF cultures may fluctuate as a function of cell density and age or passage number. Rather than investigate the details of that variability, we have tried to establish a uniform protocol for cell plating and growth, conducting short-term experiments with cultures that are 70-80% confluent. However, in an experiment in which, for instance, collagen synthesis is assessed over long times (24 to 72 hours), significant cell growth occurs during the experiment and it is necessary to begin with less dense cultures. All cells used in our experiments are from passages 2 through 4. In our experience, the rate of growth of CFs slows as the passage number increases beyond 4 or 5. In addition, we begin to observe enhanced levels of staining for α-smooth muscle actin in these older cells, as though either a small population of smooth muscle cells is expanding or the CFs are adopting a myo-fibroblast like phenotype.

CFs are generally grown in Dulbecco's modified Eagle's medium with 5% or 10% fetal calf serum; in other words, a carefully defined medium is perverted by the addition of serum, with all the growth factors and metabolites, vitamins and amino acids that serum contains. As vital as the serum is for growth, it can confound studies of hormone action, since serum contains traces of many hormones (e.g., lysophosphatidic acid, cytokines, vasoactive peptides, insulin, catecholamines, nucleotides) that can activate and down-regulate signaling pathways. Both serum and common growth media contain small molecules (e.g., myo-inositol) that can dilute the specific radioactivity of compounds with which CFs may be pulsed. We generally precede short term experiments with a 2 hour incubation in serum-free conditions but the literature is not consistent on this point and it is not always possible to judge this detail from the *Methods* sections of published work. Some workers refer to serum starvation but actually include 0.5% heat inactivated serum or bovine serum albumin; indeed, in long-term assessments such as collagen synthesis it is probably necessary to include limited amounts of serum.

Thus, we cannot be sure that the purity, age, density and conditions of assessment of CFs in the cited reports are similar; these are variables that users of primary cultures of CFs should seek to control and make explicit in research papers. In addition, we should realize that CFs in culture may not perfectly represent CFs *in situ* and that CFs in diseased heart may exhibit different phenotypes under the influence of mechanical, endocrine, paracrine and autocrine factors relating to the disease process.

3. Receptor-Transducer-Effector Coupling in Intact Cardiac Fibroblasts

The classification of receptors that couple to G_s is generally based on the functional assessment of cyclic AMP accumulation in response to a short exposure to an agonist. The extent of cyclic AMP accumulation is markedly attenuated by the high capacity of the cyclic AMP degradative enzymes (phosphodiesterases or PDEs) in CFs. To be certain that an agonist acts to increase cyclic AMP synthesis and does not simply reduce degradation, we recommend conducting a series of experiments in the presence of an inhibitor of phosphodiesterases, such as isobutylmethylxanthine (IBMX, a non-specific inhibitor). We find that 1 mM IBMX is needed to provide maximal cyclic AMP accumulation in response to isoproterenol in CFs (Figure 1). Similar experiments comparing the efficacies of isoform-specific PDE inhibitors suggest the approximate distribution of PDE isoforms in rat CFs (Table I): the functionally predominant PDE in rat CFs is PDE4 (inhibited by rolipram). Adenosine and its analogues are important exceptions: many PDE inhibitors are methylxantine congeners and may also antagonize the interaction of adenosine with its membrane receptors; thus, a modest concentration of non-methylxanthine inhibitor is preferred in experiments with adenosine; the optimal concentration of PDE inhibitor can be determined from a concentration-dependence curve at constant adenosine (minimal concentration that is maximally effective).

For a quantitative assessment of cyclic AMP production, it is necessary to account for the fraction of cellular cyclic AMP (up to 20% in normal cells) that is extruded and recovered in the extracellular medium [12-14]. PDE inhibitors and a variety of drugs will inhibit this extrusion [12].

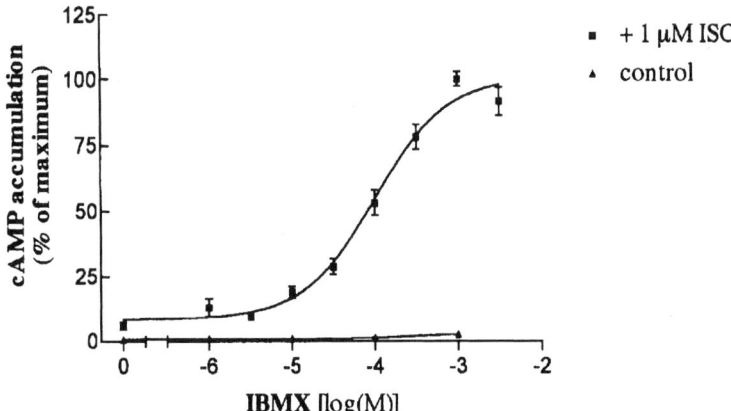

Figure 1. Concentration-dependent increase of isoproterenol-stimulated cyclic AMP accumulation by IBMX. Cardiac fibroblasts were treated with IBMX for 15 minutes and then stimulated with 1 μM isoproterenol (ISO) for 5 minutes. The cyclic AMP content was determined as described (5, 6). Data are mean±SEM of three experiments and are expressed as a percentage of maximal cAMP accumulation, ~1250 pmol/mg protein.

Table I. Distribution of PDE activities in rat cardiac fibroblasts

ISOFORM	INHIBITOR	FRACTIONAL EFFECT
	(concentration)	(vs. 1 mM IBMX)
PDE1	vinpocentine (25 μM)	0
PDE2	EHNA (200 μM)	0.08
PDE3	milrinone (10 μM)	0.01
PDE4	rolipram (10 μM)	0.62
PDE5	MBCQ (10 μM)	0

The effect of each of the indicated inhibitors was compared to the effect (cyclic AMP accumulation in a 5 minute treatment with 1 mM isoproterenol) of IBMX as described in legend to figure 1. RT-PCR analysis demonstrated transcripts for PDE isoforms 2A, 3A, 3B, 4A, 4B and 4D (8).

In CFs as in many cells, extruded cAMP may be converted to adenosine by the sequential actions of extracellular PDEs and 5'-nucleotidase [15]. The adenosine thus produced can act as an autocrine/paracrine modulator of adenosine receptors, several of which (A_{2a} and A_{2b}) may couple to G_s. For an idea of the physiologic consequences of receptor-G_s activation, no PDE inhibitor should be used. Given some of the confounding features involving PDEs and autocrine/paracrine factors, we recommend that receptor-G_s-adenylyl cyclase coupling be verified in an assay of adenylyl cyclase activity in isolated plasma membranes.

The presence of receptors coupling to G_i is readily judged by the effect of an agonist to reduce the activity of stimulated adenylyl cyclase (hormone or forskolin); such an effect should be sensitive to prior treatment of the cells with pertussis toxin, which prevents activation of G_i. Activation of G_i can also lead to activation of MAPK and ERK1/2 through pertussis toxin-sensitive and -insensitive pathways not yet fully defined; pertussis toxin-sensitive activation of these enzymes can be used as evidence of the involvement of $G_{i/o}$. There seem to be no receptors on CFs that couple to G_i to inhibit adenylyl cyclase although several will activate MAPK/ERK family members through G_i/G_o (pertussis toxin-sensitive). However, other agents, such as adenosine, appear to activate ERK1/2 via a cyclic AMP-dependent mechanism [16].

The existence of receptors that couple to G_q is generally based on the functional assessment of [^{3}H] inositol phosphate accumulation in response to an agonist added to [^{3}H]inositol-labeled cells, ascribed to receptor-G_q-phospholipase C (PLC) coupling. The resulting mobilization of intracellular Ca^{++} can be assessed, secondarily, by fluorescent techniques (see, for instance, references 5 and 6). For a sensitive assessment of PLC activation, it is important to include 10 mM LiCl to reduce the dephosphorylation of inositol monophosphate. The activation of receptors coupling to the G_q-PLC pathway can also result in the production and/or release of a variety of autocrine/paracrine factors such as eicosanoids and ATP. For instance, the diacylglycerol produced when PLC removes inositol polyphosphate from phosphatidylinositol is a good substrate for PLA_2 (that may be activated by Ca^{++}), as is phosphatidic acid. Any C_{20} fatty acids released by PLA_2 may become substrates for eicosanoid synthesis. The resulting PGE_2 and PGI_2 can activate receptors linked to G_s, stimulating cyclic AMP synthesis.

4. G-protein Coupled Receptors on Cardiac Fibroblasts

Applying these techniques, a number of laboratories have defined receptor-G-protein coupling in the adult rat cardiac fibroblast in culture. Table II summarizes much of this information and compares it with receptor distribution and coupling in adult rat cardiac myocytes. CFs have G_s-linked

Table II. Functional second messenger signaling: rat cardiac fibroblasts vs. myocytes

FUNCTIONAL SECOND MESSENGER SIGNALING:
RAT CARDIAC FIBROBLASTS vs MYOCYTES

Ligand/Receptor	Fibroblasts	Myocytes	References
G_q signaling (IP synthesis)			
BK	moderate	large	5,6,40
ET-1	negligible	large (ET_A)	34-36,39,#
AT1	large	negligible	5,6,36
P2Y (ATP, UTP)	large	negligible	5,17-20,#
Muscarinic	negligible	moderate	5,42
α_1-adrenergic	negligible	large	5,41-43
LPA	--	moderate	16,#
$PGF_{2\alpha}$	moderate	--	#
Thrombin	moderate	moderate	45
G_s signaling (cyclic AMP synthesis)			
β-adrenergic	large (β_2)	moderate (β_1)	5-7,44
PGE_2	moderate	negligible	5-7,#
Adenosine (A_2)	moderate	--	16
CGRP	moderate	small	#
G_i signaling (reduced cyclic AMP synthesis)			
Muscarinic	negligible	large	5,42,43
α_2-adrenergic	negligible	negligible	5,41-43
ET-1	negligible	large	16,34-36,#
LPA	negligible	moderate	16,46,#

The terms used, in order of size of effect are: negligible, small, moderate and large.
A line (--) indicates no data cited at this time. (#, unpublished observations, Brunton et al.)

receptors for catecholamines (β_2), prostaglandins (PGE_2), adenosine (A_{2a} and A_{2b}) and calcitonin gene-related peptide (CGRP). In relation to catecholamines, the main distinctions between CFs and myocytes are the type of β receptor and the negligible effects of PGEs on cyclic AMP production in the rat myocyte. If this difference in β receptor distribution is also true in humans, it could provide for substantial activation of cyclic AMP production (and its sequelae, reduced synthesis of collagen and reduced cell growth) [10] in CFs and concomitant antagonism of the β_1 positive inotropic effects of catecholamines in patients treated with β_1-antagonists (atenolol, metoprolol), a potentially useful combination of effects in heart failure and hypertension.

We have detected no receptors on CFs that couple to G_i to reduce the rate of stimulated cyclic AMP production [5,16]. This is in contrast to myocytes

Table III. Activation of MAP Kinases in Rat Cardiac Fibroblasts

Agonist	Extent	References
Pertussis toxin-sensitive activation of MAPK or ERK 1/2		
LPA	large	16
ET-1	moderate	#
muscarinic	moderate	#
adenosine	negligible	16
Pertussis toxin-insensitive activation of MAPK or ERK1/2		
LPA	negligible	16
ET-1	moderate	#
muscarinic	moderate	#
adenosine	large	16

The terms used, in order of size of effect are: negligible, small, moderate and large.
(#, unpublished observations, Brunton et al.)

Table IV. Cyclic GMP and NO Synthesis: Cardiac Fibroblasts vs. Myocytes

Ligand	Fibroblast	Myocyte	Reference
Nitric oxide production			
G_q-PLC-IP-Ca^{++} coupled	negligible	small	3,4,#
Ionomycin	negligible	small	3,#
Interleukin-1β (24 h induction of iNOS)	large	moderate	4,7,8
Cyclic GMP production			
ANP	large	small	3,4,#
Guanylin	small	small	3,4,#
Nitroprusside	large	moderate	3,4,#

The terms used, in order of size of effect are: negligible, small, moderate and large.
(#, unpublished observations, Brunton et al.)

where endothelin-1 (ET-1), lysophosphatidic acid (LPA) and muscarinic cholinergic agonists couple to G_i-linked receptors and thence to adenylyl cyclase in a pertussis toxin-sensitive manner. Nonetheless, there is evidence of functional $G_{i/o}$ in CFs as summarized in Table III. LPA, ET-1 and carbachol enhance the phosphorylation of ERK 1/2 and/or MAPK. Prior treatment with pertussis toxin abolishes the effect of LPA and reduces the effects of endothelin and carbachol, suggesting that activation of ERK/MAPK occur all (LPA) or in part (ET-1, carbachol) via $G_{i/o}$. ET-1 and carbachol also activate ERK/MAPK by a pertussis toxin-insensitive mechanism. Adenosine, acting via A_2 receptors that couple to G_s-adenylyl cyclase, leads to enhanced cyclic AMP synthesis and activates MAPK in CFs; 8-Br-cyclic AMP and forskolin mimic this effect of adenosine [16].

With respect to G_q-PLC-IP-Ca^{++} signaling in CFs, there are no effects in response to neurotransmitters of the autonomic nervous system (norepinephrine and acetylcholine) that affect myocytes and pacemaker cells in the heart so dramatically. However, the peptides bradykinin and angiotensin II activate prominent IP-Ca^{++} responses in CFs, as do ATP and UTP [5]. ATP is stored and released with norepinephrine from adrenergic storage granules; in addition, there is evidence that under certain conditions many types of cells, including endothelial, will secrete ATP; thus, these P2Y receptors on CFs probably do not lack ligand, physiologically. In our hands, BK and Ang II do not activate G_q-linked signaling in cardiac myocytes. Although extracellular ATP can increase intracellular Ca^{++} in ventricular myocytes, this effect seems to be independent of G_q-PLC signaling [17-20]. A prominent feature of the G_q-linked responses in CFs is the speed with which some of them are down-regulated. For example, the data of Meszaros et al. [6] demonstrate that the G_q-linked responses to Ang II fully desensitize within 30 seconds, much more rapidly than G_q-linked responses to agonists of P2Y receptors. Pharmacologic evidence indicates that this rapid desensitization is mediated by the activity of protein kinase C.

5. Nitric Oxide and Cyclic GMP

Rat CFs in culture express functional receptors for activating ligands of both membrane-bound and soluble guanylyl cyclases (see Table IV) [3, 4]. Atrial natriuretic peptide stimulates a large accumulation of cyclic GMP; guanylin stimulates a smaller accumulation, demonstrating the presence of the membrane monospanning guanlylyl cyclase. There is no response to muscarinic agonists although NO donors do cause cyclic GMP accumulation, indicating the presence of the soluble guanylyl cyclase. CFs also turn out to be able producers of NO, although they differ from the textbook example whereby Ca^{++} activates a Ca^{++}/CaM sensitive form of eNOS. In rat CFs, (as opposed to myocytes) elevation of cellular Ca^{++} (with ionomycin or in response to a G_q-linked agonist such as ATP, Ang II or BK) is ineffective until the enzyme has been phosphorylated by a cyclic AMP-dependent means (unpublished observations). Interleukin-1β (IL-1β) and the combination of tumor necrosis factor- α (TNF-α) and interferon-γ (IFN-γ) are efficacious inducers of iNOS in rat cardiac fibroblasts, a process that β-agonists and other stimulators of cyclic AMP accumulation enhance, via a PKA-dependent stabilization of the mRNA for iNOS [7]. The interactions of the cyclic AMP and NO/cyclic GMP pathways occur at several levels, an immediate activation by cyclic GMP of PDE2, that lowers cyclic AMP accumulation, and a long term enhancement of iNOS induction by cyclic AMP-PKA (see Figure 2). The capacity of CFs to make NO is quite large and may effect neighboring cells such as myocytes.

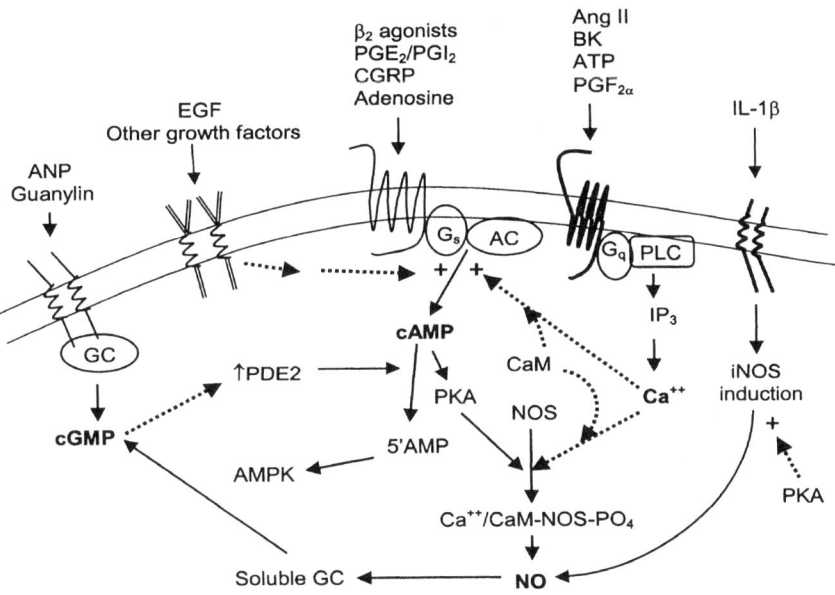

Figure 2. Signaling crosstalk in rat cardiac fibroblasts. The figure shows examples of crosstalk amongst pathways in cardiac fibroblasts (CFs). In addition, conditions that activate phospholipase C often activate phospholipase A_2, with concomitant production of eicosanoids, opening the possibility of local autocrine loops. Furthermore, many $\beta\gamma$ heterodimers of G-proteins transactivate the EGF receptor, which may enhance receptor-activated G_s adenyl cyclase activity in addition to initializing P-tyr/SH2-dependent signaling. See text and Figure 3 for information on AMPK.

6. Crosstalk

Mammalian cells are not stimulated by a single hormone except in carefully contrived experiments *in vitro*. When we have attempted to simulate the simultaneous activation of receptors linked to G_s and G_q, we have found a web of interactions (Figure 2). Activation of the G_q pathway (for instance, with

UTP, ATP or angiotensin II) enhances activation of the G_s-AC pathway (doubling cyclic AMP accumulation), at least partially via an effect of Ca^{++} on adenylyl cyclase-3, the Ca^{++}/CaM-sensitive form of adenylyl cyclase [10]. The converse is without effect: activation of the G_s pathway does not alter G_q-PLC activation and IP production. The combination of elevated intracellular cyclic AMP and mobilization of intracellular Ca^{++} leads to activation of eNOS. The resultant NO activates the soluble guanylyl cyclase, and the cyclic GMP produced activates PDE2, reducing cellular accumulation of cyclic AMP. Induction of iNOS achieves the same effect and cyclic AMP, acting via PKA, enhances iNOS induction by stabilizing iNOS mRNA (doubles half life) [7].

It is worth speculating further on the role of cyclic AMP breakdown by PDE activities. As Figure 1 demonstrates, the production of cyclic AMP in maximally stimulated CFs is quite large, around 1250 pmol/mg (in a 5 minute treatment with a β-agonist) but accumulation is normally limited by PDE activities to about 48 pmol/mg; the rest is metabolized to 5'AMP. It seems possible that this large quantity of AMP (~1202 pmol/mg), could, despite further metabolism of AMP, achieve sufficient concentration to activate the AMP-activated protein kinase, AMPK (see Figure 2 and reference 39). Indeed, assuming that cells are 70% water and 10% protein, we calculated that 1200 pmol/mg of 5'AMP is roughly 171 μM. The K_{act} for AMPK by 5'AMP is 1-10 μM. Thus, a fraction of the metabolized cyclic AMP (about 5%) could half-maximally activate AMPK. This would provide for cyclic AMP-dependent, PKA-independent actions of cyclic AMP (see Figure 3), challenging the dogma that PKA is the exclusive mediator of the actions of cyclic AMP.

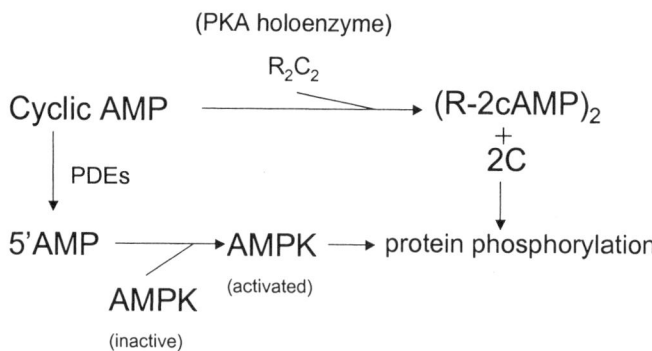

Figure 3. PKA-dependent and PKA-independent actions of cAMP: Potential role of AMP kinase. In cardiac fibroblasts and other cells with a large capacity to synthesize and destroy cAMP, it seems plausible that cAMP could activate both PKA and AMPK. See text for numerical estimates.

7. Regulation of the Cardiac Fibroblast Function by G-protein Linked Receptors

Following a myocardial infarction, the activation of CFs to replace the dead myocytes with scar tissue is a positive feature of the healing process. Similarly, if the heart is remodeling in response to increased demand, it is appropriate for the CFs to keep up. Unfortunately, the increase in CF cell division and the net deposition of extracellular matrix (ECM) may occur in excess, leading to conditions where CFs and ECM interrupt appropriate electrical and mechanical connections between cardiac myocytes. It seems likely that many of the beneficial and adverse properties of CFs could be therapeutically regulated by stimulating and inhibiting G-protein linked receptors.

Angiotensin II is known as a prominent stimulant of CF functions [2]. Signaling is mediated through AT_1 receptors linked to G_q, and stimulates JAK-STAT, p38 MAPK and ERK 1/2 activities, proliferation, motility, secretion of growth factors and net accumulation of ECM [21-29,47,48,50-53,55,57,59]. Ang II enhances synthesis of mRNAs and/or production of growth factors, collagens I and III, and other ECM proteins. Blockade of AT_1 receptors suppresses CF migration, proliferation and ECM synthesis in response to Ang II. Blockade of AT_2 receptors is reported to have minimal effects or to further enhance the AT_1 effect. An unresolved issue is why bradykinin, which shares G_q coupling with Ang II, is not also pro-fibrotic; rather it is generally reported as anti-fibrotic [30, 31].

β-adrenergic stimulation may be hypertrophic or hypotrophic, pro-fibrotic or anti-fibrotic. Repeated exposure of animal models to a β-agonist such as isoproterenol leads to cardiac hypertrophy. However, in studies using isolated rat CFs, we found that the elevation of cyclic AMP by isoproterenol reduces net collagen production [10]. We expect that this apparent paradox is explained on the basis of multiple sites of isoproterenol action *in vivo*. *In vitro*, with a culture of CFs, the action of isoproterenol is limited to the stimulation of β_2 receptors with reduction in net collagen synthesis [10]. *In vivo*, not only are the β_2 receptors of the CFs stimulated but so are receptors found on endothelial and smooth muscle cells, potentially giving rise to autocrine and paracrine factors that could influence CF activities. Furthermore, β_1 stimulation of myocytes may produce locally active factors and stimulation of β_1 receptors on juxtaglomerular cells in the kidney will enhance renin secretion. Increased plasma renin activity would lead to increased hydrolysis of angiotensinogen and increased levels of Ang II, which would lead to the anticipated effects of Ang II on CF functions.

A number of other hormones affect CF functions. Some effects of IL-1β on iNOS induction have been noted above. Transforming growth factor-β1, markedly over-expressed under conditions of cardiac remodeling, increases net collagen types I and III accumulation [32,49]. Endothelin-1 receptors (ET_A and ET_B) have been detected on CFs [33] although as noted above we do not detect coupling to G_q-PLC or G_i-AC, as we do in rat cardiac myocytes [34-36]. ET-1 activates ERK 1/2 and MAPK in CFs (through both pertussis toxin-sensitive and pertussis toxin-insensitive mechanisms, Table III), yielding enhanced production of collagens I and III, DNA synthesis and cell proliferation [51,54,56,59]. Atrial natriuretic peptide reportedly inhibits the proliferative response to Ang II via inhibition of ET_1 gene expression [53]. A number of laboratories report that CFs produce factors that have trophic effects on myocytes [57,58]. Activation of P2Y receptors on CFs activates c-fos gene expression and reduces overall DNA synthesis [60].

The effects of activating these signaling pathways are further discussed elsewhere in this volume, and some of those effects may seem disparate, possibly reflecting variable cross-talk. For instance, Ostrom and colleagues report that $β_2$-adrenergic activation reduces collagen synthesis in serum activated CFs [10]. On the other hand, norepinephrine reportedly activates STAT activity and proliferation of CFs, possibly via induction of IL-6 [37]. Koch and colleagues have demonstrated that the addition of a β-agonist to serum-starved cells stimulates ERK phosphorylation and DNA synthesis in a manner that inhibitors of EGFR activity and pp60 src attenuate [38]; ERK activation seems to occur via a PI3 kinase pathway and the adrenergic stimulation of mitogenesis in CFs apparently occurs via $β_2$-mediated transactivation of the EGFR tyrosine kinase pathway.

Additional hormones that may regulate CF function via G-protein linked receptors include vasopressin and adrenomedullin [61,62]. In neonatal rat CFs, adrenomedullin seems to act by elevating cAMP, thereby antagonizing responses to Ang II [62].

6. Conclusions

The responses of CFs to individual hormones are beginning to be cataloged (Tables II, III and IV). It is not yet known whether CFs from subjects with cardiovascular diseases differ from "normal" CFs. Actions of hormones are likely much more complex *in vivo* than *in vitro*, as the example of β-adrenergic agonists makes clear. In addition, cross-talk amongst signaling pathways when cells are stimulated by multiple agonists complicates the picture and the overall responses. Moreover, although elevated cyclic AMP reduces collagen synthesis by CFs in culture, other components of the ECM, such as fibronectin, seem to be derived from genes with cyclic AMP response elements (CREs). Despite these complexities we believe that activation and blockade of cell surface receptors, especially G-protein linked, could be fruitful way of manipulating CF function *in vivo*.

References

1.	Eghbali, M., Cardiac fibroblasts: function, regulation of gene expression, and phenotypic modulation. Basic Res Cardiol, 1992. 87: p. 183-9.
2.	Villarreal, F. and N. Kim, Regulation of myocardial extracellular matrix components by mechanical and chemical growth factors. Cardiovasc Pathol, 1998. 7: p. 145-151.
3.	Villegas, S. and L.L. Brunton, Separation of cyclic GMP and cyclic AMP. Anal Biochem, 1996. 235: p. 102-3.
4.	Villegas, S. and L.L. Brunton, Cellular localization and characterization of cyclic GMP production in rat heart. Cardiovasc Pathobiol, 1996. 1: p. 5-12.
5.	Meszaros, J.G., et al., Identification of G protein-coupled signaling pathways in cardiac fibroblasts: cross talk between G_q and G_s. Am J Physiol, 2000. 278: p. C154-62.
6.	Meszaros, J.G., et al., Protein kinase C contributes to desensitization of angiotensin II signaling in adult rat cardiac fibroblasts. Am J Physiol, 2000. 279: p. C1978-85.
7.	Gustafsson, Å.B. and L.L. Brunton, Regulation of the expression of inducible nitric oxide synthesa in adult rat cardiac fibroblasts by IL-1β and β-adrenergic stimulation. Mol Pharmacol, 2000. 58: p. 1470-1478.
8.	Gustafsson, Å.B. and L.L. Brunton, Attenuation of cAMP accumulation in adult rat cardiac fibroblasts by IL-1β and NO: role of cGMP-stimulated PDE2. Am J Physiol, 2002. 283p. C463-71.
9.	Gustafsson, Å.B., Interactions of the cyclic AMP and nitric oxide pathways in cardiac fibroblasts. Dissertation, 2001.
10.	Ostrom, R.S., et al., Angiotensin II enhances adenylyl cyclase signaling via Ca^{2+}/calmodulin. G_q-G_s cross-talk regulates collagen production in cardiac fibroblasts. J Biol Chem, 2003. 278: p. 24461-8.
11.	Klett, C.P., et al., Evidence for differences in cultured left ventricular fibroblast populations isolated from spontaneously hypertensive and Wistar-Kyoto rats. J Hypertens, 1995. 13: p. 1421-31.
12.	Brunton, L.L. and S.E. Mayer, Extrusion of cyclic AMP from pigeon erythrocytes. J Biol Chem, 1979. 254: p. 9714-20.
13.	Brunton, L.L., and Buss J.E., Export of cyclic AMP by mammalian reticulocytes. J Cyclic Nucleotide Res, 1980. 6: p. 369-77.
14.	Heasley, L.E. and L.L. Brunton, Prostaglandin A_1 metabolism and inhibition of cyclic AMP extrusion by avian erythrocytes. J Biol Chem, 1985. 260: p. 11514-9.
15.	Dubey, R.K., et al., Cardiac fibroblasts express the cAMP-adenosine pathway. Hypertension, 2000. 36: p. 337-42.
16.	Epperson, S., F. Villarreal, and L.L. Brunton, Adenosine receptor signaling in cardiac fibroblasts. J Mol Cell Cardiol, 2004. 36: p. A628.
17.	Zheng, J.S., et al., Ca^{2+} mobilization by extracellular ATP in rat cardiac myocytes: regulation by protein kinase C and A. Am J Physiol, 1992. 263: p. C933-40.
18.	Puceat, M., et al., Extracellular ATP-induced acidification leads to cytosolic calcium transient rise in single rat cardiac myocytes. Biochem J, 1991. 274: p. 55-62.
19.	Song, Y. and L. Belardinelli, ATP promotes development of after-depolarizations and triggered activity in cardiac myocytes. Am J Physiol, 1994. 267: p. H2005-11.
20.	Yang, Z. and D.S. Steele, Effects of cytosolic ATP on Ca^{2+} sparks and SR Ca^{2+} content

in permeabilized cardiac myocytes. Circ Res, 2001. 89: p. 526-33.

21. Booz, G.W., et al., Involvement of protein kinase C and Ca^{2+} in angiotensin II-induced mitogenesis of cardiac fibroblasts. Am J Physiol, 1994. 267: p. C1308-18.

22. Schorb, W., et al., Angiotensin II is a potent stimulator of MAP-kinase activity in neonatal rat cardiac fibroblasts. J Mol Cell Cardiol, 1995. 27: p. 1151-60.

23. Booz, G.W. and K.M. Baker, Molecular signalling mechanisms controlling growth and function of cardiac fibroblasts. Cardiovasc Res, 1995. 30: p. 537-43.

24. Dostal, D.E., et al., Molecular mechanisms of angiotensin II in modulating cardiac function: intracardiac effects and signal transduction pathways. J Mol Cell Cardiol, 1997. 29: p. 2893-902.

25. Villarreal, F.J., et al., Identification of functional angiotensin II receptors on rat cardiac fibroblasts. Circulation, 1993. 88: p. 2849-61.

26. Campbell, S.E. and L.C. Katwa, Angiotensin II stimulated expression of transforming growth factor-beta1 in cardiac fibroblasts and myofibroblasts. J Mol Cell Cardiol, 1997. 29: p. 1947-58.

27. Ahmed, M.S., et al., Connective tissue growth factor--a novel mediator of angiotensin II-stimulated cardiac fibroblast activation in heart failure in rats. J Mol Cell Cardiol, 2004. 36: p. 393-404.

28. van Kesteren, C.A., et al., Angiotensin II-mediated growth and antigrowth effects in cultured neonatal rat cardiac myocytes and fibroblasts. J Mol Cell Cardiol, 1997. 29: p. 2147-57.

29. Lijnen, P.J., V.V. Petrov, and R.H. Fagard, Angiotensin II-induced stimulation of collagen secretion and production in cardiac fibroblasts is mediated via angiotensin II subtype 1 receptors. J Renin Angiotensin Aldosterone Syst, 2001. 2: p. 117-22.

30. Kim, N.N., et al., Regulation of cardiac fibroblast extracellular matrix production by bradykinin and nitric oxide. J Mol Cell Cardiol, 1999. 31: p. 457-66.

31. Villarreal, F.J., T. Bahnson, and N.N. Kim, Human cardiac fibroblasts and receptors for angiotensin II and bradykinin: a potential role for bradykinin in the modulation of cardiac extracellular matrix. Basic Res Cardiol, 1998. 93: p. 4-7.

32. Villarreal, F.J. and W.H. Dillmann, Cardiac hypertrophy-induced changes in mRNA levels for TGF-beta 1, fibronectin, and collagen. Am J Physiol, 1992. 262: p. H1861-6.

33. Fareh, J., et al., Endothelin-1 and angiotensin II receptors in cells from rat hypertrophied heart. Receptor regulation and intracellular Ca^{2+} modulation. Circ Res, 1996. 78: p. 302-11.

34. Hilal-Dandan, R., et al., Coupling of the type A endothelin receptor to multiple responses in adult rat cardiac myocytes. Mol Pharmacol, 1994. 45: p. 1183-90.

35. Hilal-Dandan, R., et al., Endothelin ET_A receptor regulates signaling and ANF gene expression via multiple G protein-linked pathways. Am J Physiol, 1997. 272: p. H130-7.

36. Hilal-Dandan, R., K. Urasawa, and L.L. Brunton, Endothelin inhibits adenylate cyclase and stimulates phosphoinositide hydrolysis in adult cardiac myocytes. J Biol Chem, 1992. 267: p. 10620-4.

37. Yin, F., et al., Interleukin-6 family of cytokines mediates isoproterenol-induced delayed STAT3 activation in mouse heart. J Biol Chem, 2003. 278: p. 21070-5.

38. Kim, J., et al., Beta-adrenergic receptor-mediated DNA synthesis in cardiac fibroblasts is dependent on transactivation of the epidermal growth factor receptor and subsequent activation of extracellular signal-regulated kinases. J Biol Chem, 2002. 277: p. 32116-23.

39. Kemp, B.E., et al., Dealing with energy demand: the AMP-activated protein kinase. Trends Biochem Sci, 1999. 24: p. 22-5.

40. Minshall, R.D., et al., Characterization of bradykinin B2 receptors in adult myocardium and neonatal rat cardiomyocytes. Circ Res, 1995. 76: p. 773-80.

41.   Buxton, I.L. and L.L. Brunton, Alpha-adrenergic receptors on rat ventricular myocytes: characteristics and linkage to cAMP metabolism. Am J Physiol, 1986. 251: p. H307-13.

42.   Buxton, I.L. and L.L. Brunton, Action of the cardiac alpha 1-adrenergic receptor. Activation of cyclic AMP degradation. J Biol Chem, 1985. 260: p. 6733-7.

43.   Brown, J.H., I.L. Buxton, and L.L. Brunton, Alpha 1-adrenergic and muscarinic cholinergic stimulation of phosphoinositide hydrolysis in adult rat cardiomyocytes. Circ Res, 1985. 57: p. 532-7.

44.   Buxton, I.L. and L.L. Brunton, Direct analysis of beta-adrenergic receptor subtypes on intact adult ventricular myocytes of the rat. Circ Res, 1985. 56: p. 126-32.

45.   Steinberg, S.F., Protease activated receptors in heart. Mol Pharmacol, 2004. in press.

46.   Hilal-Dandan, R., et al., Lysophosphatidic acid induces hypertrophy of neonatal cardiac myocytes via activation of G_i and Rho. J Mol Cell Cardiol, 2004. 36: p. 481-93.

47.   Crabos, M., et al., Characterization of angiotensin II receptors in cultured adult rat cardiac fibroblasts. Coupling to signaling systems and gene expression. J Clin Invest, 1994. 93: p. 2372-8.

48.   Dostal, D.E., et al., Detection of angiotensin I and II in cultured rat cardiac myocytes and fibroblasts. Am J Physiol, 1992. 263: p. C851-63.

49.   Eghbali, M., et al., Differential effects of transforming growth factor-beta 1 and phorbol myristate acetate on cardiac fibroblasts. Regulation of fibrillar collagen mRNAs and expression of early transcription factors. Circ Res, 1991. 69: p. 483-90.

50.   Everett, A.D., F. Heller, and A. Fisher, AT_1 receptor gene regulation in cardiac myocytes and fibroblasts. J Mol Cell Cardiol, 1996. 28: p. 1727-36.

51.   Fareh, J., et al., Endothelin-1 and angiotensin II receptors in cells from rat hypertrophied heart. Receptor regulation and intracellular Ca^{2+} modulation. Circ Res, 1996. 78: p. 302-11.

52.   Fischer, T.A., et al., Role of AT_1 and AT_2 receptors in regulation of MAPKs and MKP-1 by ANG II in adult cardiac myocytes. Am J Physiol, 1998. 275: p. H906-16.

53.   Fujisaki, H., et al., Natriuretic peptides inhibit angiotensin II-induced proliferation of rat cardiac fibroblasts by blocking endothelin-1 gene expression. J Clin Invest, 1995. 96: p. 1059-65.

54.   Guarda, E., et al., Effects of endothelins on collagen turnover in cardiac fibroblasts. Cardiovasc Res, 1993. 27: p. 2130-4.

55.   Ishihata, A. and M. Endoh, Species-related differences in inotropic effects of angiotensin II in mammalian ventricular muscle: receptors, subtypes and phosphoinositide hydrolysis. Br J Pharmacol, 1995. 114: p. 447-53.

56.   Katwa, L.C., E. Guarda, and K.T. Weber, Endothelin receptors in cultured adult rat cardiac fibroblasts. Cardiovasc Res, 1993. 27: p. 2125-9.

57.   Kim, N.N., et al., Trophic effects of angiotensin II on neonatal rat cardiac myocytes are mediated by cardiac fibroblasts. Am J Physiol, 1995. 269: p. E426-37.

58.   Long, C.S., C.J. Henrich, and P.C. Simpson, A growth factor for cardiac myocytes is produced by cardiac nonmyocytes. Cell Res, 1991. 2: p. 1081-95.

59.   Touyz, R.M., et al., Intracellular Ca^{2+} modulation by angiotensin II and endothelin-1 in cardiomyocytes and fibroblasts from hypertrophied hearts of spontaneously hypertensive rats. Hypertension, 1996. 28: p. 797-805.

60.   Zheng, J.S., et al., Stimulation of P2Y receptors activates c-fos gene expression and inhibits DNA synthesis in cultured cardiac fibroblasts. Cardiovasc Res, 1998. 37: p. 718-28.

61.   Yang, X.D., et al., Effects of arginine vasopressin on growth of rat cardiac fibroblasts: role of V1 receptor. J Cardiovasc Pharmacol, 2003. 42: p. 132-5.

62.   Autelitano, D.J., et al., Adrenomedullin inhibits angiotensin AT1A receptor expression and function in cardiac fibroblasts. Regul Pept, 2003. 112: p. 131-7.

II. *DIAGNOSTIC TOOLS FOR THE IDENTIFICATION OF MYOCARDIAL FIBROSIS AND CHANGES IN VENTRICULAR MECHANICS*

Chapter 5

Serum Markers of Fibrillar Collagen Metabolism in Cardiac Diseases

Javier Diez
Universidad de Navarra, Pamplona, Spain

1. Introduction

A substantial increase in fibrillar collagen types I and III deposition, that leads to interstitial and perivascular fibrosis, is an integral feature of the detrimental structural remodeling of cardiac tissue seen in cardiac diseases, namely hypertensive heart disease [1] and ischemic heart disease [2]. Fibrous tissue accumulation adversely affects ventricular and diastolic function, coronary vasomotor reactivity, and electrical stability of the myocardium [3, 4]. Thus, given the importance of fibrosis in leading to myocardial dysfunction and failure, noninvasive monitoring of myocardial fibrosis could prove a clinically useful tool, particularly given the potential for cardioprotective and cardioreparative pharmacological strategies. An emerging experimental and clinical experience holds promise for the use of radioimmunoassays of various serological markers of fibrillar collagen types I and III metabolism in cardiac disease. This approach represents an exciting and innovative strategy and available data set the stage for larger trials, where biochemical monitoring of fibrosis in cardiac diseases could prove useful.

2. Serum Markers of Collagen Types I and III Metabolism

2.1 Metabolism of fibrillar collagen

In physiological conditions, a metabolic equilibrium exists between the synthesis and degradation of fibrillar collagen molecules [5]. Fibrillar collagen is synthesized in the fibroblasts as procollagen containing an amino-terminal and a carboxy-terminal propeptide. Figure 1 illustrates the structure of a procollagen type I molecule [6].

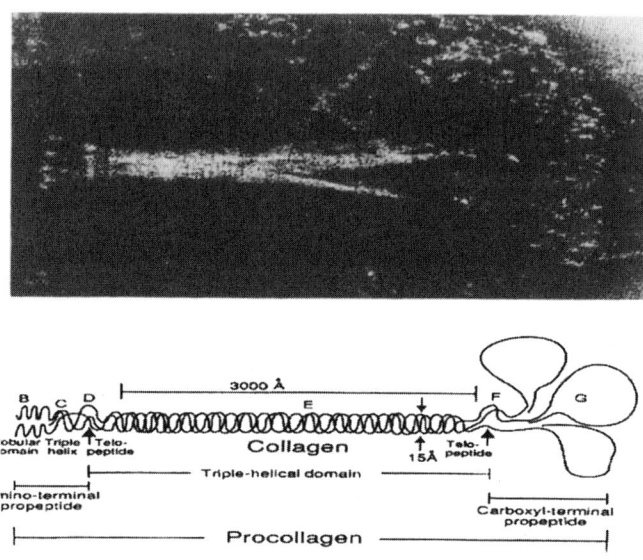

Figure 1. Electron micrograph of segment-long-spacing aggregates of procollagen type I (top) and a model of procollagen type I molecule (bottom). [B, amino-terminal propeptide extension; C, minor triple helix of the amino-terminal protease cleavage site; D, amino-terminal telopeptide; E, triple helix; F, carboxy-terminal telopeptide; G, carboxy-terminal propeptide extension].

After procollagen has been secreted into the extracellular space, the propeptides are removed by specific proteinases, allowing integration of the rigid collagen triple helix into the growing fibril (Figure 2) [6]. As discussed below, these propeptides are released into the bloodstream and can be used of markers of collagen synthesis.

The rate-limiting step in the degradation of collagen fibrils is catalytic cleavage by interstitial collagenase (Figure 2) [7]. This enzyme cleaves all three α chains of collagen at a single, specific locus located at a distance of three-fourths from the amino-terminal. The 2 resulting telopeptides (collagen degradation fragments) maintain their helical structure and are resistant to further proteolytic degradation. Whereas the small telopeptide will be released into the blood stream, the big telopeptide spontaneously denatures into non-helical gelatin derivatives, which in turn, are completely degraded by interstitial gelatinases (Figure 2).

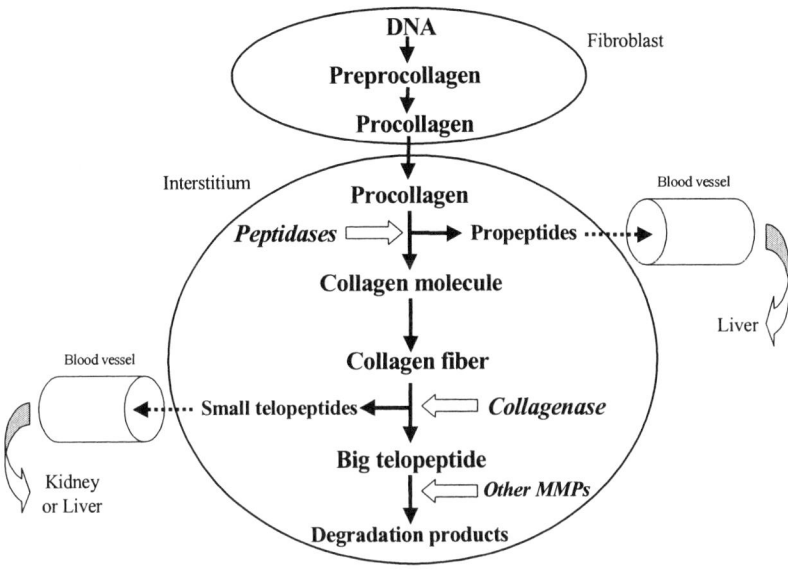

Figure 2. Diagrammatic depiction of the different compartments of fibrillar collagen metabolism. The origin and destination of the serum markers of collagen types I and III synthesis (propeptides) and degradation (telopeptides) are indicated. [MMPs, matrix metalloproteinases]. (Adapted from reference [13]).

2.2. Fibrillar collagen-derived serum peptides

Cumulative evidence suggest that some of these collagen-derived peptides may be of interest for *in vivo* monitoring of collagen types I and type III metabolism [8]. The 100-kDa procollagen type I carboxy-terminal propeptide (PIP) is cleaved from procollagen type I during the synthesis of fibril-forming collagen type I and is released into the blood stream (Figure 2). A stoichiometric ratio of 1:1 exists between the number of collagen type I molecules produced and that of PIP released. The small 12-kDa pyridinoline cross-linked carboxy-terminal telopeptide resulting from the cleavage of collagen type I (CITP) is found in an immunochemically intact form in blood, where it appears to be derived from tissues with a stoichiometric ratio of 1:1 between the number of collagen type I molecules degraded and that of CITP released (Figure 2). The

42-kDa procollagen type III amino-terminal propeptide (PIIIP) is formed during the conversion of procollagen type III into the fibril-forming molecule collagen type III and is also released into the blood stream with a stoichiometry of 1:1. Whereas terminal propeptides of procollagen type I are cleaved off before fibrils are formed, collagen type III molecules may retain the amino-terminal propeptide. Collagenase and other matrix metalloproteinases cleave this propeptide, which is then released into the blood stream with a stoichiometric ratio of 1:1.

2.3. Interpretation of serum concentrations of collagen-derived peptides

Serum concentrations of these peptides depend of several factors, including their rate of release, how they reach the blood (eg, cardiac lymph versus venous drainage) and their volume of distribution [8]. Elimination of a peptide, its potential uptake, and its metabolism are other factors that need to be considered.

Circulating PIP is cleared from the blood by the liver (Figure 2) [9]. Therefore, the serum concentration of PIP can be considered as marker of collagen type I synthesis in conditions of preserved liver function. On the other hand, CITP appears to be cleared from the circulation via glomerular filtration (Figure 2) [10]. Therefore, the serum concentration of CITP can be interpreted as reflecting collagen type I degradation when renal function is normal.

PIIIP is cleared from blood by the liver, thus its serum concentration should be interpreted as expressing both the combination of synthesis and degradation of collagen type III in conditions of normal liver function[11, 12].

Several clinical observations have demonstrated that high serum concentrations of PIP measured by specific radioimmunoassay reflect ongoing fibrosis in various organs [8]. Similarly, serum concentrations of CITP measured by radioimmunoassy have been found to be related to the intensity of the degradation of collagen type I fibrils in patients with different diseases [8]. A number of experimental and clinical observations demonstrate that high serum concentrations of PIIIP as assessed by specific radioimmunoassay are associated with stimulated fibrogenesis in several fibrotic diseases [8].

3. Serum Markers in Hypertensive Heart Disease

3.1. Experimental studies

In recent studies [14-16] we measured serum PIP and CITP concentrations by specific radioimmunoassays in adult normotensive Wistar-

Kyoto rats (WKY), adult rats with spontaneous hypertension (SHR), adult SHR chronically treated with the angiotensin converting enzyme (ACE) inhibitor, quinapril, and adult SHR chronically treated with the angiotensin AT_1 receptor antagonist (ARA), losartan. In untreated SHR, compared with WKY, we found more extensive interstitial and perivascular fibrosis, more marked myocardial accumulation of collagen type I, increased serum PIP concentrations and similar serum CITP concentrations [14-16]. Interestingly, a direct correlation was found between the extent of myocardial fibrosis and serum PIP in untreated SHR [14]. Similar results have been reported recently by Camilión de Hurtado et al. in young SHR compared with young WKY [17].

In quinapril-treated SHR [14] and losartan-treated SHR [11, 16], compared with untreated SHR, we found a marked reduction in myocardial fibrosis, diminished accumulation of collagen type I, decreased PIP concentrations and a tendency to increased CITP concentrations. In losartan-treated SHR these effects were observed at doses that did not normalize blood pressure [16]. The ratio between PIP and CITP, an index of the degree of coupling between the synthesis and degradation of collagen type I [13, 18], was abnormally increased in untreated SHR and became normalized after treatment in SHR with both agents (Figure 3) [14-16].

3.2. Clinical studies

We measured serum PIP and CITP concentrations by specific radioimmunoassays in patients with essential hypertension who had never been treated and in normotensive individuals who acted as controls [19-21]. Measurements were repeated in hypertensive patients after chronic treatment with either the ACE inhibitor lisinopril [19, 20] or the ARA losartan [21]. Baseline serum PIP concentrations were increased in hypertensive patients, compared with normotensive individuals [19-21]. No significant differences were found between baseline serum CITP concentrations in hypertensive patients compared with normotensive individuals [21]. The ratio between PIP and CITP was higher in hypertensive patients than in normotensive individuals (Figure 3) [21]. Serum PIP concentrations correlated directly with the left ventricular mass index in the hypertensive group [19]. In addition, serum PIP related directly to the severity of ventricular arrhythmias in the hypertensive group [19]. Patients treated with lisinopril [19, 20] and those treated with losartan [21] had normalization in blood pressure, regression of left ventricular mass index, improvement of diastolic dysfunction and a diminution in the number of ventricular extrasystoles. Serum PIP concentration decreased to normal values and CITP concentration tended to be increased in these patients [19-21]. The ratio between PIP and CITP was normalized in losartan-treated patients (Figure 3) [21].

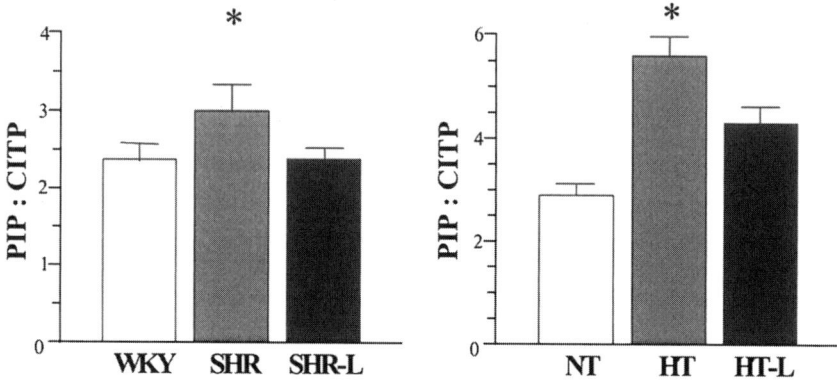

Figure 3. Ratio between procollagen type I carboxy-terminal propeptide (PIP) and collagen type I pyridoline cross-linked carboxy-terminal telopeptide (CITP) in rats (left panel) and humans (right panel). [WKY, normotensive Wistar-Kyoto rats; SHR, spontaneously hypertensive rats; SHR-L, losartan-treated SHR; NT, normotensive subjects; HT, patients with essential hypertension; HT-L, losartan-treated HT]. (Adapted from references [15, 16, 21]).

In three recent studies [22-24] transvenous endomyocardial biopsies were performed in patients with essential hypertension and collagen volume fraction was determined on picrosirius red-stained septal tissue sections with an automated image analysis system. Hearts from autopsies performed in normotensive subjects were used as controls in these studies [22-24]. In the first study we found that serum PIP concentrations correlated directly with collagen volume fraction in hypertensive patients [22] (Figure 4). Furthermore, using receiver operating characteristic curves, we observed that a cutoff of 127 μg/L for PIP provided 78% specificity and 75% sensitivity for predicting severe myocardial fibrosis with a relative risk of 4.80 (95% CI, 1.19 to 10.30) [22]. In the two other studies we observed a strong association between treatment-induced changes in tissue collagen content and treatment-induced changes in serum PIP in hypertensive patients [23, 24] (Figure 4). Furthermore, the ability of antihypertensive treatment to reduce blood pressure did not predict its capacity to either regress myocardial fibrosis or normalize collagen type I synthesis in these patients. Thus, chronic AT_1 blockade with losartan, but not chronic calcium channel blockade with amlodipine, resulted in a decrease of both collagen volume fraction and PIP in hypertensive patients [24]. Interestingly, the ability of losartan to induce regression of fibrosis and reduction in serum PIP was associated with diminution of myocardial stiffness in hypertensive patients [24].

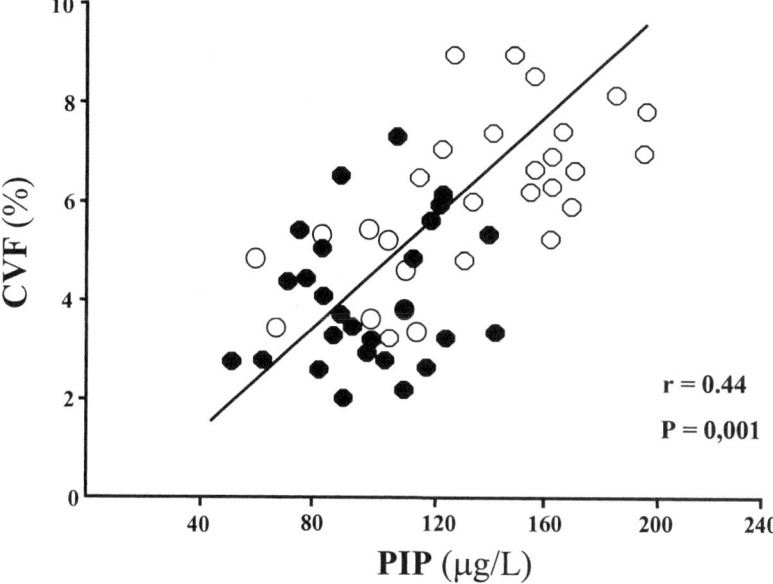

Figure 4. Direct correlation (y = 1.66 + 0.03x) between serum concentrations of procollagen type I carboxy-terminal propeptide (PIP) and collagen volume fraction (CVF) in patients with essential hypertension before (open circles) and after (filled circles) antihypertensive treatment. (Adapted from references [22-24]).

In a previous study we measured serum PIIIP concentrations in never-treated patients with essential hypertension before and after treatment with the ACE inhibitor lisinopril [25]. Serum PIIIP was higher in hypertensive patients than in normotensive controls. Serum PIIIP was directly correlated with plasma renin activity and inversely correlated with Doppler echocardiographic parameters assessing diastolic filling in hypertensives. The serum PIIIP concentrations decreased significantly after treatment with lisinopril in hypertensives. In a recent study, McLean Timms et al. [26] reported that there was no difference in serum PIIIP concentrations between patients with essential hypertension and normotensive controls. Nevertheless, it should be noted that a significant fraction of the patients were under antihypertensive treatment until 1 month before the study was performed. Thus it cannot be excluded that the lack of variation in PIIIP observed in this study is due to the pharmacological interference with the turnover of collagen type III. Clearly, more studies are necessary to ascertain the metabolism of collagen type III in arterial hypertension.

3.3. Interpretation of results from studies in hypertension

If an equilibrium is to exist between collagen synthesis and degradation, as proposed by Laurent [5], the above data suggest that the metabolism of collagen type I is abnormal in rats and humans with primary hypertension and myocardial fibrosis and that pharmacological interference of the renin-angiotensin system that results in regression of myocardial fibrosis also normalizes collagen type I metabolism in primary hypertension.

It is clear that PIP detectable in serum is not exclusively heart-specific. Nevertheless, we have calculated that in the SHR [14-16], changes in the cardiac compartment of collagen type I can alter concentrations of PIP in the circulation and that other extracardiac sources able to elevate the serum concentrations of PIP can be excluded. Whether this is also the case in hypertensive patients, we propose that measurement of serum PIP concentrations may provide indirect diagnostic information on the development of collagen type I-dependent myocardial fibrosis in arterial hypertension in the absence of liver or renal disease. This is further supported by the relations here reported between serum PIP and the extent of fibrosis in the free left ventricular wall of SHR and the interventricular septum of hypertensive patients. On the other hand, our findings that pharmacological interventions that regress myocardial fibrosis also decrease serum PIP suggest that determination of this peptide may be useful to assess the cardioreparative properties of antihypertensive treatment in hypertensives.

4. Serum Markers in Ischemic Heart Disease

4.1. Studies in patients with acute myocardial infarction

Jensen et al. [11] and Host et al. [27] sequentially monitored serum PIIIP and PIP concentrations in a group of patients with acute myocardial infarction and compared these findings with patients without enzymatic or electrocardiographic evidence of infarction. Concentrations of each propeptide rose on day 2 to 3 after infarction and remained elevated at 4 to 6 months. Individual peak changes in PIIIP correlated with the extent of infarction as gauged by the release of cardiac creatine kinase and lactic dehydrogenase isoenzymes. Peuhkurinen et al. [28] also found serum PIIIP concentrations to remain unchanged for 2 days after infarction in patients with documented myocardial infarction not treated with thrombolytic agents because of contraindications. Beyond this point in time, serum PIIIP levels rose over the course of 5 days. A rise in serum PIP was also observed beyond day 2 after infarction. Host et al. [29] found the extent to which serum PIIIP rose on days 1 to 5 after myocardial infarction to correlate with peak creatine kinase response and found that it was higher on days 0 to 2 in nonsurvivors than in survivors. Serum PIIIP was no longer increased 12 months after infarction.

4.2. Studies in patients with myocardial infarction subjected to thrombolysis

Peuhkurinen et al. [28] and Host et al. [27] each found that the pattern of PIIIP release after myocardial infarction was altered by streptokinase. This serological marker rose rapidly within the first several hours after myocardial infarction in a group of patients with probable reperfusion as contrasted to those with nonprobable reperfusion, in whom it did not. Two days later, there was a second increase in serum PIIIP observed in both groups, which coincided with a similar increase in PIP. No correlation, however, was now found between the rise in these propeptides and the extent of creatine kinase release. Host et al. [30] likewise found an earlier and more marked release of PIIIP in patients receiving tissue plasminogen activator, which activates plasmin only in the presence of fibrin, as contrasted to streptokinase, which indiscriminately activates plasmin. Uusimaa et al. [31] monitored the serum concentration of PIIIP in their patients with acute myocardial infarction who were treated with either tissue plaminogen activator or streptokinase, within 6 hours of the onset of their symptoms. PIIIP rose by 40% in patients with enzymatic evidence of definite infarction. The integral of the PIIIP response with respect to time (area under the curve above basal levels during the first 10 days after infarction) was greater in patients with acute left ventricular systolic dysfunction (ejection fraction <40%) and clinical evidence of heart failure than in their counterparts without evidence of failure. This also appeared to be the case when blood flow to the culprit coronary vessel was not adequately restored.

4.3. Interpretation of results from myocardial infarction studies

These studies suggest that scar formation after myocardial infarction can be quantified from serial measurements of serum PIIIP and that the rise in PIIIP over time is correlated with the extent of injury, ventricular dysfunction, and vessel patency. They further suggest that the rise in serum PIIIP that appears after day 2 after myocardial infarction, coincident with the collagenolytic phase of repair, together with the persistent elevation of PIIIP and PIP for months after infarction is in keeping with the dynamic and persistent nature of tissue repair in the infarcted human heart. The source of these propeptides during later stages of healing could include the site of myocardial infarction and fibrous tissue formation at sites remote to it.

The initial source of PIIIP that accompanies thrombolysis is uncertain.

It may include plasmin-induced collagen breakdown in the arterial vasculature at large, the atherosclerotic site associated with thrombosis, and the site of infarcted ventricular tissue. Plasmin, a nonspecific protease, induces fibrinolysis and activates latent matrix metalloproteinases that promote degradation of collagen types I and III. Reperfusion and an influx of neutrophils may likewise contribute to collagen degradation. Collectively, collagen degradation in the infarcted myocardium could predispose to infarct expansion and rupture of the ventricle.

5. Serum Markers in Chronic Cardiac Failure

5.1. Clinical studies

Klappacher et al. [32] monitored serum PIIIP and CITP in patients with chronic cardiac failure of varying severity secondary to either dilated (idiopathic) cardiomyopathy or ischemic cardiomyopathy. All were treated with an ACE inhibitor, nitrates, digitalis, and several diuretics. Increased levels of PIIIP and CITP were correlated with the clinical severity of failure, the degree of hemodynamic impairment, hyponatremia, and the need for cardiac transplantation. PIIIP and CITP each predicted mortality independently of hepatic function. In explanted hearts obtained from this same study population, PIIIP and CITP respectively correlated with accumulation of types III and I collagens in tissue, expressed as an interstitial and perivascular fibrosis of the right and left ventricles in either disease entity.

Zannad et al. [33] radomized a sample of 261 patients with congestive heart failure from the Randomized Aldactone Evaluation Study (RALES) to placebo or spironolactone. Serum PIP and PIIIP were assessed at baseline and at 6 months. High baseline serum concentrations of these peptides were significantly associated with poor outcome and levels decrease during spironolactone therapy but not during placebo therapy. The benefit from spironolactone was associated with higher levels of PIIIP and PIP.

5.2. Interpretation of results from heart failure studies

These findings raise the prospect that the exent of myocardial fibrosis in either dilated or ischemic cardiomyopathy evolving with cardiac failure can be predicted from collagen-derived serum peptides and that they have prognostic value in patients with these conditions. On the other hand, the limitation of the excessive extracellular matrix turnover may be one of the various mechanisms contributing to the beneficial effect of some pharmacological treatments (i.e., spironolactone) in patients with congestive heart failure.

6. Perspectives on the use of serum markers of collagen metabolism

The available experimental and clinical data suggest that the biochemical monitoring of fibrillar collagen metabolism may provide a potential non-invasive method of documenting both the extent and mechanisms of myocardial fibrosis in cardiac diseases. In fact, the use of cardiac biopsies is an invasive methodology not useful for wide-scale application, and beyond that may be subject to sampling error, whereas serum tests can be performed on a frequent basis.

In addition, this procedure can be of interest in assessing the usefulness of other diagnostic methods aimed to detect changes in myocardial composition. For instance, recent findings have shown that the combination of serum markers of collagen metabolism with parameters assessing ultrasonic reflectivity of the myocardium may provide an accurate estimation of severe myocardial fibrosis associated with cardiac disease (i.e., hypertensive heart disease) [34].

Finally, several trials underscore the potential for pharmacological measures designed to prevent the appearance of or even to cause the regression of myocardial fibrosis in hypertensive patients [23, 35, 36]. The studies here reviewed have set the stage for larger on-going trials wherein serological markers of fibrillar collagen turnover would be of great interest to assess the antifibrotic ability of pharmacological interventions aimed at repairing the myocardium in cardiac diseases.

References

1. Beltrami, C.A., et al., Structural basis of end-stage failure in ischemic cardiomyopathy in humans. Circulation, 1994. 89: p. 151-63.
2. Rossi, M.A., Pathologic fibrosis and connective tissue matrix in left ventricular hypertrophy due to chronic arterial hypertension in humans. J Hypertens, 1998. 16: p. 1031-41.
3. Weber, K.T., C.G. Brilla, and J.S. Janicki, Myocardial fibrosis: functional significance and regulatory factors. Cardiovasc Res, 1993. 27: p. 341-8.
4. Diez, J., et al., Clinical aspects of hypertensive myocardial fibrosis. Curr Opin Cardiol, 2001. 16: p. 328-35.
5. Laurent, G.J., Dynamic state of collagen: pathways of collagen degradation in vivo and their possible role in regulation of collagen mass. Am J Physiol, 1987. 252: p. C1-9.
6. Nimmi, N., Fibrillar Collagens: their biosynthesis, molecular structure, and mode of assembly. In Extracellular Matrix, Zern, M.A., Reid, L,M., eds. New York, NY: Marcel Dekker, 1993: p. 121-148.
7. Janicki, J., Collagen degradation in the heart. In Molecular Biology of Collagen Matrix in the Heart, Eghbali-Webb ed. Austin, TX: RG Landes, 1995: p. 61-76.
8. Risteli, L. and J. Risteli, Noninvasive methods for detection of organ fibrosis. In Focus on Connective Tissue in Health and Disease, Rojkind, M., ed. Boca Raton, FL: CRC

Press, 1990: p. 61-68.

9. Smedsrod, B., et al., Circulating C-terminal propeptide of type I procollagen is cleared mainly via the mannose receptor in liver endothelial cells. Biochem J, 1990. 271: p. 345-50.

10. Risteli, J., et al., Radioimmunoassay for the pyridinoline cross-linked carboxy-terminal telopeptide of type I collagen: a new serum marker of bone collagen degradation. Clin Chem, 1993. 39: p. 635-40.

11. Jensen, L.T. and N.B. Host, Collagen: scaffold for repair or execution. Cardiovasc Res, 1997. 33: p. 535-9.

12. Jensen, L.T., et al., Collagen metabolism during wound healing in rats. The aminoterminal propeptide of type III procollagen in serum and wound fluid in relation to formation of granulation tissue. APMIS, 1993. 101: p. 557-64.

13. Lopez, B., et al., Biochemical assessment of myocardial fibrosis in hypertensive heart disease. Hypertension, 2001. 38: p. 1222-6.

14. Diez, J., et al., Serum markers of collagen type I metabolism in spontaneously hypertensive rats: relation to myocardial fibrosis. Circulation, 1996. 93: p. 1026-32.

15. Varo, N., et al., Losartan inhibits the post-transcriptional synthesis of collagen type I and reverses left ventricular fibrosis in spontaneously hypertensive rats. J Hypertens, 1999. 17: p. 107-14.

16. Varo, N., et al., Chronic AT(1) blockade stimulates extracellular collagen type I degradation and reverses myocardial fibrosis in spontaneously hypertensive rats. Hypertension, 2000. 35: p. 1197-202.

17. Camilion de Hurtado, M.C., et al., Regression of cardiomyocyte hypertrophy in SHR following chronic inhibition of the Na+/H+ exchanger. Cardiovasc Res, 2002. 53: p. 862-8.

18. Diez, J. and C. Laviades, Monitoring fibrillar collagen turnover in hypertensive heart disease. Cardiovasc Res, 1997. 35: p. 202-5.

19. Diez, J., et al., Increased serum concentrations of procollagen peptides in essential hypertension. Relation to cardiac alterations. Circulation, 1995. 91: p. 1450-6.

20. Laviades, C., et al., Abnormalities of the extracellular degradation of collagen type I in essential hypertension. Circulation, 1998. 98: p. 535-40.

21. Laviades, C., N. Varo, and J. Diez, Transforming growth factor beta in hypertensives with cardiorenal damage. Hypertension, 2000. 36: p. 517-22.

22. Querejeta, R., et al., Serum carboxy-terminal propeptide of procollagen type I is a marker of myocardial fibrosis in hypertensive heart disease. Circulation, 2000. 101: p. 1729-35.

23. Lopez, B., et al., Usefulness of serum carboxy-terminal propeptide of procollagen type I in assessment of the cardioreparative ability of antihypertensive treatment in hypertensive patients. Circulation, 2001. 104: p. 286-91.

24. Diez, J., et al., Losartan-dependent regression of myocardial fibrosis is associated with reduction of left ventricular chamber stiffness in hypertensive patients. Circulation, 2002. 105: p. 2512-7.

25. Laviades, C., G. Mayor, and J. Diez, Treatment with lisinopril normalizes serum concentrations of procollagen type III amino-terminal peptide in patients with essential hypertension. Am J Hypertens, 1994. 7: p. 52-8.

26. Timms, P.M., et al., Plasma tissue inhibitor of metalloproteinase-1 levels are elevated in essential hypertension and related to left ventricular hypertrophy. Am J Hypertens, 2002. 15: p. 269-72.

27. Host, N.B., et al., Thrombolytic therapy of acute myocardial infarction alters collagen metabolism. Cardiology, 1994. 85: p. 323-33.

28. Peuhkurinen, K.J., et al., Thrombolytic therapy with streptokinase stimulates collagen

breakdown. Circulation, 1991. 83: p. 1969-75.

29. Host, N.B., et al., The aminoterminal propeptide of type III procollagen provides new information on prognosis after acute myocardial infarction. Am J Cardiol, 1995. 76: p. 869-73.

30. Host, N.B., et al., Effect on collagen metabolism of thrombolytic therapy with tissue-plasminogen activator. A randomized, placebo-controlled study. Eur J Clin Invest, 1995. 25: p. 15-8.

31. Uusimaa, P., et al., Collagen scar formation after acute myocardial infarction: relationships to infarct size, left ventricular function, and coronary artery patency. Circulation, 1997. 96: p. 2565-72.

32. Klappacher, G., et al., Measuring extracellular matrix turnover in the serum of patients with idiopathic or ischemic dilated cardiomyopathy and impact on diagnosis and prognosis. Am J Cardiol, 1995. 75: p. 913-8.

33. Zannad, F., et al., Limitation of excessive extracellular matrix turnover may contribute to survival benefit of spironolactone therapy in patients with congestive heart failure: insights from the randomized aldactone evaluation study (RALES). Rales Investigators. Circulation, 2000. 102: p. 2700-6.

34. Maceira, A.M., et al., Ultrasonic backscatter and serum marker of cardiac fibrosis in hypertensives. Hypertension, 2002. 39: p. 923-8.

35. Schwartzkopff, B., et al., Repair of coronary arterioles after treatment with perindopril in hypertensive heart disease. Hypertension, 2000. 36: p. 220-5.

36. Brilla, C.G., R.C. Funck, and H. Rupp, Lisinopril-mediated regression of myocardial fibrosis in patients with hypertensive heart disease. Circulation, 2000. 102: p. 1388-93.

Chapter 6

Ultrasonic Characterization of the Myocardium

Mark R. Holland, and Samuel A. Wickline
Washington University in St. Louis, St. Louis, MO, U.S.A.

1. Introduction

In this chapter we review the use of ultrasound as a method for obtaining noninvasive measurements of the intrinsic properties of myocardium. Our goal is to illustrate how the measurement of fundamental ultrasonic parameters is related to the presence of myocardial fibrosis. These measurements lay the foundation for the development of clinically applicable noninvasive ultrasonic assessment techniques. Although echocardiographic imaging is routinely employed to assess systolic and diastolic cardiac function [1, 2], we limit our discussion to the use of ultrasound as a method for obtaining direct measurements of intrinsic myocardial properties (i.e., ultrasonic tissue characterization techniques). Ultrasonic tissue characterization is well-suited for the direct measurement of myocardial properties because the determinants of ultrasonic propagation and scattering are directly related to fundamental physical properties such as mass density, compressibility, cellular geometry and organization [3-7].

This chapter is not intended to be a comprehensive review of the published literature. Instead we provide several examples of specific studies that illustrate the use of ultrasonic tissue characterization techniques in the assessment of myocardial fibrosis. More comprehensive reviews of the application of ultrasonic tissue characterization techniques to the quantitative assessment of a wide variety of cardiovascular pathologies can be found in the published literature [8-10].

1.1. Relationship between ultrasonic measurements and intrinsic myocardial properties.

The behavior of ultrasonic waves as they propagate, reflect, and scatter within tissue is, to a large extent, governed by the inherent material and structural properties of the tissue. Hence, estimates of specific properties of myocardium can be obtained by measuring the propagation and scattering characteristics of an ultrasonic field. Three ultrasonic parameters are typically used to characterize tissue: ultrasonic velocity (*speed of sound*), *attenuation*, and scatter (*backscatter*).

1.1.1. Speed of sound.

The *speed* at which ultrasonic waves propagate is related to the inherent elastic properties of the material and the mass density. Often these elastic properties are expressed in terms of the *bulk modulus,* which relates the hydrostatic pressure on a volume of material (i.e., a specific stress applied to all surfaces of the volume) to the fractional change in its volume. The speed of sound for a longitudinal (i.e., compressional ultrasonic waves with material displacement parallel to the direction of propagation – this mode of propagation is the predominant mode for ultrasonic waves in tissue) ultrasonic wave is given by the expression

$$c = \sqrt{B/\rho}$$

(Eq. 1)

where c is the speed of propagation, B is the bulk modulus, and ρ is the mass density. Measurements of the speed of sound can provide estimates of the inherent elastic properties of myocardium.

1.1.2. Attenuation.

As an ultrasonic wave propagates in tissue it will diminish in amplitude with distance traveled. This decrease in amplitude, or *attenuation*, is the result of both scattering and absorption mechanisms [11] that are related to the inherent viscoelastic and structural properties of the tissue. The amplitude of an ultrasonic wave decreases exponentially with distance traveled, x, and is expressed as

$$A(x) = A_0 e^{-\alpha x}$$

(Eq. 2)

where $A(x)$ is the amplitude after propagating a distance x, A_0 is the initial amplitude, and α is the *attenuation coefficient* expressed in units of cm^{-1} or dB/cm. The attenuation coefficient, α, also depends on the frequency of the ultrasonic wave. For many tissues, it can be expressed as approximately linearly increasing with frequency, i.e.,

$$\alpha = \beta \cdot f \qquad\qquad (Eq.\ 3)$$

where the linear frequency dependence of the attenuation coefficient, β, or "slope of attenuation" is often expressed in units of dB/(cm·MHz). Because analyses of the frequency dependence of attenuation provide very useful information for the characterization of myocardial tissue, ultrasonic pulses with a relatively broad spectrum of frequency components are often used. Figure 1 illustrates the effect of the frequency dependence of attenuation on a propagating ultrasonic pulse. Both the time-domain and frequency-domain representations of the pulse are illustrated. This figure shows how the frequency-dependent effects of attenuation affect the spectral content of the pulse: higher frequencies are more severely attenuated than are lower frequencies.

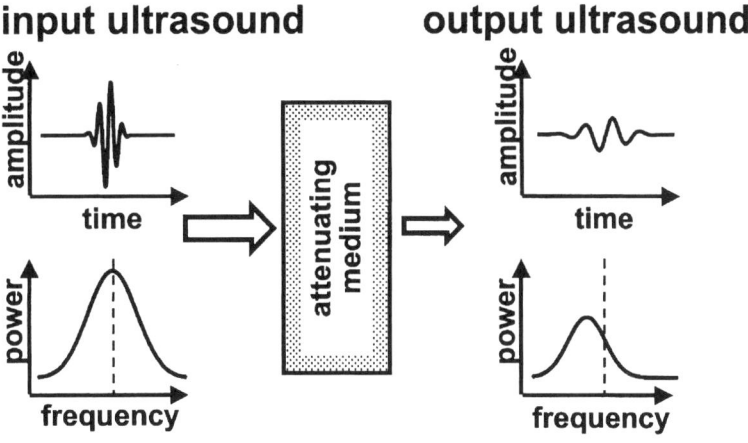

Figure 1. Effect of attenuation on an ultrasonic pulse transmitted through tissue

1.1.3. Backscatter.

When a propagating ultrasonic wave encounters a boundary between two media with different acoustic properties the amount of the incident ultrasonic intensity that is reflected is related to the difference in *acoustic impedance* properties of the two media. The characteristic *acoustic impedance, Z, of a medium is related to the inherent mass density, ρ, and sound speed, c,* properties and is expressed as $Z = \rho c$. Because the sound speed is related to the inherent elastic properties of the tissue (Eq. 1) so too is the acoustic impedance.

When a propagating ultrasonic wave encounters an object (or a distribution of objects) with acoustic impedance different from that in which it is embedded and whose dimensions are *small* compared with the wavelength of the insonifying field, the wave will be *scattered*. Unlike specular reflection where the reflected ultrasonic wave propagates in a specific direction, scattered ultrasound tends to travel in all directions. Such is the case for scattered ultrasound arising from within tissue.

Scattered ultrasonic waves travelling back to the insonifying source, which now acts as a receiver, are referred to as *backscattered* ultrasonic waves. This concept is illustrated in Figure 2. The inherent backscatter efficiency of a tissue is expressed in terms of its *backscatter coefficient,* which typically manifests a strong frequency dependence on the order of frequency to the fourth power. The inherent scattering efficiency of tissue and blood tends to be relatively small. Backscatter depends on the incident ultrasonic frequency, and typical values for the intensity of ultrasonic backscatter might be on the order of 100,000 times less than the amount transmitted (i.e., 50 dB less) [12-16]. Although the backscatter coefficient represents the inherent backscatter efficiency of the tissue, the intensity of ultrasonic backscatter measured by an ultrasonic imaging system is affected by other factors as well (e.g., attenuation, ultrasonic beam propagation effects, etc.) and hence the uncompensated measured level of backscatter is often referred to as the *apparent backscatter* [17, 18].

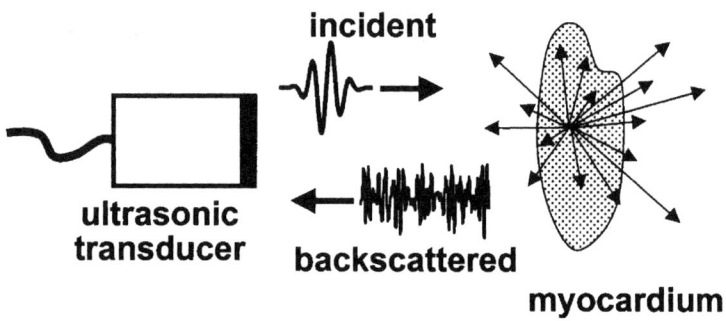

Figure 2. Backscattered ultrasound from myocardium.

Of the three ultrasonic parameters that are typically measured and used to characterize tissue (speed of sound, attenuation, and backscatter), backscatter measurements are often the most readily obtainable, especially in a clinical setting. The measured backscattered signal, $S(f)$ (expressed in terms of its frequency domain representation), comprises a number of individual components [18, 19]:

$$S(f) = E(f) \cdot V(f) \cdot A(f) \cdot \eta(f) \qquad \text{(Eq. 4)}$$

where $E(f)$ is the electrical/mechanical response of the measurement system, $V(f)$ is the effects of ultrasonic beam propagation and volume of tissue insonified, $A(f)$ is the effect of attenuation, and $\eta(f)$ is the *inherent* backscatter properties of tissue.

One of the challenges associated with obtaining quantitative estimates of the inherent backscatter properties of myocardial tissue (i.e., the backscatter coefficient) is to identify appropriate methods for compensating for the effects of the other contributions to the measured signal. This is accomplished readily in measurements of excised myocardial tissue but is much more difficult in clinical studies. These issues will be further discussed in the following sections of this chapter.

1.2. Myocardial properties and ultrasonic parameters

Both the intrinsic material properties (e.g., viscoelasticity, density) as well as the geometrical properties of myocardial fibrosis will affect the measured ultrasonic parameters. This has been incorporated into specific models proposed to explain the observed ultrasonic backscatter from myocardium [6, 7]. In the models developed by Wickline et al. [3] and Rose et al. [7], the relationship between myocardial intra- and extracellular properties and ultrasonic backscatter measurements is explicitly addressed. It is hypothesized that the difference in acoustic impedance responsible for the scattering is that between the extracellular collagen network that surrounds each myocyte (or myocyte bundle) and the rest of the tissue (the myocytes' intracellular contents). An elementary myocardial scatterer is modeled as an ellipsoidal shell, having the material properties of collagen, imbedded in a host medium having the average properties of myocardium. Figure 3 illustrates the structure of the elementary myocardial scatterer assumed in this model. In this model the first Born approximation to elastic scattering [20] is used to calculate the frequency-dependent scattering from a single scatterer. Backscatter from a distribution of scatterers is calculated as the sum of the power scattered in the direction of the transducer by each individual scatterer located in the active volume of the ultrasonic beam (an independent-scatterer approximation). The resultant characteristics of ultrasonic backscatter predicted by this model are consistent with published experimental measurements.

The effects of the geometrical properties of myocardium on the measured ultrasonic parameters are manifest in their angle of insonification dependent (*anisotropic*) nature. These anisotropic properties have been observed in measurements of ultrasonic backscatter [21-23], attenuation [24, 25], and speed of sound [26-28].

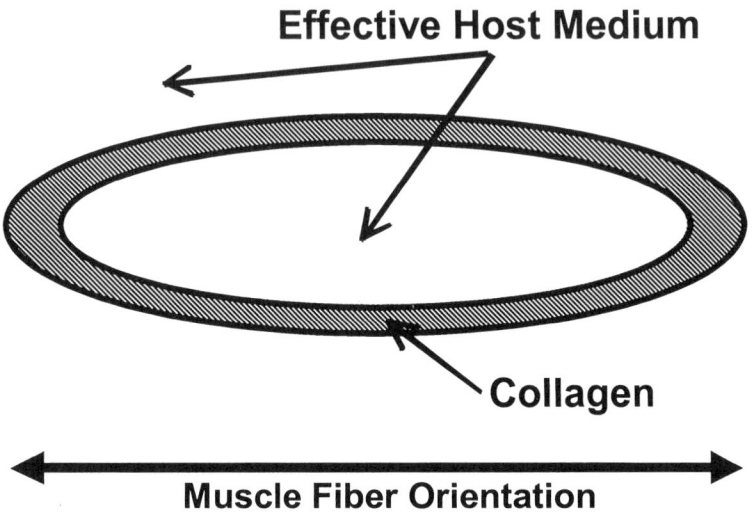

Effective Host Medium

Collagen

Muscle Fiber Orientation

Figure 3. Model of elementary myocardial scatterer (ellipsoidal shell) proposed by Rose et al. [7].

Figure 4 shows the arrangement for acquiring the anisotropic measurement of ultrasonic parameters. Ultrasonic backscatter is greatest for perpendicular insonification with respect to the predominate myofiber direction and least for parallel insonification. Conversely, ultrasonic attenuation and velocity are a minimum for perpendicular insonification and at a maximum for parallel insonification.

2. Ultrasonic Characterization of Excised Myocardium

Several studies have been published illustrating the relationship between measured ultrasound and collagen content in excised specimens of myocardial tissue. These studies provide a more complete understanding of the relationship between myocardial fibrosis and ultrasonic parameters. Utilization of excised tissue specimens permits robust estimates of the absolute intrinsic ultrasonic properties of myocardium to be obtained because compensation for the effects of the measurement system and ultrasonic beam propagation are more readily implemented. This is typically accomplished by comparing measurements of the myocardial specimens to those of a well-characterized reference material. These studies form the basis for further development of more clinically applicable ultrasonic tissue characterization techniques.

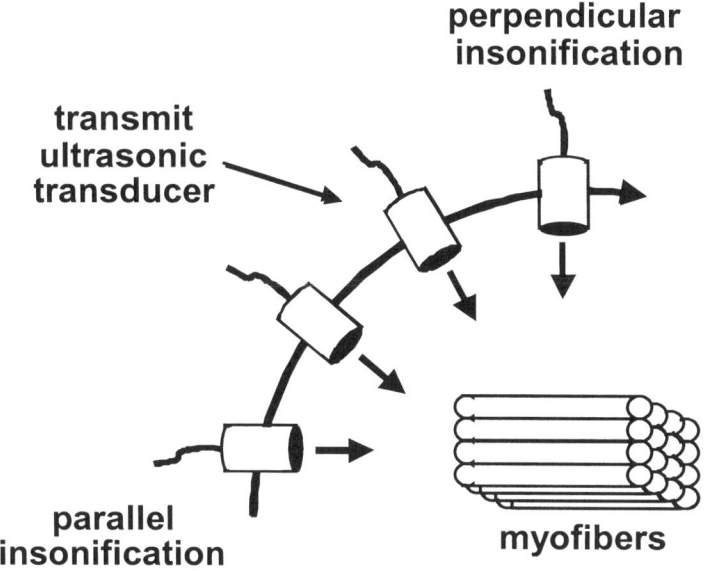

Figure 4. Measurement of anisotropic ultrasonic properties.

The nature of the observed anisotropy in these ultrasonic parameters is illustrated in Figure 5. Measurements of the anisotropic properties of myocardium provide a means for developing insights into the geometrical and structural attributes of myocardial fibrosis. These anisotropic effects can be readily observed in clinical echocardiographic images [29, 30].

3. Relationships Between Collagen and Ultrasonic Attenuation

Assessment of the inherent attenuation properties of excised myocardium are typically obtained by measuring the ultrasonic signal after it has propagated through the tissue specimen and compensating this measurement by a reference measurement obtained without the intervening specimen. This analysis is most often performed on the frequency-domain representations of the through-specimen and reference pulses. If the frequency power spectra are expressed in the logarithmic domain (i.e., in decibels), the frequency-dependent attenuation properties can be obtain by subtracting the through-specimen power spectrum from the reference power spectrum, compensating for any transmission losses occurring at the surface interfaces, and dividing by the specimen thickness. If a line is fit to the attenuation versus frequency data the resultant

122

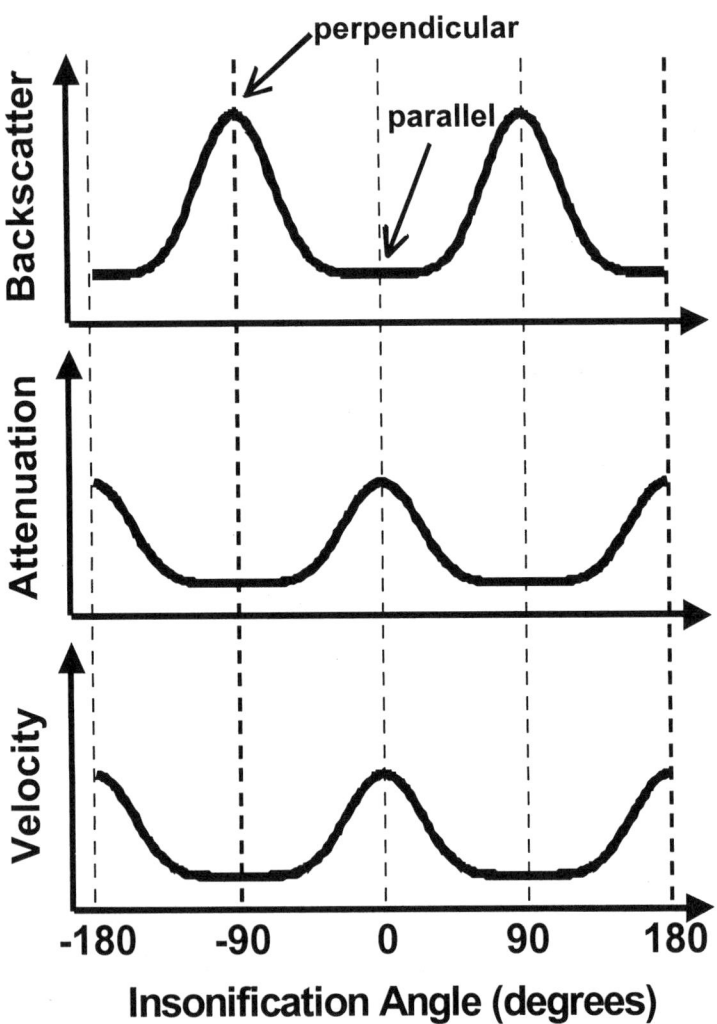

Figure 5. Anisotropic of ultrasonic backscatter, attenuation, and velocity.

frequency dependence of attenuation is often referred to as the "slope of attenuation." This process is illustrated in Figure 6.

In a series of studies published in the late 1970's and early 1980's, Miller's group investigated the relationship between collagen content and ultrasonic attenuation in myocardium [31-34]. In these studies, hearts from animals subjected to ischemic injury by coronary occlusion were used to elucidate the relationship between ultrasonic attenuation and collagen content. Increased ultrasonic attenuation correlated well with increased collagen content

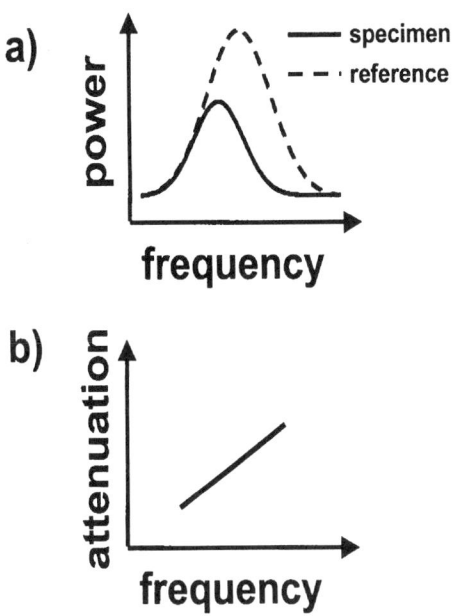

Figure 6. a) Typical nature of the power spectra corresponding to the reference and after propagating through tissue. b) attenuation as a function of frequency derived from the measured power spectra.

determined biochemically in regions of several-week old ischemic injury. Figure 7 shows the relationship between the ultrasonic attenuation properties and collagen content based on results from measurements on the hearts from normal dogs and dogs subjected to ischemic injury as reported by O'Donnell et al. [32]. This figure shows the frequency dependence of the attenuation coefficient expressed as the slope of a linear fit to the measured attenuation values over the frequency range from 2 to 10 MHz for three levels of collagen content.

Infarcted human myocardial tissue was characterized ultrasonically at the microscopic level by Saijo et al. in a paper published in 1997 [35]. Specimens of infarcted human myocardium were obtained at autopsy, formalin fixed, and sectioned in a plane parallel to the normal muscle fiber orientation. Regions in which the myocytes appeared normal by optical microscopy were defined as "normal" and regions that exhibited collagen fibers were defined as "fibrosis". A specially developed scanning acoustic microscope system, operating in the 100- to 200-MHz frequency range, was used to obtain the measurements. Measured values of the slope of attenuation were found to be 9.4±0.4 in "normal" regions and 17.5±1.1 dB/(cm·MHz) in "fibrosis".

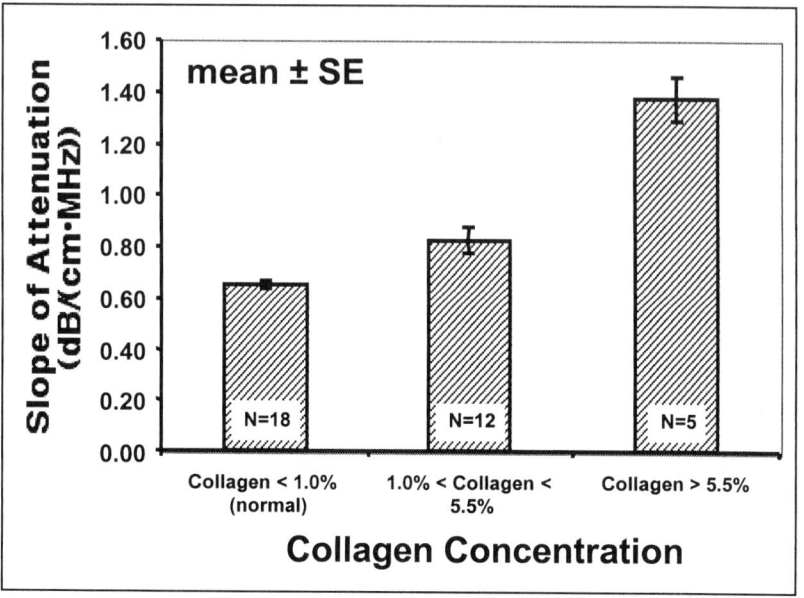

Figure 7. Measured myocardial slope of attenuation values for three levels of collagen concentration. Adapted from data published by O'Donnell et al. [32].

These results indicate that the slope of attenuation associated with collagen in regions of "fibrosis" is almost a factor of two greater than the values observed in the regions of "normal" tissue for the relatively high ultrasonic frequencies used in this study. It is interesting to note that the values of the slope of attenuation appear to be significantly higher for these specific myocardial regions at these relatively high frequencies when compared with the values found for myocardium as a whole measured at lower frequencies.

In a study designed to investigate age-related alterations of cardiac tissue microstructure and material properties, Nguyen et al. reported measurements of ultrasonic properties in Fischer 344 rats as they aged [36]. Age-related changes in myocardial properties were delineated using high-frequency ultrasound (over a bandwidth ranging from 30 to 44 MHz) to interrogate the hearts from Fischer 344 rats at 6, 18, and 24 months of age. For each age group, the excised lateral wall of the left ventricle from each rat was insonified using a 50-MHz nominal-frequency acoustic microscope for determination of the ultrasonic properties. In addition, histological and biochemical analyses for collagen content and cardiac myocyte diameter were performed on each specimen. The slope of attenuation increased with age for the 18 month-old rats compared with 6 month-old rats, but relatively little additional change was observed between the 18 month-old and

24 month-old rats (1.23 ± 0.83, 2.26 ± 0.33, and 2.23 ± 0.66 dB/(cm·MHz) for 6, 18, and 24 months, respectively). These attenuation data appear to be consistent with the trend observed in the data from the biochemical analyses. Collagen concentration was observed to increase progressively with age, with the greatest increment occurring from 6 to 18 months (38.0 ± 6.3 to 53.0 ± 7.1 mg/g dry wt), and leveling off at 24 months (60.0 ± 7.4 mg/g dry wt).

Because the progressive increase in stiffening of the myocardium may be related, in part, to an excessive number of collagen cross-links within the myocardial extracellular matrix [37, 38], the effects of collagen cross-linking on measured ultrasonic attenuation in myocardial tissue was investigated by Hall et al. [39]. A model was employed in which increased cross-linking was induced by treating excised rat myocardial tissue with chemical fixatives. Groups of 4 to 6 month old, male Sprague-Dawley rat hearts were arrested at end-diastole, excised, and sectioned. These specimens were interrogated with an acoustic microscope with a bandwidth ranging from 30 to 50 MHz. Initial measurements were obtained on the fresh tissue specimens within a few minutes of excision and then groups of these specimens were placed in a fixative of either 10% formalin or 2.5% glutaraldehyde. A separate group of specimens were placed in Hanks Buffered Saline Solution and served as a control. Each group of specimens was serially interrogated at specific time intervals thereafter for 24 hours. Results of this study showed that the frequency dependence of the attenuation coefficient (slope of attenuation) increased as a function of the extent of collagen cross-links in 10% formalin with a maximal change of 0.8 ± 0.3 dB/(cm·MHz) and in 2.5% glutaraldehyde with a maximal change of 0.9 ± 0.6 dB/(cm·MHz) over a 24-hour period. The control group of specimens exhibited a decline in the slope of attenuation of 0.7 ± 0.5 dB/(cm·MHz) over the same time period. This study suggests that measurements of ultrasonic attenuation may provide a method for quantifying the progression of collagen cross-linking in myocardium.

4. Relationship Between Collagen and Ultrasonic Velocity and Impedance

Direct measurements of the relationship between the level of myocardial collagen content and the ultrasonic velocity and acoustic impedance have not been extensively performed. However, studies have shown that collagen plays a dominant role in the determination of ultrasonic velocity and impedance properties in biological tissues [40-42].

In the same 1997 paper detailed above describing attenuation measurements [35], Saijo et al. reported measured values of the sound speed to be 1620.2 ± 8.2 m/sec in the "normal" tissue regions and 1690.3 ± 9.1 m/sec in "fibrosis" for acoustic microscope measurements operating in the 100- to 200-MHz frequency range. Again, it is interesting that these reported values of velocity appear to be significantly higher in these specific myocardial regions at these relatively high frequencies when compared with the mean value of 1540 m/sec value typically found for normal myocardium as a whole measured at lower frequencies. In an additional study [43], Saijo et al. reported the acoustic impedance of the "normal" and "fibrosis" tissue regions in infarcted myocardium to be 1.75 x 10^5 rayl and 1.85 x 10^5 rayl, respectively (1 rayl (in cgs units) = 1 dyne·sec/cm^3, a unit of acoustic impedance) for measurements in the same frequency range. Similarly, in an acoustic microscopy study at 600 MHz, Chandraratna et al. [44] showed that the intensity of reflected ultrasonic echoes from collagen depends upon collagen fiber morphologic characteristics in thin sections of excised myocardium.

In a study designed to quantify the effect of the structural organization of collagen on ultrasonic velocity, Hoffmeister et al. measured the anisotropy of velocity in fixed bovine Achilles tendon [26]. The authors chose tendon for this study because it possesses a high content of collagen and has a well-defined unidirectional arrangement of fibers. Ultrasound velocity exhibited a substantial angular dependence: the larger velocities were associated with propagation parallel to the collagen fiber axis (2024 ± 53 m/sec) and the smaller velocities corresponded with propagation perpendicular to the fiber axis (1672 ± 18 m/sec). By modeling the tendon as an anisotropic elastic medium with five independent elastic constants, and measuring the mass density of the tendon specimens, the authors estimated the elastic properties of the tendon perpendicular and parallel to the fiber axis based on the velocity measurements. The elastic stiffness coefficients C_{11} (perpendicular to the fibers) and C_{33} (parallel to the fibers) were computed to be (mean ± SD) 3.08 ± 0.06 GPa and 4.51 ± 0.23 GPa, respectively.

Similar measurements of the anisotropy of ultrasonic velocity and elastic properties in fixed normal human myocardium have been reported by Verdonk et al. [28]. Results from this study yielded information about ultrasound velocity in-line and perpendicular to cardiac myofiber direction. Ultrasound velocity exhibited an angular dependence with a value of 1550 ± 9 m/sec for propagation parallel to the myofibers and 1530 ± 5 m/sec for propagation perpendicular to the myofibers. This resulted in estimates of the elastic constant coefficients to be 2.462 ± 0.004 GPa for C_{11} (perpendicular to the myofibers) and 2.527 ± 0.004 GPa for C_{33} (parallel to the myofibers). In a comparison of the results obtained from myocardium with the corresponding values from tendon, Hoffmeister et al. [26] noted that the angular dependence of velocity between myocardium and tendon are qualitatively similar, both exhibiting a maximum velocity (and stiffness) parallel to the inherent fiber orientation.

5. Relationship Between Collagen and Ultrasonic Backscatter

A large number of studies have explored the relationship between collagen content and properties of ultrasonic backscatter from myocardium. However, the use of the measured backscattered signals to provide a quantitative assessment of the inherent backscatter properties of tissue is a relatively complex task because the measured backscattered signal consists of contributions from a number of individual components as indicated in Eq. 4. Assessment of the inherent backscatter properties of excised myocardium are accomplished by measuring the backscattered ultrasonic signal from the tissue specimen and compensating this measurement by a reference measurement obtained from a medium with well-known properties. Often the specular reflection from a stainless-steel plate serves as the reference signal. This analysis is typically performed using the frequency-domain representations of the specimen and reference backscattered signals. If the frequency power spectra are expressed in the logarithmic domain (i.e., in decibels), the backscatter properties of myocardium can be obtain by subtracting the reference power spectrum from the specimen backscatter power spectrum to give the *apparent backscatter transfer function*. This process is illustrated in Figure 8. Further compensation of the *apparent backscatter transfer function* for the effects of attenuation, transmission losses occurring at the surface interfaces, and beam volume will provide a measure of the inherent *backscatter coefficient* properties of myocardium. The frequency dependence of the *backscatter coefficient,* and to a lessor extent, the *apparent backscatter transfer function,* can provide additional information regarding the nature of the scattering media within the myocardium. Another very useful parameter used to characterize the backscatter from myocardium is the *integrated backscatter* value. This can be obtained by determining the average value of the backscatter transfer function over a specific frequency range (Figure 8b). In addition, time-domain methods of estimating integrated backscatter values [45] are also widely implemented.

6. Importance of the Extracellular Matrix in Ultrasonic Backscatter

The importance of the extracellular matrix in determining the properties of ultrasonic backscatter from myocardium was demonstrated by Hall et al. [46]. Their approach was to determine whether measurements of backscatter from the isolated extracellular matrix could reproduce the measured anisotropy of backscatter obtained from intact cardiac tissues. Specimens of the left ventricular free wall from formalin fixed porcine hearts were insonified using a 50 MHz center-frequency acoustic microscope system and the anisotropy of

128

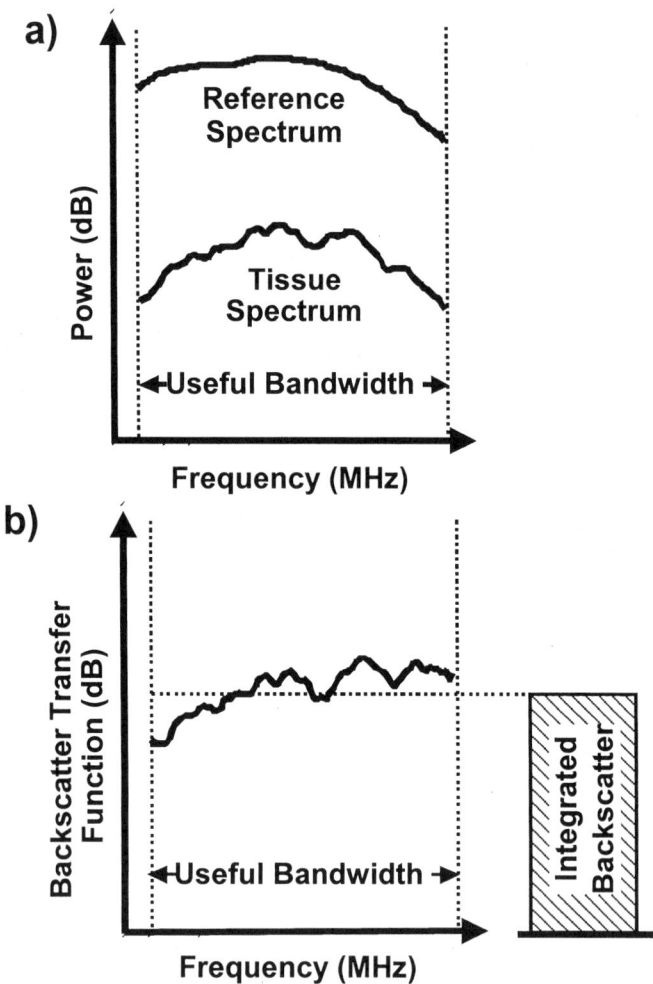

Figure 8. a) Example spectra from myocardial backscatter and reference reflection. b) Backscatter Transfer Function obtained from measured power spectra. The integrated backscatter value represents the mean value of the Backscatter Transfer Function over the useful bandwidth of the measurement system.

myocardial backscatter, defined as the difference between maximum (perpendicular to fibers) and minimum (parallel to fibers) backscatter amplitude was determined. These same tissue specimens were subsequently treated with 10% NaOH for 7 to 10 days to dissolve all of the cellular components, leaving only the intact extracellular matrix and reinsonified after treatment to compare the anisotropy of integrated backscatter. Results showed the mean magnitude of the anisotropy of integrated backscatter of these specimens decreased from 15.4 ± 0.8 dB prior to digestion to 12.6 ± 1.1dB after digestion. Because the digestion of the myocardium leaves only the extracellular matrix as a source of ultrasonic scattering, and because backscatter measurements of the isolated extracellular matrix exhibited a similar ultrasonic anisotropy as that observed for the intact myocardium, it was concluded that there appears to be a direct association between the extracellular matrix and the anisotropy of backscatter within intact tissue. Hence this study suggests that the structure and function of the extracellular matrix may well be characterized using ultrasonic backscatter measurement techniques.

In an earlier study, Hoyt et al. [47] demonstrated the importance of collagen content in normal myocardium on measurements of ultrasonic backscatter. Measurements of integrated ultrasonic backscatter from the right and left ventricles of freshly excised normal canine hearts and formalin-fixed normal human hearts were obtained using a 2.25 MHz center-frequency ultrasonic measurement system. The segments of myocardium corresponding to the integrated ultrasonic backscatter sample volumes were assayed for hydroxyproline as a marker for collagen. Results from this study showed that, in freshly excised canine hearts, the integrated ultrasonic backscatter from right ventricle was significantly larger than that from left ventricle (-60.4 ± 1.6 vs. -66.9 ± 1.0 dB, respectively). This finding correlated with the hydroxyproline measurements that showed the collagen content of right ventricle to be significantly higher than that of left ventricle (4.40 ± 0.26 vs. 3.58 ± 0.13 micrograms/mg dry weight, respectively). Similar results were obtained for the measurements of human hearts. This study illustrates the relationship between collagen content and the level of ultrasonic backscatter in normal myocardium and is consistent with earlier papers [13, 34], which demonstrated this relationship in infarcted myocardial tissue, and confirmed the relationship between backscatter and collagen content in myocardium relative to other tissues with different collagen contents [48].

7. Characterization of Myocardial Infarction with Ultrasonic Backscatter

In 1981 O'Donnell et al. reported the relationship between ultrasonic backscatter and collagen concentration estimated from hydroxyproline concentration in hearts from dogs subjected to ischemic injury by coronary occlusion [13]. The ultrasonic backscatter was found to increase significantly in regions of ischemic injury studied several weeks following occlusion and the average backscatter coefficients were approximately six times larger in regions of infarct studied at 8 to 10 weeks after occlusion than the values obtained from regions of infarct studied 5 to 6 weeks after occlusion as illustrated in Figure 9.

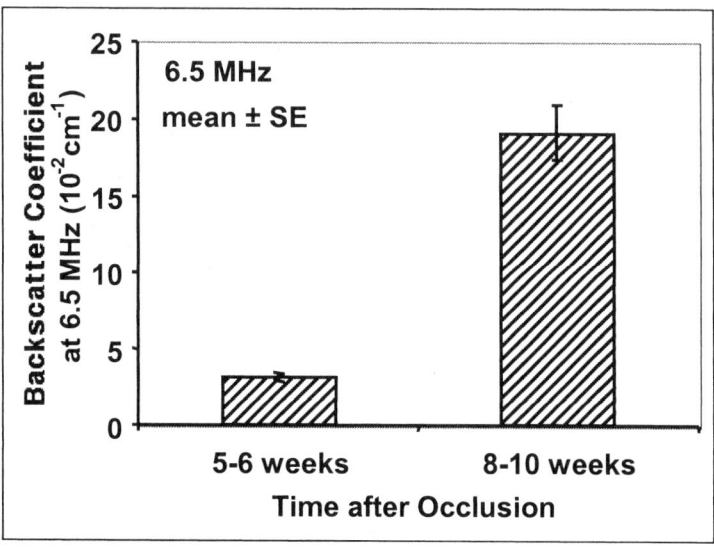

Figure 9. Measured backscatter coefficients at 6.5MHz from infarcted myocardium at specific times post-coronary occlusion. Adapted from data published by O'Donnell et al. [13].

However, it was observed that the average hydroxyproline concentration for regions of infarct (used as a quantitative index of molecular collagen content) was only approximately 25% larger for infarcted regions at 8 to 10 weeks compared with that for infarcted regions at 5 to 6 weeks after occlusion. These results suggest that the magnitude of ultrasonic scattering in regions of mature myocardial infarct is sensitive not only to the concentration of molecular collagen but, perhaps, to the organizational state of this structural protein as well.

These results appear consistent with a published investigation in which Mimbs et al. reported the dependence of ultrasonic backscatter on collagen content in rabbit hearts studied 5 to 7 weeks after coronary occlusion [34]. In this study it was found that both the ultrasonic backscatter coefficient (at 2.25 MHz) and collagen content within zones of infarction were increased significantly when compared with the values from nonischemic regions in the same rabbit hearts. In order to determine whether the increased ultrasonic backscatter depended on the content of intact collagen in regions of infarction, isolated rabbit hearts were perfused with either a modified Krebs-Henseleit solution or a modified Krebs-Henseleit solution containing a potent clostridial collagenase. Measurements showed a significant reduction of the backscatter coefficient in the infarcted regions of those hearts perfused with the modified Krebs-Henseleit solution with collagenase when compared with those hearts perfused with only the modified Krebs-Henseleit solution. However, measurements of the average hydroxyproline concentration did not change significantly in the infarcted regions of these two groups of hearts. This may not be surprising considering the protocol used in perfusing the hearts with collagenase resulted in a modification in the structure of collagen but not the complete removal of the hydroxyproline-rich peptide fragments. It does, however, suggest that the structural organization of the collagen may play an important role in determining the extent of backscattered ultrasound.

In a study designed to investigate the relationship between ultrasonic backscatter and collagen deposition in human hearts, Hoyt et. al published the results of measurements of excised human hearts with old infarcts.[49] They reported measurements of the backscattered ultrasonic energy from infarcted myocardial regions, expressed in terms of integrated backscatter values referenced to the reflection from a stainless-steel plate. Ultrasonic measurements were obtained utilizing a 2.25 MHz ultrasonic transducer. The interrogated myocardium was excised and portions were either assayed for hydroxyproline or stained with Masson's trichrome to provide an assessment of the transmural distribution of collagen. Results showed a linear correlation between the magnitude of integrated backscatter and myocardial collagen content estimated by hydroxyproline assay. However, quantitative histologic analysis revealed a somewhat variable relationship between the transmural distribution of collagen and the corresponding transmural pattern of the backscatter signal. Only in those specimens that exhibited a discrete layer of subendocardial fibrosis did the transmural collagen and ultrasonic backscatter profiles bear a close resemblance. In specimens with other patterns of fibrosis, the local backscatter amplitude did not correspond to the transmural pattern of collagen distribution. The authors speculated that this may be due to the limitations of the axial resolution of the ultrasonic backscatter measurement technique employed.

Measurements of the frequency content of backscatter and the magnitude of the backscattered signal from infarcted human myocardial specimens were reported by Wickline et al. [50]. In this study, cylindrical biopsy specimens from normal and infarcted regions of fixed, explanted human hearts were investigated. Backscatter data was obtained using a 5 MHz broadband transducer for specific angles of incidence with respect to the predominant myofiber orientation at specific transmural levels for each specimen. The dependence of the *apparent backscatter* (i.e., uncompensated for attenuation or beam width factors) on frequency was computed from spectral analyses of the backscattered signals. The frequency (f) dependence of the backscattered ultrasound was determined by fitting the backscattered data to a functional form of f^n over the frequency range of 3 to 7 MHz. Values for the *apparent integrated backscatter* were also determined over this same frequency bandwidth. Averaging the measurements over all angles of insonification and transmural levels resulted in average values for the *apparent frequency dependence, n,* to be 0.9 ± 0.1 for normal tissue and 0.6 ± 0.1 for infarcted tissue ($p < 0.05$). Further analyses showed that infarct manifested a significant decrease of n from epicardial to endocardial levels (epi: $n = 0.9$, mid: $n = 0.7$, endo: $n = 0.2$; $p < 0.05$) whereas normal tissue manifested similar values for n at each transmural level (epi: $n = 0.8$, mid: $n = 1.1$, endo: $n = 0.9$; $p = $ NS). The average value of integrated backscatter across all transmural levels for infarct (-48.3 ± 0.5 dB) was found to be significantly greater than the value for normal tissue (-53.4 ± 0.4 dB) ($p < 0.05$). Comparison of the ultrasonic measurements with histologic analyses revealed that the presence of fibrosis in the myocardial specimens was associated with smaller values of n and larger values of integrated backscatter. These results appear to be consistent with those reported by O'Donnell et al. [13] and by Wear et al. [51] both of whom reported that the frequency dependence of the backscatter coefficient was less in regions of infarct compared with the values obtained from regions of normal myocardium.

In a follow-on study, Hall et al. [17] demonstrated that both noninfarcted and infarcted myocardium exhibited anisotropy of the *apparent* frequency dependence of backscatter, with maxima occurring at angles of insonification perpendicular to the predominant myofiber direction and minima when parallel to the fibers. Backscattered ultrasonic data were obtained from specimens taken from noninfarcted and infarcted regions of the left ventricular free wall of formalin-fixed human hearts explanted because of ischemic cardiomyopathy. The *apparent* frequency dependence of ultrasonic backscatter (n) was determined over the frequency range of 3-7 MHz. Perpendicular insonification yielded results for n of 1.8 ± 0.1 for noninfarcted myocardium and 1.2 ± 0.1 for infarcted myocardium while parallel insonification yielded results of 0.4 ± 0.1 for noninfarcted and 0.0 ± 0.1 for infarcted myocardium.

8. Characterization of Cardiomyopathy and Aging with Ultrasonic Backscatter

Characterization of cardiomyopathy with ultrasonic backscatter was reported by Mimbs et al. [52] In this study, cardiomyopathy was induced by prolonged administration of doxorubicin (Adriamycin) in rabbits and results were compared with those obtained from normal control rabbits. The hearts from treated animals were excised at selected intervals 10 to 18 weeks after initiation of drug administration and integrated backscatter measurements were obtained along with analysis of collagen content based on hydroxyproline concentration.

Results showed a marked increase in the integrated backscatter values from regions of fibrosis (a 7 dB or approximately 500% increase) and a significant, although less marked, increase in myopathic regions without marked collagen deposition.

The use of ultrasonic backscatter measurements to characterize specific alterations in the three-dimensional transmural architecture of idiopathic dilated cardiomyopathy was reported by Wong et al. [53] In this study the frequency-dependent backscatter from specimens obtained from explanted fixed hearts of patients who underwent heart transplantation for idiopathic cardiomyopathy along with specimens of normal myocardial regions from and additional set of explanted fixed human hearts were investigated. Consecutive transmural levels from each specimen were insonified using a 5MHz broadband transducer and the *apparent* frequency dependence of backscatter was measured. Histologic analyses of myofiber diameter and percentage fibrosis were determined at each transmural level for each specimen. For regions of cardiomyopathic tissue, the *apparent* frequency dependence of backscatter (n) was found to increase progressively from the epicardial to endocardial levels (0.02 ± 0.37 to 1.01 ± 0.12, $p < 0.05$) in conjunction with a progressive decrease in myofiber diameter (29.5 ± 0.9 to 21.4 ± 0.6 microns, $p < 0.0001$). In contrast, in tissue from areas of infarction, the apparent frequency dependence decreased progressively from epicardium to endocardium (0.91 ± 0.20 to 0.23 ± 0.21, $p < 0.05$) in conjunction with a progressive increase in the percentage of fibrosis ($23.5 \pm 9.4\%$ to $54.5 \pm 4.9\%$, $p < 0.005$). Specimens of normal tissue exhibited no significant transmural trend for the apparent frequency dependence, myofiber diameter, or percentage fibrosis. These data indicate the presence of a heterogeneous transmural distribution of scattering structures associated with human idiopathic cardiomyopathy and myocardial infarction that may be detected by ultrasonic backscatter.

In a study designed to assess changes in myocardial architecture associated with cardiomyopathy characterized by diffuse interstitial fibrosis, Wong et al. [54] reported measurements of ultrasonic backscatter from the hearts of 4- to 6-month-old *tight-skin mice*, which exhibit a cardiomyopathy analogous to cardiac scleroderma with excessive interstitial and perivascular fibrosis. Ultrasonic backscatter was measured from excised segments of left ventricular free walls of tight-skin mice and sex- and age-matched normal controls using a 50 MHz center frequency, broadband ultrasonic measurement system. Integrated backscatter from tight-skin mice (-53.6 ± 0.6 dB) was larger than the value obtained from the control group (-56.6 ± 0.7 dB; $p < 0.02$). The myocardial collagen content determined by hydroxyproline assay was found to be 11% larger in the tight-skin mice compared with the controls. A significant linear relationship was observed between myocardial hydroxyproline concentration and integrated backscatter values ($r = 0.74$; $p < 0.02$). Thus, ultrasonic tissue characterization appears to permit sensitive detection of modest changes in the extent of interstitial fibrosis that accompany tissue remodeling in the early stages of cardiomyopathy.

The use of backscatter measurements as a means to assess the effects of aging and cardiomyopathy was reported by Perez et al. [55] where excised hearts from Syrian hamsters and normal hamsters (controls) were measured at specific ages. Integrated backscatter measurements, obtained using a broadband 25 MHz center-frequency data acquisition system, showed young Syrian hamsters to exhibit average values (-53.87 ± 0.26 dB) significantly larger than the values from age-matched control hamsters (-58.07 ± 0.08 dB). Cardiomyopathic hearts from older animals also exhibited average backscatter values (-50.87 ± 0.22 dB) significantly larger than the values from age-matched control hamsters (-55.91 ± 0.11 dB). These results correlated with calcification and fibrosis characteristics assessed histopathologically.

In a related study addressing the effects of angiotensin-converting enzyme (ACE) inhibitors on the measured backscatter from scar tissue and grossly normal myocardium in experimental dilated cardiomyopathy, Davison et al. [56] reported ultrasonic measurements from cardiomyopathic Syrian hamsters treated with captopril compared with those from untreated cardiomyopathic Syrian hamsters. Ultrasonic integrated backscatter measurements were obtained from the excised hearts of the animals using a broad bandwidth 50-MHz acoustic microscope system. Results showed the integrated backscatter values from the hearts of treated cardiomyopathic hamsters to be significantly less in both grossly normal regions of myocardium and scar tissue regions when compared with integrated backscatter values from hearts of the untreated hamsters. This observed reduction in integrated backscatter from hearts of treated cardiomyopathic hamsters suggests direct alterations in the material properties of cardiomyopathic hearts after captopril therapy.

Effects of age-related alterations of cardiac tissue on backscatter measurements were reported by Nguyen et al.[36] Using a 50 MHz nominal-frequency acoustic microscope system, measurements of integrated backscatter from aging Fischer 344 rats was performed. Results demonstrated a progressive increase in integrated backscatter values (from -44.7 ± 1.8 to -40.8 ± 1.9 dB) and backscatter coefficients (from $0.73 \pm 0.16 \times 10^{-5}$ to $3.76 \pm 1.6 \times 10^{-5}$ cm^{-1}), from 6 to 24 months of age. These data correlate with the trend observed in the data from the biochemical analyses showing a progressive increase in collagen concentration with age. These results suggest that measurements of ultrasonic backscatter may prove to be a sensitive tool to monitor changes in the cardiac microstructure, such as increased collagen deposition, that occur within age-related diastolic dysfunction.

9. Noninvasive Clinical Ultrasonic Characterization of Myocardium

From the studies described above it is evident that measurements of ultrasonic parameters can provide a quantitative assessment of the nature of myocardial fibrosis. Thus the continuing development and utilization of noninvasive ultrasonic tissue characterization techniques in the clinical setting remains a worthy goal. Perhaps the biggest challenge associated with transferring these ultrasonic measurement techniques from the assessment of excised tissue specimens to the clinical arena is the identification of appropriate and reliable reference sources and techniques. This is necessary in order to compensate the measured ultrasonic signals from patients for the effects of the ultrasonic imaging system and thus provide measurements of the inherent properties of the myocardium. Several approaches have been proposed and implemented in order to obtain "absolute" measurements of ultrasonic parameters (most often backscatter) from patients. These include measurements of backscattered energy from myocardium referenced to the specular echo from the pericardial region [57-64] or referenced to the backscattered energy from blood obtained from the Doppler power signal [65-69]. Several studies have been performed which show the relationship between the magnitude of the backscatter signal and degree of myocardial fibrosis measured in patients [60, 62, 70-74].

Perhaps one of the most successful approaches to clinical ultrasonic tissue characterization is the measurement of the systematic variation of backscattered energy from the myocardium over the heart cycle. This measurement of the "cyclic variation" of backscattered energy was first reported by Madaras et al. [75] and has been successfully applied to characterize a large spectrum of specific cardiac pathologies [8, 9, 76, 77]. The relative success cyclic variation-based methods have enjoyed in diagnosing specific cardiac pathologies in a clinical environment is due, in part, to the self-reference nature of these measurements. That is, the quantification of the cyclic variation of backscatter from the myocardium over the heart cycle does not require the acquisition or estimation of separate reference backscatter measurements as would be required for "absolute" measurements of myocardial tissue properties. This self-reference approach enables cyclic variation to be a very robust approach to characterize myocardial tissue and relatively easy to implement clinically.

Measurements of the cyclic variation of backscattered energy from specific myocardial regions are obtained by placing a region-of-interest in the mid-myocardium of two-dimensional or M-mode integrated backscatter images and adjusting its position over the heart cycle so that approximately the same region of myocardium is investigated. The average integrated backscatter value within the region-of-interest is determined subsequently yielding a data trace of mean backscatter values as a function of time. This is illustrated in Figure 10.

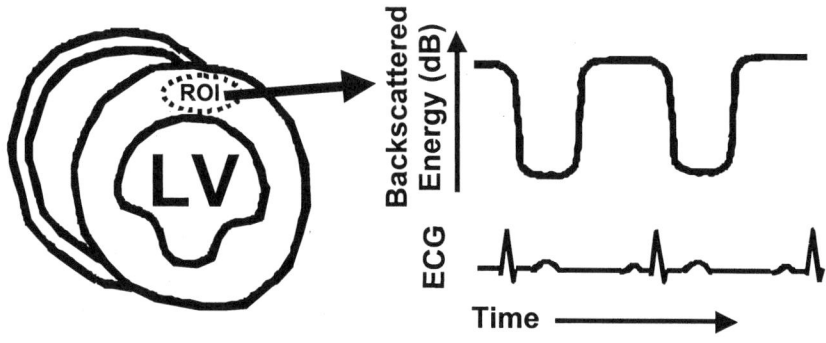

Figure 10. Systematic variation of backscattered energy over the heart cycle (cyclic variation).

Cyclic variation of backscatter data is often characterized in terms of its *magnitude* and *normalized time delay* [78, 79]. The magnitude of cyclic variation is defined as the difference in backscatter between the *average peak* and *average nadir* values. The corresponding normalized time delay of cyclic variation is expressed in terms of a dimensionless ratio, obtained by dividing the time interval from end-diastole to the nadir of the integrated backscatter trace (Δt_n) by the systolic interval (Δt_s). This parameterization of the cyclic variation data is depicted in Figure 11.

Although the specific mechanisms responsible for the observed cyclic variation of backscattered energy over the heart cycle remain unspecified, one model proposed by Wickline et al. [3, 4] illustrates the specific role the inherent properties of the extracellular matrix may contribute. In this model, contributions to the observed cyclic variation of backscattered energy are due, in part, to a systematic variation in the relative acoustic impedance differences between intracellular and extracellular elastic properties. Because differences in acoustic impedance are determined partly by differences in elastic moduli, changes in local elastic moduli resulting from the non-Hookian behavior of myocardial elastic elements may alter the degree of local acoustic impedance mismatch over the heart cycle and hence concomitant changes in backscatter.

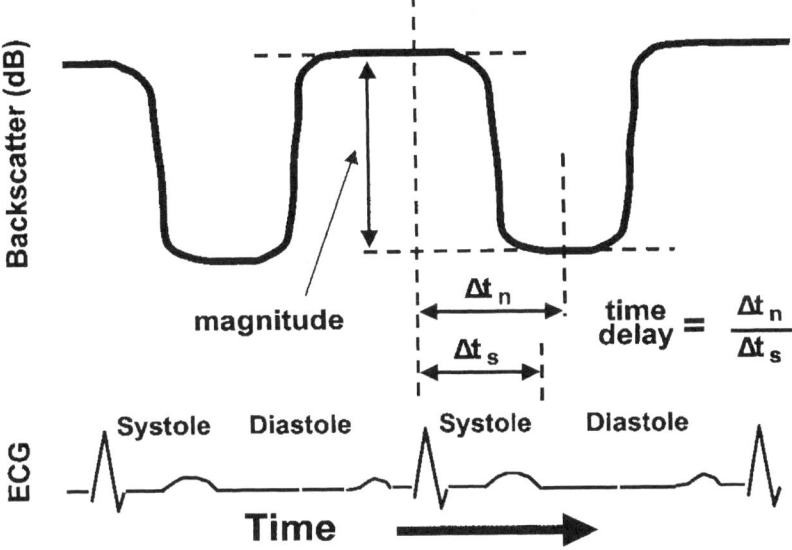

Figure 11. Magnitude and normalized time delay of cyclic variation.

Figure 12 depicts the three-component Maxwell-type model of muscle mechanics used. In this model the intracellular contractile element shortens during systole and hence stretches the series-elastic element. The series-elastic element demonstrates non-Hookian elastic behavior with elongation and therefore the elastic modulus (E_{se}) increases leading to increased intracellular stiffness and, hence, acoustic impedance during contraction. Thus, the intracellular impedance will become closer in value to the relatively larger impedance of the stiffer extracellular impedance during systole resulting in concomitant reduction in acoustic impedance mismatch.

However, other sources may also contribute significantly to the observed cyclic variation of backscattered energy. These include changes in scatterer (myocyte) size and number density as well as geometric alterations induced by myofiber strain over the heart cycle. Glueck et al. showed that superfused stimulated frog gastrocnemius preparations manifested changes in backscatter from the relaxed to the tetanized state [80]. Wear et al. reported that isotonic but not isometric contractions induced altered backscatter in stimulated superfused papillary muscle preparations [81]. The Rose model [7] allows for changes in geometry as a source of cyclic variation, since the backscatter coefficients depend on shape and orientation of the principal scattering unit, which can

138

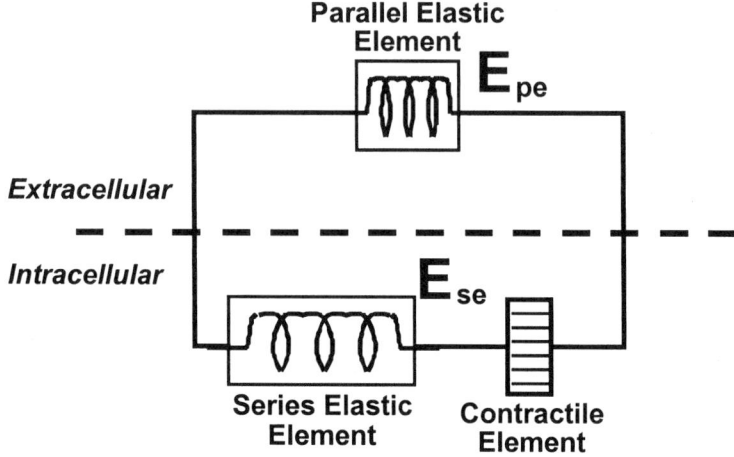

Figure 12. Model incorporating contributions of intra- and extracellular elastic properties to the observed cyclic variation of backscatter. Adapted from Wickline et al. [3].

change across the heart cycle as myofibers contract (by shortening and thickening) and also alter their principal orientation through shear mechanisms. The effects of myocardial fiber orientation can be readily observed and quantified in echocardiographic images of humans [30] and may play a significant role in the observed echocardiographic view-dependence of cyclic variation [82-85].

Regardless of the mechanism of cyclic variation, a large body of literature attests to its potential clinical utility for diagnosing cardiac pathology. A sampling of the published literature shows how this approach has been applied successfully in the characterization of ischemia and infarction [83, 86-95], cardiomyopathy [69, 96-106], diabetes [107-112], hypertension [97, 109, 110, 113-121], hypertrophy [67, 113, 114, 122], heart failure [123, 124], and aging [125] among other pathologies. More comprehensive reviews of cyclic variation in myocardial tissue characterization can be found in the published literature [9, 10, 126, 127]. In the following paragraphs, we offer several specific examples which help illustrate the potential of cyclic variation measurements as a method of myocardial assessment in the clinical setting.

9.1. Clinical Utility of Cyclic Variation Measurements.

In a study designed to investigate the use of cyclic variation measurements for detection of acute myocardial infarction and reperfusion,

Milunski et al. reported measurements on patients studied within the first 24 hours after the onset of symptoms indicative of acute myocardial infarction [86]. The measured mean magnitude of cyclic variation was found to be significantly larger in normal myocardial regions (4.8 ± 0.5 dB) remote from the infarct zone compared with the value from acute infarct regions (0.8 ± 0.3 dB) within the first 24 hours after the onset of symptoms. In subsequent studies after therapeutic intervention, those patients exhibiting vessel patency (by coronary arteriography) demonstrated a significant increase in the magnitude of cyclic variation from the infarct regions. However, those patients with occluded infarct-related arteries exhibited no significant recovery of cyclic variation. Blinded analyses of standard two-dimensional echocardiographic images revealed no significant recovery of wall thickening in either group over the same time intervals. Hence this study suggests that cyclic variation measurements can promptly detect acute myocardial infarction and may delineate the beneficial effects of coronary artery reperfusion manifest by restoration of cyclic variation in the continued presence of severe wall motion abnormalities.

Masuyama et al. published a study in which cyclic variation measurements were used to assess acute cardiac allograft rejection [128]. Measurements of the magnitude of cyclic variation from cardiac allograft recipients, performed within 24 hours of right ventricular endomyocardial biopsy acquired for rejection surveillance, were compared with the cyclic variation measurements from normal subjects. Results indicated no significant difference in the measured magnitude of cyclic variation between the normal subjects and the cardiac allograft recipients without previous or current histological evidence of acute rejection. However, a significant decrease in the cyclic variation from the septum was observed in cardiac allograft recipients exhibiting left-ventricular hypertrophy compared with the normal subjects. Studies performed on those recipients exhibiting moderate acute rejection showed a significant decrease in the magnitude of the cyclic variation when values were compared before and during rejection. These data suggest acute cardiac rejection is accompanied by an alteration in the magnitude of cyclic variation properties of the myocardium which is detectable by serial measurements.

Koyama et al. demonstrated the prognostic significance of cyclic variation measurements in patients with cardiac amyloidosis [129]. In a study that prospectively examined 208 consecutive biopsy-proven patients with primary amyloidosis, the magnitude of cyclic variation was analyzed and its prognostic value was compared with standard Doppler measurements using the Tei index (isovolumic contraction time plus isovolumic relaxation time divided by ejection time). Univariate analysis showed that cyclic variation was the best predictor of cardiac death and all-cause death and that the Tei index did not identify those patients at risk of death. Multivariate analysis showed that measurement of the magnitude of cyclic variation from the posterior wall was the only independent predictor of both cardiac and overall deaths. This study shows that the measurement of cyclic variation is a powerful predictor of clinical outcome and appears superior to standard echocardiographic indexes among patients with cardiac amyloidosis.

A study which addressed the measurement of cyclic variation in diabetic patients and its correlation with the severity of disease was published in a 1992 paper by Perez et al. [107]. The objective of this paper was to determine if cyclic variation measurements could detect changes potentially indicative of occult cardiomyopathy in patients with insulin-dependent diabetes exhibiting no overt cardiac disease. Measurements of the magnitude of cyclic variation were found to be significantly reduced and the normalized time delay was found to be significantly increased in diabetic patients when compared with the corresponding values from previous measurements of age-matched patients without diabetes. The reduction of magnitude was greatest in patients with diabetes who had neuropathy, as was the increase in delay. Retinopathy and nephropathy were associated with abnormal cyclic variation as well. In a follow-on paper distinguishing diabetic patients from normal control subjects, Wagner et al. [130] showed specific combinations of the cyclic variation parameters yielded areas under the receiver operating characteristic (ROC) curves of approximately 0.80 (0.90 for such patients with retinopathy). This is comparable to the performance of many commonly used diagnostic tests. These studies suggest occult cardiomyopathic changes in diabetic patients without overt heart disease are readily detectable by cyclic variation measurements and the results parallel the severity of noncardiac diabetic complications.

9.2. Correlation Between Cyclic Variation Measurements and Degree of Fibrosis.

The relationship between measurements of cyclic variation and degree of fibrosis in human subjects was demonstrated in a series of recent papers by Maceira et al. [120, 121]. In a study designed to assess whether the magnitude of cyclic variation was related to the severity of myocardial fibrosis as estimated by serum concentration of the carboxy-terminal propeptide of procollagen type I (PIP) [120], subjects were divided into 3 groups: normotensives with PIP <127 μg/L (group 1), hypertensives with PIP <127 mg/L (group 2), and hypertensives with PIP >127 mg/L (group 3). The magnitude of cyclic variation was found to be the largest in group 1 and the least in group 3 with an intermediate value for those subjects in group 2. Analyses utilizing receiver operating characteristic curves yielded 75% sensitivity and 63% specificity for predicting PIP >127 mg/L in hypertensives for the magnitude of cyclic variation decision threshold set at a value of 2.90 dB for measurements in the apex. These results suggest an association between diminished magnitude of cyclic variation of backscatter and increased serum concentration of PIP in hypertension.

One hundred nine subjects were included in a study designed to assess whether cyclic variation measurements were related to the severity of diastolic dysfunction as assessed by Doppler echocardiography in patients with essential hypertension [121]. The largest magnitude of cyclic variation was found for normotensives with normal diastolic function and hypertensives with normal diastolic function and the smallest values were associated with hypertensives with a pseudonormal filling pattern and hypertensives with a restrictive filling pattern. Intermediate values of the magnitude were found for hypertensives with a delayed relaxation pattern. The magnitude of cyclic variation was found to be inversely correlated with left ventricular chamber stiffness and directly correlated with midwall fractional shortening in all hypertensives. Hence measurements of this parameter may be useful for the assessment of diastolic dysfunction in hypertension.

The relationship between right-ventricular endomyocardial biopsy and measurement of cyclic variation in patients with dilated cardiomyopathy was reported by Fujimoto et al. [101]. Results indicated that the magnitude of cyclic variation was significantly lower in patients with dilated cardiomyopathy when compared with normal controls. Biopsy findings showed the decrease in cyclic variation measurements to be correlated with the extent of fibrosis in myocardial tissue and the myocyte diameter suggesting that the magnitude of cyclic variation is a good indicator of the severity of fibrosis and myocyte atrophy in patients with dilated cardiomyopathy. Similarly, in a study designed to compare cyclic variation measurements with endomyocardial biopsy findings in patients with hypertrophic cardiomyopathy, Mizuno et al. [70] found the magnitude of cyclic variation to be significantly smaller in patients with hypertrophic cardiomyopathy compared with normal subjects. Biopsy results from patients with hypertrophic cardiomyopathy showed the degree of myocardial disarray, interstitial fibrosis, and nonhomogeneity of myocyte size correlated with a decrease in the magnitude of cyclic variation. These results suggest that cyclic variation measurements enables the noninvasive evaluation of myocardial histological abnormalities in patients with hypertrophic cardiomyopathy.

10. Summary

Hopefully this chapter has provided the reader with insights regarding how measurements of fundamental ultrasonic parameters are related to the presence of myocardial fibrosis and how these measurements lay the foundation for the continuing development of clinically applicable noninvasive ultrasonic assessment techniques. Several studies were cited which show how ultrasonic propagation and scattering are directly related to fundamental physical properties of myocardium and hence ultrasonic tissue characterization techniques are well-suited for the direct measurement of specific myocardial properties. The clinical utility of ultrasound-based noninvasive assessment techniques, such as measurements of the cyclic variation of backscattered energy over the heart cycle, have been clearly demonstrated. Nonetheless significant challenges remain before these techniques can be routinely applied in clinical practice. Enhanced measurement techniques which increase the reliability and decrease the uncertainty associated with clinical estimates of ultrasonic parameters will help foster the continued acceptance of these methods. Development of additional robust reference measurement methods that permit compensation of the measured ultrasonic backscatter for the effects of the imaging system will provide absolute measures of inherent ultrasonic properties of myocardium. A better understanding of the underlying mechanisms associated with the observed ultrasonic properties (e.g., the role of scatterer (myocyte) size, number density, myocardial fiber orientation, inherent viscoelastic properties of the intra- and extracellular components) will enhance the interpretation of ultrasonic measurements and permit direct assessment of specific myocardial properties. The future for ultrasonic-based methods of directly characterizing myocardium continues to look promising and these techniques will continue to be useful tools for the assessment of properties such as the degree and nature of myocardial fibrosis.

References

1. Feigenbaum, H., Echocardiography. Lea & Febiger, 1994. 5th Ed.
2. Weyman, A., Principles and Practice of Echocardiography. Lea & Febiger, 1994. 2nd Ed.
3. Wickline, S.A., et al., A relationship between ultrasonic integrated backscatter and myocardial contractile function. J Clin Invest, 1985. 76: p. 2151-60.
4. Wickline, S.A., et al., The dependence of myocardial ultrasonic integrated backscatter on contractile performance. Circulation, 1985. 72: p. 183-92.
5. Wickline, S., J. Perez, and J. Miller, Cardiovascular Tissue Characterization *in vivo*. In: Ultrasonic Scattering in Biological Tissue. Shung, KK, Thieme, G, eds. Boca Raton, FL: CRC Press, 1993: p. 313-345.
6. Kumar, K. and J. Mottley, Quantitative Modeling of the Anisotropy of Ultrasonic Backscatter from Canine Myocardium. IEEE Trans Ultrasonics Ferroelect Freq Contr., 1994. 41: p. 441-450.
7. Rose, J.H., et al., A proposed microscopic elastic wave theory for ultrasonic backscatter from myocardial tissue. J Acoust Soc Am, 1995. 97: p. 656-68.
8. Miller, J., et al., Backscatter Imaging and Myocardial Tissue Characterization. In: Proc IEEE Ultrasonic Symposium. Symposium- Sendai,Japan, 1998: p. 1373-1383.
9. Perez, J., et al., Ultrasonic Characterization of Cardiovascular Tissue. In: Cardiac Imaging - A companion to Brauwald's Heart Disease. 2nd Ed. Skorton DJ, Seelbert HR, Wolf GL, Brundage BH, eds. W.B. Saunders Co., 1996: p. 606-622.

10. Miller, J.G., J.E. Perez, and B.E. Sobel, Ultrasonic characterization of myocardium. Prog Cardiovasc Dis, 1985. 28: p. 85-110.

11. Wells, P., Biomedical Ultrasonics. London: Academic Press, 1977.

12. Fei, D.Y. and K.K. Shung, Ultrasonic backscatter from mammalian tissues. J Acoust Soc Am, 1985. 78: p. 871-6.

13. O'Donnell, M., J.W. Mimbs, and J.G. Miller, Relationship between collagen and ultrasonic backscatter in myocardial tissue. J Acoust Soc Am, 1981. 69: p. 580-8.

14. Shung, K., Ultrasound Scattering from Blood as a Function of Hemocrit. IEEE Transactions on Sonics and Ultrasonics, 1982. SU-29: p. 327-331.

15. Shung, K., Physics of Blood Echogenicity. J Cardiovascular Ultrasonogr, 1983. 2: p. 401-406.

16. Waag, R.C., A review of tissue characterization from ultrasonic scattering. IEEE Trans Biomed Eng, 1984. 31: p. 884-93.

17. Hall, C.S., et al., Anisotropy of the apparent frequency dependence of backscatter in formalin fixed human myocardium. J Acoust Soc Am, 1997. 101: p. 563-8.

18. Miller, J., et al., Myocardial Tissue Characterization: An Approach Based on Quantitative Backscatter and Attenuation. Proc. IEEE Ultrasonics Symposium-Atlanta, 1983. CH 1947: p. 782-793.

19. O'Donnell, M. and J. Miller, Quantitative Braodband Ultrasonic Backscatter: An Approach to Non-Destructive Evaluation in Acoustically Inhomogeneous Materials. J Appl Physics, 1981. 52: p. 1056-1065.

20. Rose, J. and J. Richardson, Time Domain Born Approximation. J Appl Physics, 1982. 3: p. 45-53.

21. Hoffmeister, B.K., et al., Comparison of the anisotropy of apparent integrated ultrasonic backscatter from fixed human tendon and fixed human myocardium. J Acoust Soc Am, 1995. 97: p. 1307-13.

22. Madaras, E.I., et al., Anisotropy of the ultrasonic backscatter of myocardial tissue: II. Measurements in vivo. J Acoust Soc Am, 1988. 83: p. 762-9.

23. Mottley, J.G. and J.G. Miller, Anisotropy of the ultrasonic backscatter of myocardial tissue: I. Theory and measurements in vitro. J Acoust Soc Am, 1988. 83: p. 755-61.

24. Mottley, J.G. and J.G. Miller, Anisotropy of the ultrasonic attenuation in soft tissues: measurements in vitro. J Acoust Soc Am, 1990. 88: p. 1203-10.

25. Verdonk, E.D., et al., Anisotropy of the slope of ultrasonic attenuation in formalin fixed human myocardium. J Acoust Soc Am, 1996. 99: p. 3837-43.

26. Hoffmeister, B.K., et al., Effect of collagen on the anisotropy of quasi-longitudinal mode ultrasonic velocity in fibrous soft tissues: a comparison of fixed tendon and fixed myocardium. J Acoust Soc Am, 1994. 96: p. 1957-64.

27. Hoffmeister, B.K., et al., Ultrasonic determination of the anisotropy of Young's modulus of fixed tendon and fixed myocardium. J Acoust Soc Am, 1996. 100: p. 3933-40.

28. Verdonk, E.D., S.A. Wickline, and J.G. Miller, Anisotropy of ultrasonic velocity and elastic properties in normal human myocardium. J Acoust Soc Am, 1992. 92: p. 3039-50.

29. Recchia, D., J.G. Miller, and S.A. Wickline, Quantification of ultrasonic anisotropy in normal myocardium with lateral gain compensation of two-dimensional integrated backscatter images. Ultrasound Med Biol, 1993. 19: p. 497-505.

30. Holland, M.R., et al., Effects of myocardial fiber orientation in echocardiography: quantitative measurements and computer simulation of the regional dependence of backscattered ultrasound in the parasternal short-axis view. J Am Soc Echocardiogr, 1998. 11: p. 929-37.

31. O'Donnell, M., et al., Collagen as a Determinant of Ultrasonic Attenuation in Myocardial Infarcts. Ultrasound in Medicine., 1978. 4: p. 503-513.

32. O'Donnell, M., J.W. Mimbs, and J.G. Miller, The relationship between collagen and ultrasonic attenuation in myocardial tissue. J Acoust Soc Am, 1979. 65: p. 512-7.

33. Mimbs, J.W., et al., Changes in ultrasonic attenuation indicative of early myocardial ischemic injury. Am J Physiol, 1979. 236: p. H340-4.

34. Mimbs, J.W., et al., The dependence of ultrasonic attenuation and backscatter on collagen content in dog and rabbit hearts. Circ Res, 1980. 47: p. 49-58.

35. Saijo, Y., et al., Ultrasonic tissue characterization of infarcted myocardium by scanning acoustic microscopy. Ultrasound Med Biol, 1997. 23: p. 77-85.

36. Nguyen, C.T., et al., Age-related alterations of cardiac tissue microstructure and material properties in Fischer 344 rats. Ultrasound Med Biol, 2001. 27: p. 611-9.

37. Weber, K.T., Cardiac interstitium in health and disease: the fibrillar collagen network. J Am Coll Cardiol, 1989. 13: p. 1637-52.

38. Spector, K.S., Diabetic cardiomyopathy. Clin Cardiol, 1998. 21: p. 885-7.

39. Hall, C., et al., High Frequency Ultrasound Detection of the Temporal Evolution of Protein Cross-Linking in Myocardial Tissue. IEEE Trans Ultrasound Ferroelec Freq Contr., 2000. 47: p. 1051-1058.

40. Fields, S. and F. Dunn, Letter: Correlation of echographic visualizability of tissue with biological composition and physiological state. J Acoust Soc Am, 1973. 54: p. 809-12.

41. O'Brien, W., The Role of Collagenn in Determining Ultrasonic Propagation Properties in Tissue. In: Acoustical Holography. Kessler, L ed. New York, NY : Plenum, 1977: p. 37-50.

42. Shung, K.K. and J.M. Reid, Ultrasound velocity in major bovine blood vessel walls. J Acoust Soc Am, 1978. 64: p. 692-4.

43. Saijo, Y., et al., Ultrasonic tissue characterization of diseased myocardium by scanning acoustic microscopy. J Cardiol, 1995. 25: p. 127-32.

44. Chandraratna, P.A., et al., Characterization of collagen by high-frequency ultrasound: evidence for different acoustic properties based on collagen fiber morphologic characteristics. Am Heart J, 1997. 133: p. 364-8.

45. Thomas, L.J III., et al., Quantitative Real-Time Imaging of Myocardium Based on Ultrasonic Integrated Backscatter. IEEE Trans Ultrason Ferroelect Freq Contr., 1989. UFFC-36: p. 466-470.

46. Hall, C.S., et al., The extracellular matrix is an important source of ultrasound backscatter from myocardium. J Acoust Soc Am, 2000. 107: p. 612-9.

47. Hoyt, R.M., et al., Ultrasonic backscatter and collagen in normal ventricular myocardium. Circulation, 1984. 69: p. 775-82.

48. Pohlhammer, J. and W.D. O'Brien, Jr., Dependence of the ultrasonic scatter coefficient on collagen concentration in mammalian tissues. J Acoust Soc Am, 1981. 69: p. 283-5.

49. Hoyt, R.H., et al., Assessment of fibrosis in infarcted human hearts by analysis of ultrasonic backscatter. Circulation, 1985. 71: p. 740-4.

50. Wickline, S.A., et al., Identification of human myocardial infarction in vitro based on the frequency dependence of ultrasonic backscatter. J Acoust Soc Am, 1992. 91: p. 3018-25.

51. Wear, K.A., et al., Differentiation between acutely ischemic myocardium and zones of completed infarction in dogs on the basis of frequency-dependent backscatter. J Acoust Soc Am, 1989. 85: p. 2634-41.

52. Mimbs, J.W., et al., Detection of cardiomyopathic changes induced by doxorubicin based on quantitative analysis of ultrasonic backscatter. Am J Cardiol, 1981. 47: p. 1056-60.

53. Wong, A.K., et al., Detection of unique transmural architecture of human idiopathic cardiomyopathy by ultrasonic tissue characterization. Circulation, 1992. 86: p. 1108-15.

54.   Wong, A.K., et al., Quantification of ventricular remodeling in the tight-skin mouse cardiomyopathy with acoustic microscopy. Ultrasound Med Biol, 1993. 19: p. 365-74.

55.   Perez, J.E., et al., Applicability of ultrasonic tissue characterization for longitudinal assessment and differentiation of calcification and fibrosis in cardiomyopathy. J Am Coll Cardiol, 1984. 4: p. 88-95.

56.   Davison, G., et al., Ultrasonic tissue characterization of end-stage dilated cardiomyopathy. Ultrasound Med Biol, 1995. 21: p. 853-60.

57.   Tanaka, M., Y. Teresawa, and H. Hikichi, Quantitative Evaluation of the Heart Tissue by Ultrasound. Journal of Cardiography, 1977. 7: p. 515-530.

58.   Hikichi, H. and M. Tanaka, Ultrasono-cardiotomographic evaluation of histological changes in myocardial infarction. Jpn Heart J, 1981. 22: p. 287-98.

59.   Lattanzi, F., et al., Normal ultrasonic myocardial reflectivity in athletes with increased left ventricular mass. A tissue characterization study. Circulation, 1992. 85: p. 1828-34.

60.   Gigli, G., et al., Normal ultrasonic myocardial reflectivity in hypertensive patients. A tissue characterization study. Hypertension, 1993. 21: p. 329-34.

61.   Lucarini, A.R., et al., Regression of hypertensive myocardial hypertrophy does not affect ultrasonic myocardial reflectivity: a tissue characterization study. J Hypertens, 1994. 12: p. 73-9.

62.   Di Bello, V., et al., Increased echodensity of myocardial wall in the diabetic heart: an ultrasound tissue characterization study. J Am Coll Cardiol, 1995. 25: p. 1408-15.

63.   Di Bello, V., et al., Increased myocardial echo density in left ventricular pressure and volume overload in human aortic valvular disease: an ultrasonic tissue characterization study. J Am Soc Echocardiogr, 1997. 10: p. 320-9.

64.   Lucarini, A.R., et al., Increased myocardial ultrasonic reflectivity is associated with extreme hypertensive left ventricular hypertrophy: a tissue characterization study in humans. Am J Hypertens, 1998. 11: p. 1442-9.

65.   Nakayyama, K. and S. Yagi, In vivo Tissue Characterization Using Blood Flow Doppler Signal as a Reference. In: Proc. Japan Soc. Ultrason Med. Japan Soc Ultrason Med., 1988: p. 399-400.

66.   Shiba, A., et al., A Measerement Method for Absolute Value of Integrated Backscatter. In: IEEE 1991. Ultrasonics Symposium, 1991: p. 1089-1092.

67.   Naito, J., et al., Analysis of transmural trend of myocardial integrated ultrasound backscatter for differentiation of hypertrophic cardiomyopathy and ventricular hypertrophy due to hypertension. J Am Coll Cardiol, 1994. 24: p. 517-24.

68.   Naito, J., et al., Validation of transthoracic myocardial ultrasonic tissue characterization: comparison of transthoracic and open-chest measurements of integrated backscatter. Ultrasound Med Biol, 1995. 21: p. 33-40.

69.   Naito, J., et al., Dobutamine stress ultrasonic myocardial tissue characterization in patients with dilated cardiomyopathy. J Am Soc Echocardiogr, 1996. 9: p. 470-9.

70.   Mizuno, R., et al., Myocardial ultrasonic tissue characterization for estimating histological abnormalities in hypertrophic cardiomyopathy: comparison with endomyocardial biopsy findings. Cardiology, 2001. 96: p. 16-23.

71.   Naito, J., et al., Ultrasonic myocardial tissue characterization in patients with dilated cardiomyopathy: value in noninvasive assessment of myocardial fibrosis. Am Heart J, 1996. 131: p. 115-21.

72.   Lythall, D.A., et al., Relationship between myocardial collagen and echo amplitude in non-fibrotic hearts. Eur Heart J, 1993. 14: p. 344-50.

73.   Picano, E., et al., In vivo quantitative ultrasonic evaluation of myocardial fibrosis in humans. Circulation, 1990. 81: p. 58-64.

74.   Shaw, T.R., et al., Relation between regional echo intensity and myocardial connective tissue in chronic left ventricular disease. Br Heart J, 1984. 51: p. 46-53.

75.   Madaras, E.I., et al., Changes in myocardial backscatter throughout the cardiac cycle. Ultrason Imaging, 1983. 5: p. 229-39.

76.   Tamirisa, P.K., et al., Ultrasonic tissue characterization: review of an approach to assess hypertrophic myocardium. Echocardiography, 2001. 18: p. 593-7.

77.   Donal, E., et al., Tissue characterization by study of myocardial response to ultrasound. Review of the literature and perspectives. Arch Mal Coeur Vaiss, 2000. 93: p. 301-8.

78.   Mobley, J., et al., Clinical Tissue Characterization: Online Determination of Magnitude and Time Delay Myocardial Backscatter. Video J Echocardiogr., 1995. 5: p. 40-48.

79.   Mohr, G.A., et al., Automated determination of the magnitude and time delay ("phase") of the cardiac cycle dependent variation of myocardial ultrasonic integrated backscatter. Ultrason Imaging, 1989. 11: p. 245-59.

80.   Glueck, R.M., et al., Changes in ultrasonic attenuation and backscatter of muscle with state of contraction. Ultrasound Med Biol, 1985. 11: p. 605-10.

81.   Wear, K., T. Shoup, and R. Popp, Ultrasonic Characterization of Canine Myocardium Contraction. IEEE Trans Ultrasonics Ferroelect Freq Contr., 1986. UFFC-33: p. 347-353.

82.   Vandenberg, B.F., et al., Cyclic variation of ultrasound backscatter in normal myocardium is view dependent: clinical studies with a real-time backscatter imaging system. J Am Soc Echocardiogr, 1989. 2: p. 308-14.

83.   Vandenberg, B.F., et al., Diagnosis of recent myocardial infarction with quantitative backscatter imaging: preliminary studies. J Am Soc Echocardiogr, 1991. 4: p. 10-8.

84.   Lange, A., et al., The variation of integrated backscatter in human hearts in differing ultrasonic transthoracic views. J Am Soc Echocardiogr, 1995. 8: p. 830-8.

85.   Bouki, K.P., et al., Regional variations of ultrasonic integrated backscatter in normal and myopathic left ventricles. A new multi-view approach. Eur Heart J, 1996. 17: p. 1747-55.

86.   Milunski, M.R., et al., Ultrasonic tissue characterization with integrated backscatter. Acute myocardial ischemia, reperfusion, and stunned myocardium in patients. Circulation, 1989. 80: p. 491-503.

87.   Vered, Z., et al., Ultrasound integrated backscatter tissue characterization of remote myocardial infarction in human subjects. J Am Coll Cardiol, 1989. 13: p. 84-91.

88.   Waggoner, A.D., et al., Differentiation of normal and ischemic right ventricular myocardium with quantitative two-dimensional integrated backscatter imaging. Ultrasound Med Biol, 1992. 18: p. 249-53.

89.   Naito, J., et al., Analysis of transmural trend of myocardial integrated ultrasonic backscatter in patients with old myocardial infarction. Ultrasound Med Biol, 1996. 22: p. 807-14.

90.   Lin, L.C., et al., Ultrasonic tissue characterization for coronary care unit patients with acute myocardial infarction. Ultrasound Med Biol, 1998. 24: p. 187-96.

91.   Pasquet, A., et al., Relation of ultrasonic tissue characterization with integrated backscatter to contractile reserve in chronic left ventricular ischemic dysfunction. Am J Cardiol, 1998. 81: p. 68-74.

92.   Colonna, P., et al., Effects of acute myocardial ischemia on intramyocardial contraction heterogeneity: A study performed with ultrasound integrated backscatter during transesophageal atrial pacing. Circulation, 1999. 100: p. 1770-6.

93.   Hirata, N., et al., Real time assessment of myocardial revascularization during coronary artery bypass surgery by means of ultrasonic integrated backscatter. Eur J Cardiothorac Surg, 1999. 16: p. 156-9.

94.   Liu, Y.B., et al., Alterations in ultrasonic backscatter during intra-aortic balloon counterpulsation support in patients with acute myocardial infarction. Ultrasound Med Biol, 1999. 25: p. 1185-93.

95. Maeda, S., et al., Ultrasonic integrated backscatter in early assessment of myocardial injury during open heart surgery. Jpn J Thorac Cardiovasc Surg, 2001. 49: p. 431-7.

96. Loomis, J.F., Jr., et al., Ultrasonic integrated backscatter two-dimensional imaging: evaluation of M-mode guided acquisition and immediate analysis in 55 consecutive patients. J Am Soc Echocardiogr, 1990. 3: p. 255-65.

97. Lythall, D.A., et al., Relation between cyclic variation in echo amplitude and segmental contraction in normal and abnormal hearts. Br Heart J, 1991. 66: p. 268-76.

98. Baysal, K., S. Uysal, and A. Bilgic, Diagnostic value of integrated ultrasonic backscatter in congestive cardiomyopathy. Jpn Heart J, 1991. 32: p. 621-5.

99. Popp, R.L., Recent experience with ultrasonic tissue characterization. Am J Cardiol, 1992. 69: p. 112H-116H.

100. Di Bello, V., et al., Ultrasonic videodensitometric analysis of myocardium in end-stage renal disease treated with haemodialysis. Nephrol Dial Transplant, 1999. 14: p. 2184-91.

101. Fujimoto, S., et al., Ultrasonic tissue characterization in patients with dilated cardiomyopathy: comparison with findings from right ventricular endomyocardial biopsy. Int J Card Imaging, 1999. 15: p. 391-6.

102. Suwa, M., et al., Myocardial integrated ultrasonic backscatter in patients with dilated cardiomyopathy: prediction of response to beta-blocker therapy. Am Heart J, 2000. 139: p. 905-12.

103. Kondo, I., et al., Effect of cibenzoline on regional left ventricular function in hypertrophic obstructive cardiomyopathy. Clin Cardiol, 2000. 23: p. 689-96.

104. Muro, T., et al., Prediction of contractile reserve by cyclic variation of integrated backscatter of the myocardium in patients with chronic left ventricular dysfunction. Heart, 2001. 85: p. 165-70.

105. Dagdeviren, B., et al., Prognostic implication of myocardial texture analysis in idiopathic dilated cardiomyopathy. Eur J Heart Fail, 2002. 4: p. 41-8.

106. Dagdeviren, B., et al., Myocardial texture analysis in idiopathic dilated cardiomyopathy: prediction of contractile reserve on dobutamine echocardiography. J Am Soc Echocardiogr, 2002. 15: p. 36-42.

107. Perez, J.E., et al., Abnormal myocardial acoustic properties in diabetic patients and their correlation with the severity of disease. J Am Coll Cardiol, 1992. 19: p. 1154-62.

108. Di Bello, V., et al., Ultrasonic videodensitometric analysis in type 1 diabetic myocardium. Coron Artery Dis, 1996. 7: p. 895-901.

109. Di Bello, V., et al., Ultrasonic tissue characterization analysis in type 1 diabetes: a very early index of diabetic cardiomyopathy?. G Ital Cardiol, 1998. 28: p. 1128-37.

110. Ferri, C., et al., Heart involvement in systemic sclerosis: an ultrasonic tissue characterisation study. Ann Rheum Dis, 1998. 57: p. 296-302.

111. Di Bello, V., et al., Ultrasonic videodensitometric analysis in scleroderma heart disease. Coron Artery Dis, 1999. 10: p. 103-10.

112. Renzi, M., et al., Blunted cyclic variation of ultrasonic integrated backscatter in asymptomatic patients with type 1 diabetes and normal ejection fraction. European Heart Journal, 2000. 21: p. P1812.

113. Di Bello, V., et al., Ultrasonic videodensitometric analysis of two different models of left ventricular hypertrophy. Athlete's heart and hypertension. Hypertension, 1997. 29: p. 937-44.

114. Di Bello, V., et al., Ultrasonic myocardial texture in hypertensive mild-to-moderate left ventricular hypertrophy: a videodensitometric study. Am J Hypertens, 1998. 11: p. 155-64.

115. Coucelo, J., et al., [The cyclic variation of the 2-dimensional echocardiographic densitometry spectrum as a function of the phase of the cardiac cycle. Experimental

work and its clinical application in arterial hypertension]. Rev Port Cardiol, 1997. 16: p. 63-7.

116. Di Bello, V., et al., Ultrasonic myocardial texture versus Doppler analysis in hypertensive heart: a preliminary study. Hypertension, 1999. 33: p. 66-73.

117. Di Bello, V., et al., Ultrasonic myocardial textural parameters and midwall left ventricular mechanics in essential arterial hypertension. J Hum Hypertens, 2000. 14: p. 9-16.

118. Sutton, M.S.J. and T. Plappert, Myocardial texture in hypertrophic remodelling: new insight into ventricular load and function? J Hum Hypertens, 2000. 14: p. 7-8.

119. Di Bello, V., et al., Cyclic variation of the myocardial integrated backscatter signal in hypertensive cardiopathy: a preliminary study. Coron Artery Dis, 2001. 12: p. 267-75.

120. Maceira, A.M., et al., Ultrasonic backscatter and serum marker of cardiac fibrosis in hypertensives. Hypertension, 2002. 39: p. 923-8.

121. Maceira, A.M., et al., Ultrasonic backscatter and diastolic function in hypertensive patients. Hypertension, 2002. 40: p. 239-43.

122. Zoni, A., et al., Myocardial ultrasonic tissue characterization in patients with different types of left ventricular hypertrophy: a videodensitometric approach. J Am Soc Echocardiogr, 1997. 10: p. 74-82.

123. Nozaki, S., et al., Detection of regional left ventricular dysfunction in early pacing-induced heart failure using ultrasonic integrated backscatter. Circulation, 1995. 92: p. 2676-82.

124. Zuber, M., K. Gerber, and P. Erne, Myocardial tissue characterization in heart failure by real-time integrated backscatter. Eur J Ultrasound, 1999. 9: p. 135-43.

125. Masuyama, T., et al., Ultrasonic tissue characterization with a real time integrated backscatter imaging system in normal and aging human hearts. J Am Coll Cardiol, 1989. 14: p. 1702-8.

126. Di Bello, V., et al., Ultrasonic myocardial tissue characterization: a methodological review. Ital Heart J, 2001. 2: p. 333-43.

127. Perez, J.E., et al., Ultrasonic tissue characterization: integrated backscatter imaging for detecting myocardial structural properties and on-line quantitation of cardiac function. Am J Card Imaging, 1994. 8: p. 106-12.

128. Masuyama, T., et al., Serial measurement of integrated ultrasonic backscatter in human cardiac allografts for the recognition of acute rejection. Circulation, 1990. 81: p. 829-39.

129. Koyama, J., P.A. Ray-Sequin, and R.H. Falk, Prognostic significance of ultrasound myocardial tissue characterization in patients with cardiac amyloidosis. Circulation, 2002. 106: p. 556-61.

130. Wagner, R.F., et al., Quantitative assessment of myocardial ultrasound tissue characterization through receiver operating characteristic analysis of Bayesian classifiers. J Am Coll Cardiol, 1995. 25: p. 1706-11.

Chapter 7

The Mechanics of the Fibrosed/Remodeled Heart

Robert C. Gorman, Benjamin M. Jackson, Joseph H. Gorman, L. Henry Edmunds Jr.
University of Pennsylvania, Philadelphia, Pennsylvania, U.S.A.

1. Introduction

Congestive heart failure (CHF) affects 4.7 million Americans (1). Intense basic and clinical research over the past 30 years has increased our understanding of the disease and led to new treatments. The clinical impact of these strategies has, however, been disappointing. Five year mortality remains at least 50%, even with the most aggressive medical and surgical management [1]. While early recognition and improved therapy for hypertension and valvular heart disease have tended to reduce the incidence of CHF, these improvements have been overwhelmed by a dramatic increase in the incidence of CHF secondary to coronary artery disease, which is now the primary cause of CHF, accounting for nearly 70% of the cases [2].

Except for the relatively few cases of acute cardiomyopathy, CHF results from the left ventricle's (LV) response to altered loading conditions. Perturbations in loading conditions can occur in three ways: 1. pressure overload (chronic hypertension, aortic stenosis, LV out-flow tract obstruction), 2. volume overload (regurgitant valve disease, large arterio-venous fistulae),or 3. regional loss of contractile function (myocardial infarction). Although the pathology associated with these abnormal loading conditions vary greatly, they all initiate a phenomenon termed ventricular remodeling which is manifested clinically by changes in size, shape, and contractile function of the heart. Initially, the changes associated with the remodeling process appear to be compensatory and adaptive; but if the abnormal loading is not relieved, ventricular dilatation and loss of contractile function always ensue. In pressure overload states diastolic dysfunction often precedes LV dilatation and systolic failure.

The geometric and contractile alterations that clinically characterize remodeling are associated with genomic expression, molecular, cellular [3], and interstitial [4] changes that we are beginning to understand. It is clear that the remodeling process has profound effects on the cardiac myocyte [3, 5-7] and on the extracellular matrix (ECM) [8, 9], and that these effects are primarily initiated by myocardial strain (stress) that is induced by pathologic loading conditions.

2. Myocardial Stress-Strain and Ventricular Remodeling

An understanding of ventricular mechanics in normal and pathologic states has long been of interest to cardiologists, cardiac surgeons, physiologists, and engineers. This interest has become more intense over the last half decade as *in vitro* preparations of myocardial tissue have identified stress (and resulting strain) as determinants of cellular, genetic, and biochemical perturbations that are known to be associated with the heart failure phenotype [6, 9]. Stress is easily calculated only for simplified geometry by the law of LaPlace:

$$\sigma = \frac{PR}{2h}$$

(Where P is cavity pressure, R is radius of curvature, h is wall thickness and σ is wall stress.) In practice, it is very difficult to measure stress directly. Engineers typically estimate stress from measured geometric deformations (strain) and known material properties. Modern imaging techniques have allowed improved assessments of myocardial strain in clinical and experimental settings; however, the complex structure and time-varying nature of myocardial mechanics make determination of its material properties extremely difficult. Assumptions regarding global material properties allow rough estimates of ventricular stress in normal myocardium and in valvular heart disease (Figure 1) [10]. The regional and temporal heterogeneity of material properties that are precipitated by regional ischemia or infarction make a meaningful assessment of myocardial material properties in disease states difficult or impossible. In such cases we and others have found that myocardial strain (deformation), which is measurable, provides a useful surrogate for stress. Indeed, it is likely that myocardial strain may be the key factor that initiates the genetic, cellular, and biochemical changes that result in the heart failure phenotype [11, 12].

3. Concentric Hypertrophy: Abnormal Pressure Loading

The development of concentric left ventricular hypertrophy is the pathophysiologic consequence of prolonged increased pressure load. This

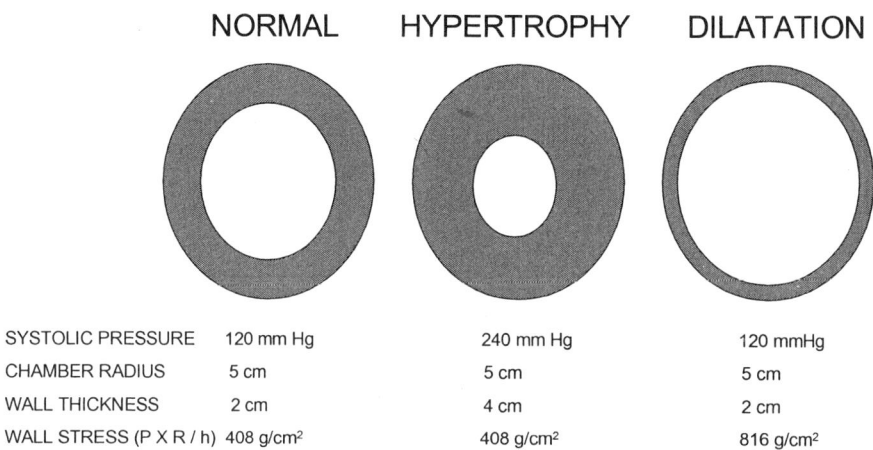

	NORMAL	HYPERTROPHY	DILATATION
SYSTOLIC PRESSURE	120 mm Hg	240 mm Hg	120 mmHg
CHAMBER RADIUS	5 cm	5 cm	5 cm
WALL THICKNESS	2 cm	4 cm	2 cm
WALL STRESS (P X R / h)	408 g/cm²	408 g/cm²	816 g/cm²

Figure 1. Schematic diagram demonstrating the dependence of wall stress on wall thickness and chamber radius. In a simple cylindrical geometry, the LaPlace formula approximates wall stress. (From Sagawa, et al. [13]).

phenomenon is seen clinically with aortic stenosis, other forms of LV outflow tract obstruction, and hypertension. Hypertrophy is an adaptive response to increased pressure loading; however, over the long term, the effects of hypertrophy on cardiac function are detrimental. Important negative effects of prolonged hypertrophic stimuli are decreased diastolic function, increased need for coronary blood flow, decreased coronary flow reserve, and biochemical changes such as ATP and ATPase depletion [14]. These changes are frankly maladaptive and contribute to the onset of angina and congestive heart failure.

3.1. Ventricular adaptations

Grossman [15] has demonstrated that increased peak systolic pressure leads to an increase in mid-myocardial wall stress. Increased wall stress induces development of concentric hypertrophy and results in increased wall thickness with near normal intraventricular volume. Consequently, the ratio of chamber radius to wall thickness is below normal (Figure 1). New myofibrils are added to individual myocytes as remodeling occurs, but the number of myocytes is not increased. Cells become thicker as myofibrils are added in parallel [16]. This remodeling has the effect of normalizing peak systolic wall stress in the short term.

How wall stress drives this remodeling process has not been fully elucidated. It is generally accepted though recently questioned [17, 18] that after birth cardiac myocytes are no longer capable of replication. Any increase in myocardial mass results from an increase in size of individual myocytes; this phenomenon is likely the result of stress-induced changes in gene expression and is initiated almost immediately after pressure load is increased. The response mechanism seems to recapitulate a fetal pattern of gene expression that alters both cellular contractile elements and the biochemistry that affects the relaxation capabilities of the ventricle. The exact mechanism by which increasing wall stress induces changes in genetic expression is currently under active investigation [14, 19-25]. Prolonged pressure overload also influences the metabolism of both matrix metalloproteinases (MMPs) and tissue-inhibitors of metalloproteinases (TIMPs), resulting in ventricular fibrosis, which further exacerbates diastolic and systolic function [26].

3.2. Ventricular mechanics and pressure load-induced remodeling

The functional response to pressure overload passes through several stages. The changes in diastolic and systolic function that result from the myocyte and interstitial changes outlined above are most clearly and concisely presented by use of the pressure volume relationship. The stages of ventricular remodeling are not always clearly differentiated clinically, but it is conceptually helpful to think in terms of a progression, as in Figure 2.

The pressure-volume curves (Figure 2A) for a normal ventricle (dashed lines) and a hypertrophied ventricle (solid lines) are shown. If no compensatory mechanisms were operative, that is if the venous filling pressure and the afterload impedance were not altered, then the hypertrophied ventricle generates a smaller stroke volume and lower pressure than the normal heart. However, compensatory mechanisms act to maintain the systolic pressure and cardiac output, most notably through an increase in venous pressure or preload. As a consequence, the mildly hypertrophied and compensated left ventricle can generate a pressure-volume loop characterized by a normal systolic pressure and stroke volume, albeit at a slightly smaller absolute ventricular volume (Figure 2B). As the hypertrophy progresses, however, the influence of stiff diastolic properties begins to dominate and stroke volume declines in relation to diastolic volume (Figure 2C). To maintain stroke volume, cardiac output, and systolic pressure, the end-diastolic pressure increases until symptomatic pulmonary congestion occurs.

Although the pump function of the pressure-overloaded hypertrophied ventricle appears to be improved in the early stages, controversy exists as to whether muscle contractile function is actually depressed from the earliest stages. Initially, as the muscle mass increases, systolic performance of the heart is improved with relatively little change in its diastolic properties.

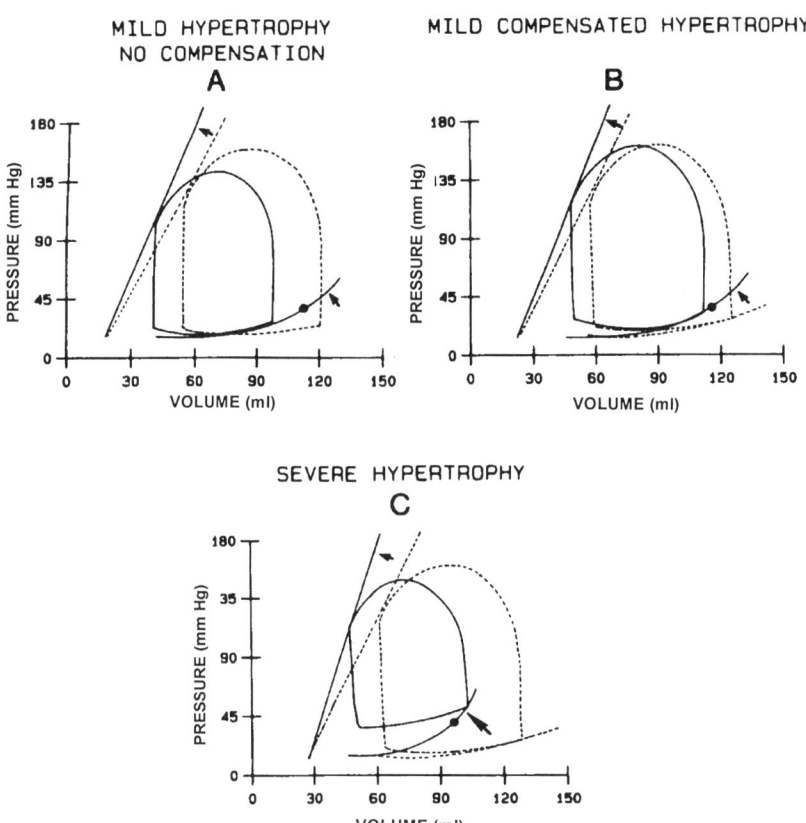

HYPERTROPHY

Figure 2. Pressure-volume (P-V) diagrams demonstrating the progression of ventricular hypertrophy. In each panel, the dashed lines represent the normal condition, for comparison. Panel A shows the leftward shift in the end-systolic and end-diastolic pressure-volume relationships (ESPVR and EDPVR, respectively) with hypertrophy. At a normal filling pressure, the ventricle would fill to a smaller end-diastolic volume, and a smaller P-V loop would result. The normal compensatory response of the circulatory system is to increase filling pressure, and as a result stroke volume returns to normal (B). As hypertrophy progresses, filling pressure exceeds the symptomatic limits (solid dot on diastolic curve), and CHF ensues. (From Sagawa, et al. [13])

As hypertrophy progresses, there is a phase in which systolic chamber function continues to improve, but significant increases in diastolic stiffness occur and a restrictive picture develops. This situation is frequently seen in patients with aortic stenosis who have small end-systolic ventricular volumes with both high systolic and diastolic ventricular pressures [13].

Finally, there is a stage of hypertrophy during which systolic pump function is clearly depressed [13, 27, 28], the heart dilates, and an end-stage dilated cardiomyopathy - similar to end-stage primary muscle failure - develops. In syndromes of left ventricular hypertrophy, the mechanism of the ultimate congestive heart failure is not known, but ischemic injury related to inadequate coronary flow has been suggested [13, 29, 30].

All of the studies cited have limitations in their methods that prevent a definitive statement about the time course of myocardial contractile function as hypertrophy develops. However, evidence indicates that the pumping ability of the left ventricle is initially improved with hypertrophy, while in advanced hypertrophy, the pumping ability declines and produces a decompensated state.

4. Eccentric Hypertrophy and Dilatation: Abnormal Volume Loading

The remodeling induced by prolonged volume overload secondary to regurgitant valvular disease is distinct from that associated with pressure overload states. As we have seen, prolonged pressure loading initiates an almost simultaneous stimulus for wall thickening and chamber dilatation. The result is a stabilization of wall stress until late in the natural history of the disease. Symptoms tend to occur earlier and are associated with diastolic dysfunction and relative ischemia.

Volume overload conditions, such as are seen with aortic insufficiency, lead to generalized chamber dilatation without significant wall thickening. The exact mechanism of dilatation in these cases remains to be elucidated; however, it is evident that any situation in which the ventricle chronically compensates using the Starling mechanism leads to a progressive rightward shift of the diastolic pressure volume relationship. Although the geometric restructuring that occurs with such eccentric hypertrophy allows the ventricle to eject a greater stoke volume, it does so at the cost of a higher wall tension (Figure 1) [13].

4.1. Ventricular adaptations

Aortic insufficiency (AI) increases left ventricular diastolic wall stress. As a result, myocytes add myofibrils in series [14]. The length- to-width ratio of each myocyte increases to accommodate increasing left ventricular volume. Because wall thickness is initially constant, systolic stress increases in accordance with LaPlace's law. Hypertrophy develops in response to the elevated systolic stress.

In addition to the myocyte changes, chronic volume overload states lead to progressive and severe changes in the ECM. Myocardial fibrosis is common in patients with AI who undergo aortic valve replacement [31-33]. Although sequential data in humans are unavailable, observational studies using biopsy material from cardiac catheterization and surgery suggest that fibrosis precedes - and may be related to - the development of CHF in AI [31-34]. Recent laboratory experiments using a rabbit model that closely replicates the pathophysiology of chronic AI in humans confirms these clinical observations: fibrosis preceded CHF and was particularly marked when CHF had developed [35]. The myocardium in these experimental animals revealed normal collagen content [36] despite histologically severe fibrosis, suggesting disproportionate accumulation of non-collagen elements within the fibrotic myocardium. Subsequent exploratory analysis with differential display polymerase chain reaction in cardiac fibroblasts from animals with chronic AI indicated upregulated expression of several genes that code for non-collagen extracellular matrix (ECM) proteins [37]. Elucidation of the ECM response to AI is potentially important: myocardial fibrosis may be involved in the pathogenesis of CHF [38, 39] or may modulate the disordered hemodynamics imposed by myocyte dysfunction [40]. In either case, knowledge of the cellular and molecular bases of fibrosis in AI may suggest treatment modalities and allow more reliable recognition of imminent LV dysfunction [40].

4.2. Ventricular mechanics and volume load-induced remodeling

The response of the heart to aortic valvular regurgitation is similar to the sequence of dilatation with volume overload created by an arterio-venous shunt or mitral regurgitation. This sequence is shown schematically in Figure 3 [13]. In acute aortic regurgitation, the heart does not have a chance to dilate except by means of the Starling mechanism: high diastolic filling pressures are required to maintain cardiac output, and heart failure symptoms are common.

In chronic aortic regurgitation, increased diastolic volumes are a prominent feature from the very earliest stages. Progressive diastolic dilatation is associated with increase in the systolic volume so that, despite marked increases in stroke volume, the ejection fraction remains in the normal range. The course of the disease is one of prolonged and progressive dilatation of the ventricle, without symptoms. This asymptomatic period appears to be associated with a progressive decrease in cardiac reserve, however. Deterioration of ventricular function, in the form of a decrease in forward cardiac output, takes place only very late in the disease and only after tremendous dilatation.

In assessing regurgitation lesions, knowledge of the pressure-volume properties of the heart may be particularly useful. The altered loading pattern of regurgitation leads to large stroke volumes and high ejection velocities. Only in the late stages of myocardial failure do most performance parameters drop below normal. Thus, ejection fraction may overestimate ventricular function

156

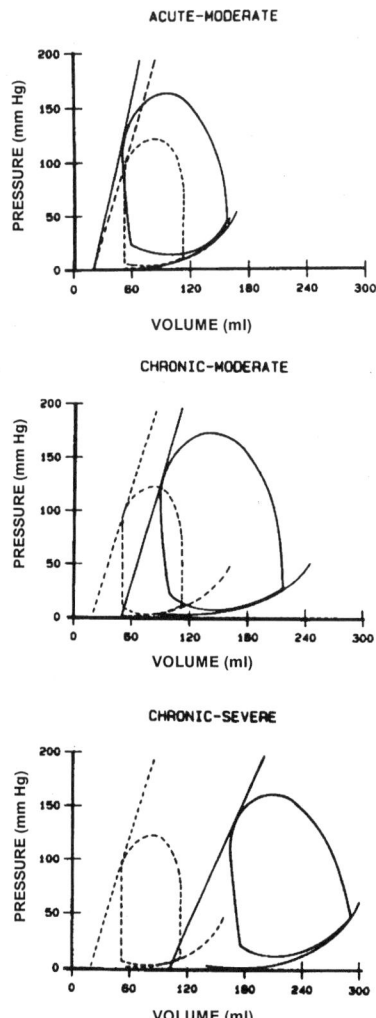

Figure 3. Schematic diagram showing changes in the P-V loops in aortic regurgitation. The dashed lines in each panel represent the normal, for comparison. In acute aortic regurgitation (top), the heart relies on both increased sympathetic tone (resulting in a steeper ESPVR line and increased heart rate) and increased ventricular filling (by the Frank-Starling mechanism) to maintain adequate forward cardiac output. With time (middle), the pressure volume relationship shifts to the right, in response to geometric restructuring of the ventricle. Because the diastolic shift is greater that the systolic shift at any given end-diastolic pressure, the stroke volume is greater than in the normal heart. With chronic severe disease (bottom), the progressive disease leads to a massively dilated heart, high wall tension, and the deterioration of systolic function (decreased slope of the ESPVR). (From Sagawa, et al. [13]).

in patients with aortic regurgitation, and the end-systolic pressure volume relationship may be especially useful [41-43].

In acute aortic regurgitation, the heart cannot immediately compensate for the load and there is an increase in end-diastolic pressure and development of congestive heart failure with little dilatation [44]. In contrast, patients may go for years without developing symptoms from chronic aortic regurgitation [45, 46]. Even severe chronic aortic regurgitation in the absence of symptoms is associated with a favorable prognosis, with 75% survival to 5 years and 50% survival to 10 years [46, 47].

5. Infarction-Induced Ventricular Remodeling

Ventricular remodeling is manifested clinically by changes in size, shape, and function of the heart as a result of a localized myocardial injury (i.e., infarction) [48, 49]. These geometric and functional alterations are associated with gene expression, molecular, cellular [3], and interstitial [4] changes that we are only beginning to understand. While the myocyte is the major cardiac cell involved in remodeling [7], the process affects other myocardial components, including the interstitium, fibroblasts, collagen, and the coronary vasculature.

It is now clear that early infarct expansion is associated with progressive ventricular dilatation, contractile dysfunction, and a poor long-term prognosis. Once initiated, the remodeling progresses inexorably even if coronary obstructions are treated and further infarctions are prevented [50].

Using an ovine infarct model and sonomicrometry array imaging, our laboratory has demonstrated that early infarct expansion is associated with progressive loss of contractile function in normally-perfused myocardium adjacent to the infarct [11]. We have also shown that this dysfunctional borderzone becomes progressively more hypocontractile and recruits additional normally-perfused myocardium as remodeling continues and CHF develops. Contractile dysfunction and geometric distortion (stretching and wall thinning) within the borderzone can be so extensive that normally-perfused borderzone myocardium cannot be distinguished from infarct without a definitive assessment of perfusion status.

Progressive enlargement of uninfarcted borderzone myocardium following a transmural myocardial infarction has been identified clinically, as well. Narula and colleagues, using a novel four-stage single-photon emission computed tomographic imaging protocol in patients with ischemic, dilated cardiomyopathy demonstrated that over 50% of the severely dysfunctional myocardium had normal blood flow [51]. These data suggest that the extension of borderzone myocardium described in our laboratory [11] contributes significantly to the development of post-infarction cardiomyopathy in patients.

Narula proposed that this hypocontractile but normally perfused myocardium be called "remodeled myocardium."

Myocardium is composed of two compartments: the myocytes and the extracellular matrix. The extracellular matrix maintains ventricular geometry and acts to harness the contractile force produced by myocytes. The remodeling process compromises the integrity of both compartments [3, 4]. Recent data from human and animal studies indicates that myocyte apoptosis is an important contributor to the contractile dysfunction seen in remodeled myocardium [3, 5, 7, 52]. Apoptosis is an energy-requiring form of programmed cell death, distinct from necrosis, which may be initiated by external stressors such as cytokines and oxygen free-radicals. Once initiated, the apoptotic pathway leads to the production of a group of enzymes called caspases; these enzymes are capable of completing the apoptotic pathway. Alternatively, new evidence suggests that even if the full apoptotic process is not completed, the caspases can destroy cytosolic contractile proteins (Figure 4), leaving the myocyte alive but unable to function [53].

Postinfarction remodeling also leads to changes in the structure of the myocardial interstitium. As noted above, the extracellular matrix is a highly organized collagen scaffolding that maintains ventricular geometry and acts to harness the contractile force of myocytes. This collagen network is maintained by a precise balance in expression and activity of two groups of enzymes: MMPs and TIMPs. Experimental studies have demonstrated that the balance between MMP and TIMP activity is profoundly disturbed in remodeled myocardium; this imbalance leads to disruption of the normal ECM structure (Figure 5) and is associated with progressive fibrosis and ventricular dilatation [4, 9].

Both myocyte apoptosis and ECM disruption have been associated with external stressors such as cytokine activation (tumor necrosis factor-α and interleukin-6) [54] and reactive oxygen species (ROS) generation. Myocardial strain (i.e., stretching) and the accompanying stress are potent stimuli for the production of such mediators, as indicated by both *in vivo* and *in vitro* data [12].

Theoretical and experimental analyses indicate that acute infarct expansion causes high stress levels in the adjacent (normally-perfused) borderzone myocardium [55]. The early contractile dysfunction seen in this borderzone region is likely due to a mechanical disadvantage associated with increased regional stress. However, using finite element analysis, Guccione and colleagues [56] have demonstrated that elevated stress levels alone account for neither the severity nor the extent of contractile dysfunction that occur as a result of chronic post-infarction remodeling: there must be additional intrinsic alterations in the biochemistry, gene expression, and ultrastructure of remodeled myocardium to account for the severe mechanical dysfunction.

These mechanical results, therefore, correlate with the cellular and biochemical findings described earlier. Remodeled myocardium functions poorly

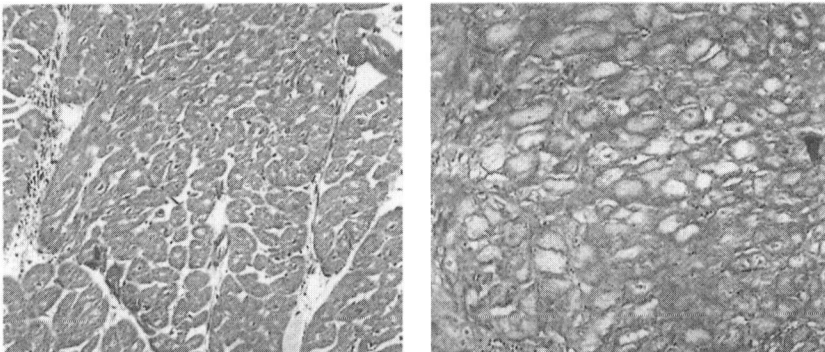

Figure 4. Left: Histological section of normal sheep myocardium. Right: Histological section of sheep myocardium adjacent to an 8-week old infarction. Note the diffuse fibrosis (upper left hand corner = low myocyte density) and the vacuolated myofibrillarlytic (MFL) myocytes in the pathologic specimen. MFL myocytes result from cytosolic caspase activation (apoptosis) that results in the destruction of myocyte contractile proteins.

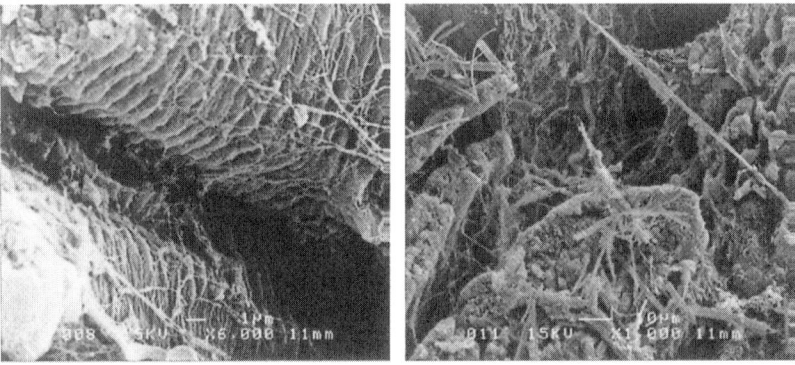

Figure 5. Left: scanning electron micrograph of normal sheep myocardium; note the well-organized extracellular matrix surrounding the myocytes. Right: scanning electron micrograph of the borderzone myocardium demonstrating disruption of the collagen network, thickening of the collagen fibers, and fragmented orientation of the collagen lattice.

not only as a result of increased mechanical stress, but also because it has been inherently and fundamentally altered at a genetic, biochemical, and cellular level. We propose a mechanism for the development of remodeled myocardium. Acute infarct expansion is the critical initiating event. Infarct expansion leads

160

to increased borderzone stress and results in stretching of this normally-perfused myocardium. Stretching induces the production of cytokines and ROS, which in turn stimulate myocyte apoptosis, disruption of the ECM, and fibrosis. It is not known whether the histologic and biochemical changes that occur in remodeled myocardium are reversible. However, given the questionable results of ventricular reshaping surgery, it is likely that these changes are difficult or impossible to reverse.

6. Conclusions

As discussed above, perturbations in loading conditions can occur as a consequence of pressure overload, volume overload or due to the loss of muscle as following myocardial infarction. Ultimately these pathologies initiate the process of ventricular remodeling which is manifested clinically by changes in size, shape, and contractile function of the heart. Initially, these changes can be considered to be compensatory and adaptive. However, if the abnormal loading is not relieved, ventricular dilatation and loss of diastolic and contractile function ensues. It is clear that the remodeling process has profound effects on the cardiac myocyte and on the extracellular matrix (excess fibrosis) and that these effects are primarily initiated by myocardial strain-stress that is induced by pathologic loading conditions. The geometric and contractile alterations that clinically characterize remodeling are associated with genomic expression, molecular, cellular, and interstitial changes that we are only beginning to understand. However, the advent of technologies such as microchip RNA and protein arrays combined with advances in imaging and biomechanical methodologies should further our understanding of the underlying molecular pathophysiology and altered mechanical behavior of the abnormally remodeled heart.

References

1.   Levy, D., et al., Long-term trends in the incidence of and survival with heart failure. N Engl J Med, 2002. 347: p. 1397-402.
2.   Gheorghiade, M. and R.O. Bonow, Chronic heart failure in the United States: a manifestation of coronary artery disease. Circulation, 1998. 97: p. 282-9.
3.   Olivetti, G., et al., Apoptosis in the failing human heart. N Engl J Med, 1997. 336: p. 1131-41.
4.   Spinale, F.G., Matrix metalloproteinases: regulation and dysregulation in the failing heart. Circ Res, 2002. 90: p. 520-30.
5.   Saraste, A., et al., Apoptosis in human acute myocardial infarction. Circulation, 1997. 95: p. 320-3.
6.   Cheng, W., et al., Programmed myocyte cell death affects the viable myocardium after infarction in rats. Exp Cell Res, 1996. 226: p. 316-27.
7.   Kang, P.M. and S. Izumo, Apoptosis and heart failure: A critical review of the literature.

Circ Res, 2000. 86: p. 1107-13.

8. Bowen, F.W., et al., Restraining acute infarct expansion decreases collagenase activity in borderzone myocardium. Ann Thorac Surg, 2001. 72: p. 1950-6.

9. Wilson, E., et al., Region and species specific induction of matrix metalloproteinases occurs with post-myocardial infarction remodeling. Circulation, In Press.

10. Grossman, W., D. Jones, and L.P. McLaurin, Wall stress and patterns of hypertrophy in the human left ventricle. J Clin Invest, 1975. 56: p. 56-64.

11. Jackson, B., et al., Progressive borderzone extension leads to heart failure after anteroapical myocardial infarction. J Amer Col Cardiol, 2002. 40: p. 1160-1167.

12. Cheng, W., et al., Stretch-induced programmed myocyte cell death. J Clin Invest, 1995. 96: p. 2247-59.

13. Sagawa, K., et al., Cardiac Contraction and the Pressure-Volume Relationship. Oxford University Press, 1988: p. Chapter 6.

14. Maier, G., A. Wechsler, and J.L. Edmunds, "Pathophysiology of aortic valve disease". Cardiac Surgery in the Adult, 1996: p. 835-58.

15. Grossman, W., Cardiac hypertrophy: useful adaptation or pathologic process? Am J Med, 1980. 69: p. 576-84.

16. Sawada, K. and K. Kawamura, Architecture of myocardial cells in human cardiac ventricles with concentric and eccentric hypertrophy as demonstrated by quantitative scanning electron microscopy. Heart Vessels, 1991. 6: p. 129-42.

17. Anversa, P., et al., Myocyte growth and cardiac repair. J Mol Cell Cardiol, 2002. 34: p. 91-105.

18. Nadal-Ginard, B., et al., Myocyte death, growth, and regeneration in cardiac hypertrophy and failure. Circ Res, 2003. 92: p. 139-50.

19. Komuro, I., et al., Stretching cardiac myocytes stimulates protooncogene expression. J Biol Chem, 1990. 265: p. 3595-8.

20. Komuro, I., et al., Mechanical loading stimulates cell hypertrophy and specific gene expression in cultured rat cardiac myocytes. Possible role of protein kinase C activation. J Biol Chem, 1991. 266: p. 1265-8.

21. Komuro, I. and Y. Yazaki, Control of cardiac gene expression by mechanical stress. Annu Rev Physiol, 1993. 55: p. 55-75.

22. Schunkert, H., et al., Localization and regulation of c-fos and c-jun protooncogene induction by systolic wall stress in normal and hypertrophied rat hearts. Proc Natl Acad Sci U S A, 1991. 88: p. 11480-4.

23. Chien, K.R., et al., Regulation of cardiac gene expression during myocardial growth and hypertrophy: molecular studies of an adaptive physiologic response. FASEB J, 1991. 5: p. 3037-46.

24. Hannan, R.D., F.A. Stennard, and A.K. West, Expression of c-fos and related genes in the rat heart in response to norepinephrine. J Mol Cell Cardiol, 1993. 25: p. 1137-48.

25. Mansier, P., et al., Membrane proteins of the myocytes in cardiac overload. Br J Clin Pharmacol, 1990. 30: p. 43S-48S.

26. Walther, T., et al., Regression of left ventricular hypertrophy after surgical therapy for aortic stenosis is associated with changes in extracellular matrix gene expression. Circulation, 2001. 104: p. I54-8.

27. Levine, H.J., et al., Force-velocity relations in failing and nonfailing hearts of subjects with aortic stenosis. Am J Med Sci, 1970. 259: p. 79-89.

28. Pantely, G., M. Morton, and S.H. Rahimtoola, Effects of successful, uncomplicated valve replacement on ventricular hypertrophy, volume, and performance in aortic stenosis and in aortic incompetence. J Thorac Cardiovasc Surg, 1978. 75: p. 383-91.

29. Bache, R.J., et al., Regional myocardial blood flow during exercise in dogs with chronic left ventricular hypertrophy. Circ Res, 1981. 48: p. 76-87.

162

30. Sink, J.D., et al., Response of hypertrophied myocardium to ischemia: correlation with biochemical and physiological parameters. J Thorac Cardiovasc Surg, 1981. 81: p. 865-72.

31. Borer, J.S., et al., Myocardial fibrosis in chronic aortic regurgitation: molecular and cellular responses to volume overload. Circulation, 2002. 105: p. 1837-42.

32. Krayenbuehl, H.P., et al., Left ventricular myocardial structure in aortic valve disease before, intermediate, and late after aortic valve replacement. Circulation, 1989. 79: p. 744-55.

33. Schwarz, F., et al., Myocardial structure and function in patients with aortic valve disease and their relation to postoperative results. Am J Cardiol, 1978. 41: p. 661-9.

34. Maron, B.J., V.J. Ferrans, and W.C. Roberts, Myocardial ultrastructure in patients with chronic aortic valve disease. Am J Cardiol, 1975. 35: p. 725-39.

35. Liu, S.K., et al., Fibrosis, myocyte degeneration and heart failure in chronic experimental aortic regurgitation. Cardiology, 1998. 90: p. 101-9.

36. Goldfine, S.M., et al., Myocardial collagen in cardiac hypertrophy resulting from chronic aortic regurgitation. Am J Ther, 1998. 5: p. 139-46.

37. Truter, S.L., et al., Abnormal gene expression of cardiac fibroblasts in experimental aortic regurgitation. Am J Ther, 2000. 7: p. 237-43.

38. Fuster, V., et al., Quantitation of left ventricular myocardial fiber hypertrophy and interstitial tissue in human hearts with chronically increased volume and pressure overload. Circulation, 1977. 55: p. 504-8.

39. Weber, K.T., C.G. Brilla, and J.S. Janicki, Myocardial fibrosis: functional significance and regulatory factors. Cardiovasc Res, 1993. 27: p. 341-8.

40. Herrold, E., P. Lu, and P.e.a. Zanzonico, Antimyosin antibody-mediated detection of myocardial injury: relation to wall stress in chronic aortic regurgitation. Comput Cardiol, 1995: p. 59-62.

41. Borow, K., L. Green, and T.e.a. Mann, End-systolic volume as a predictor of postoperative left ventricular performance in volume overload from valvular regurgitation. Am J Med, 1980. 68: p. 655-663.

42. Osbakken, M., A.A. Bove, and J.F. Spann, Left ventricular function in chronic aortic regurgitation with reference to end-systolic pressure, volume and stress relations. Am J Cardiol, 1981. 47: p. 193-98.

43. Mehmel, H., K. Von Olshausen, and G.e.a. Shuler, Overestimation of left ventricular function by ejection fraction in aortic regurgitation: proof by end-systolic pressure-volume and stress-volume relation. Circulation, 1983. 68: p. S238.

44. Wigle, E.D. and C.J. Labrosse, Sudden, severe aortic insufficiency. Circulation, 1965. 32: p. 708-20.

45. Kennedy, J.W., et al., Quantitative angiocardiography. 3. Relationships of left ventricular pressure, volume, and mass in aortic valve disease. Circulation, 1968. 38: p. 838-45.

46. Shen, W.F., et al., Evaluation of relationship between myocardial contractile state and left ventricular function in patients with aortic regurgitation. Circulation, 1985. 71: p. 31-8.

47. Spagnuolo, M., et al., Natural history of rheumatic aortic regurgitation. Criteria predictive of death, congestive heart failure, and angina in young patients. Circulation, 1971. 44: p. 368-80.

48. Pfeffer, M.A. and E. Braunwald, Ventricular remodeling after myocardial infarction. Experimental observations and clinical implications. Circulation, 1990. 81: p. 1161-72.

49. St John Sutton, M., et al., Quantitative two-dimensional echocardiographic measurements are major predictors of adverse cardiovascular events after acute myocardial infarction. The protective effects of captopril. Circulation, 1994. 89: p. 68-

75.

50. Bolognese, L., et al., Left ventricular remodeling after primary coronary angioplasty: patterns of left ventricular dilation and long-term prognostic implications. Circulation, 2002. 106: p. 2351-7.

51. Narula, J., et al., Noninvasive characterization of stunned, hibernating, remodeled and nonviable myocardium in ischemic cardiomyopathy. J Am Coll Cardiol, 2000. 36: p. 1913-9.

52. Narula, J., et al., Apoptosis in myocytes in end-stage heart failure. N Engl J Med, 1996. 335: p. 1182-9.

53. Narula, J., et al., Apoptosis and the systolic dysfunction in congestive heart failure. Story of apoptosis interruptus and zombie myocytes. Cardiol Clin, 2001. 19: p. 113-26.

54. Mann, D.L., Inflammatory mediators and the failing heart: past, present, and the foreseeable future. Circ Res, 2002. 91: p. 988-98.

55. Jackson, B.M., et al., Border zone geometry increases wall stress after myocardial infarction: contrast echocardiographic assessment. Am J Physiol, 2003. 284: p. H475-9.

56. Guccione, J.M., et al., Mechanism underlying mechanical dysfunction in the border zone of left ventricular aneurysm: a finite element model study. Ann Thorac Surg, 2001. 71: p. 654-62.

III.   *PATHOPHYSIOLOGY OF CARDIAC REMODELING AND FIBROSIS*

Chapter 8

Renin Angiotensin Aldosterone System and Cardiac Extracellular Matrix

Yao Sun, and Karl T. Weber
University of Tennessee Health Science Center, Memphis, Tennessee. U.S.A.

1. Introduction

Circulating angiotensin (Ang) II, a derivative of angiotensinogen and product of angiotensin converting enzyme (ACE)–based conversion of biologically inactive AngI, has multiple well known endocrine properties in the cardiovasculature. AngII, produced *de novo* within the heart, independent of plasma angiotensinogen, plasma renin activity, and endothelial cell bound ACE, has various autocrine and paracrine properties on resident cells that include: cardiomyocytes, representing but one-third of all cells found in the myocardium, and the remaining two-thirds which consist of fibroblasts, endothelial and vascular smooth muscle cells, and macrophages. Based on current evidence AngII type I (AT_1) receptor-ligand binding accounts for the majority of these respective endocrine and auto/paracrine actions of AngII on blood vessels and cardiac tissue.

Herein, we discuss the heart's renin-angiotensin system (RAS) and its involvement in high turnover connective tissue formation normally found in heart valve leaflets and that which appears at sites of repair (e.g., following myocardial infarction, MI). A role for AngII in regulating connective tissue formation in other tissues has been reviewed elsewhere [1, 2].

2. Tissue RAS and the Heart

2.1 Normal heart

Many tissues have the capacity to generate Ang peptides via a tissue RAS. By *in situ* hybridization and immunohistochemistry, Sun et al. [3] found renin expression (mRNA and protein) to be undetectable in the normal rat myocardium, but high within heart valve leaflets. Quantitative *in vitro* autoradiography identifies a similar heterogenous distribution to ACE and AngII receptor binding densities within the heart. Low ACE binding density, for example, is present throughout the myocardium of the right and left atria and ventricles of the adult rat heart, as is the case for AngII receptor binding. High-density binding for ACE and AngII receptors, on the other hand, is found within all four heart valve leaflets [4-6]. Autoradiography further identifies heart valve leaflets as sites of high-density receptor binding for Transforming growth factor-β_1 (TGF-β_1), a fibrogenic cytokine [6, 7]. By *in situ* hybridization marked mRNA expression of type I collagen is seen in valve leaflets where, unlike normal myocardium, collagen turnover is high [8].

Valve leaflets and their chordae tendineae represent exteriorized portions of the heart's extracellular matrix [9]. Residing within leaflet connective tissue and responsible for matrix formation are valvular interstitial cells [10]. The anatomic concordance between renin, ACE and receptors for AngII and TGF-β_1, together with marked type I collagen mRNA expression, in heart valve leaflet tissue implicates the heart's RAS and *de novo* generation of AngII in governing leaflet interstitial cell collagen turnover via autocrine-based regulation of TGF-β_1. Such has been demonstrated for cultured valve leaflet cells [11, 12] and will prove the case for the infarct scar (vida infra). Elevations in plasma AngII found in renovascular hypertension raise collagen synthesis of tricuspid, mitral and aortic valve leaflets [13].

2.2. Infarcted heart

Messenger RNA expression for renin is upregulated at the site of MI and other sites of injury in the infarcted rat heart model. High-density ACE binding is found in the infarct scar and is related to fibrous tissue formation. Temporal and spatial responses in autoradiographic ACE binding have been assessed in a rat heart model of MI. Other forms of injury involving cardiac and noncardiac tissues that appear in this model were also examined. Each served as positive controls in the analysis of ACE and its relationship to tissue repair. They included: the foreign-body fibrosis that surrounds a silk ligature placed around the left coronary artery and fibrosis of visceral pericardium that accompanies manual handling of the heart in sham-operated, noninfarcted controls; renal thromboembolic infarction; and incised/sutured skin. Serial sections involving injured and non-injured tissues, were examined together with picrosirius red histochemistry and videodensitometry, *in vitro* quantitative autoradiography and quantitative *in situ* hybridization to address the temporal and spatial appearance of RAS components that included localization of ACE, AngII and TGF-β_1 binding densities and mRNA expression of TGF-β_1 and type I collagen, together with fibrillar collagen accumulation.

Following left coronary artery ligation and the appearance of an anterior transmural MI, high-density renin mRNA expression and ACE binding density first appear at the infarct site on day 7 coincident with the initial accumulation of fibrillar collagen. As a fibrillar collagen network forms scar tissue over the course of 8 wks, the density of renin expression and ACE binding at this site increases progressively [14]. The appearance of fibrosis at sites remote to the infarct, including the noninfarcted left ventricle, interventricular septum and right ventricle, are also sites of renin expression and high-density ACE binding. The appearance of fibrosis at these remote sites is directly related to the extent of infarction [15-17]. When the transmural infarct is extensive, the entire myocardium (including infarcted and noninfarcted ventricular tissue) is involved in tissue repair and subsequent structural remodeling by fibrosis tissue.

Noninfarct-related sites of injury and repair serve to further address the relationship between the appearance of renin, ACE and fibrosis. Sham operation includes manual handling of the heart. This alone leads to inflammation and subsequent fibrosis of the visceral pericardium. Silk ligature placement around the left coronary artery or within skin to close surgical incision are each associated with a foreign-body fibrosis. The appearance of a mural thrombus in the infarcted left ventricle can be associated with subsequent endocardial fibrosis and on occasion thromboembolic renal infarction. At these sites of repair, high-density renin mRNA expression and autoradiographic ACE binding are temporally and spatially concordant with fibrous tissue formation [3, 14].

Nonischemic models of cardiac myocyte necrosis and repair have also been examined relative to ACE expression. They included: endogenous release of catecholamines that accompanies AngII infusion from implanted mini-pump [18] or administration of isoproterenol, a synthetic catecholamine [19]; and chronic (>3 wk) administration of aldosterone by mini-pump in uninephrectomized rats on a high salt diet and which is accompanied by enhanced urinary potassium excretion and subsequent cardiac myocyte potassium depletion with necrosis [20, 21]. At each site of nonischemic cardiac myocytes necrosis, and irrespective of its etiologic basis, the temporal and spatial appearance of high-density ACE binding is coincident with tissue repair and the deposition of scar tissue [22, 23] and resembles reparative responses observed with ischemic necrosis following MI.

Thus, irrespective of the etiologic basis of injury, the tissue involved, or the presence of repair in the setting of ischemic or nonischemic cardiomyocyte necrosis, a recruitable source of renin and ACE appear at sites of fibrous tissue formation. Examination of serial heart sections of the infarcted rat heart further demonstrates high-density renin expression and ACE binding to be spatially and temporally concordant with marked autoradiographic AngII and TGF-β receptor binding densities and expression of TGF-β_1 and type I collagen mRNAs (by *in situ* hybridization) at these sites of repair. Collectively, these findings in various injured tissues implicate both tissue AngII and TGF-β_1 as a common signaling pathway involved in promoting repair [24].

3. Cells and RAS in the Heart

3.1. Normal heart

Using monoclonal antibodies and immunolabeling, cells expressing renin and ACE in normal rat heart valve leaflets were identified as valvular interstitial cells residing within leaflet matrix [10, 25, 26]. Cultured valvular interstitial cells, obtained from intact rat heart valve leaflets and maintained under serum-deprived conditions, demonstrate autoradiographic ACE binding and converting enzyme (and kininase II) activities to substrates that include AngI, bradykinin, substance P and enkephalin [25].

3.2. Infarcted heart

Renin is expressed by macrophages and myofibroblasts (myoFb) found at the infarct site. ACE-positive cells are seen at and remote to the infarct site and involve endothelial cells of the neovasculature (constitutive ACE), macrophages and myofibroblasts (recruitable ACE) [27, 28]. Within 24 hrs of MI, macrophages appear at the interface between viable and necrotic myocardium; by day 3, fibroblasts co-aggregate with macrophage clusters bordering on the infarct. Thereafter, fibroblast differentiation follows resulting in the α-smooth muscle actin (α-SMA) positive myofibroblast phenotype which proliferates and migrates into the site of necrosis during the remainder of week 1. A combination of cell growth with spatial control of growth and fibrillar collagen assembly govern rebuilding of infarcted tissue. Myofibroblasts at sites of repair are aligned parallel to epi- and endocardium following transmural MI and parallel to the long axis of viable myocytes in nontransmural MI and suggest spatial alignment of myofibroblasts and their actin filaments are important to their function and serve to prevent tissue deformation. Recent evidence implicates transmission of polarity signals and homologues of *Drosophila* tissue polarity genes *frizzled 2* in the expression and alignment of α-SMA positive myofibroblasts at the infarct site [29].

By immunolabeling activated macrophages and myofibroblasts at the infarct site are each renin and ACE-positive. Beyond day 14, the gradual disappearance of macrophages from this site leaves only myofibroblasts as renin and ACE-expressing cells. Persistent high-density ACE binding and renin mRNA expression are present at the infarct site long after MI. This is based on α-SMA positive myofibroblasts, which remain in infarct scar tissue for prolonged periods of time. In the infarcted human heart these myofibroblasts persist at the site of MI for years [30]. Myofibroblasts and their cell-cell and cell-matrix connections confer contractile behavior to scar tissue and this accounts for thinning of the infarct scar. Mediators of scar tissue contraction include AngII and endothelins [31]. These contractile fibroblast-like cells are a common feature of repair seen in diverse injured tissues of rat and man [32].

Myofibroblasts have considerable phenotypic and functional diversity [33]. Immunolabeling with α-SMA, vimentin and desmin defines myofibroblast phenotype at the infarct site. Fibroblast-like cells express vimentin (V). ACE-labeled fibroblasts present in the infarct scar and involved in the expression of fibrillar collagen mRNA are also positive for α-SMA (A). These VA-positive myofibroblasts are instrumental in tissue repair, including wound contraction. They are likewise found in connective tissue that comprises endocardial fibrosis, pericardial fibrosis, renal scarring and foreign-body fibrosis [34]. Unlike incised skin, where myofibroblasts contribute to tissue repair and then progressively disappear through programmed cell death (apoptosis) coincident with wound closure and scar tissue formation at week 4 [35], VA phenotype remains at the infarct site for prolonged periods [28, 30]. An abnormal persistence of VA-positive cells in healed skin is associated with hypertrophic scarring. Whether progressive fibrosis that appears at and remote to MI found in the infarcted heart is related to its persistent myofibroblasts is presently uncertain. Nonetheless, it is clear that the infarct scar is not inert tissue. Indeed it is a dynamic tissue containing persistent, metabolically active cells nourished by a neovasculature [34].

In vitro emulsion autoradiography identifies VA-positive myofibroblasts as expressing AngII receptors [36]. Based on displacement studies using either an AT_1 receptor antagonist (losartan) or AT_2 receptor antagonist (PD123177) the great majority of AngII receptors in the infarcted rat heart are of the AT_1 subtype. Myofibroblasts found at sites of microscopic scarring in both infarcted and noninfarcted tissue express TGF-β_1 and its receptors [7].

4. Function of RAS in the Heart

4.1. Connective tissue

Renin, ACE and AngII receptors impart connective tissue cells with metabolic activity [37]. Scar tissue renin activity is independent of plasma renin activity [3]. ACE substrate utilization involves substances contributing to cell behavior, including cell growth, apoptosis and fibrillar collagen turnover [37]. ACE acting as a kininase II catabolizes AngI, bradykinin, AcSDKP, substance P and enkephalins, each of which are involved in inflammation and repair. N-acetyl-seryl-aspartyl-lysyl-proline (AcSDKP) is a recently described tetrapeptide found in a variety of tissues [38]. It suppresses cell growth by preventing the recruitment of pluripotent cells into the S-phase of the cell cycle; instead, they remain quiescent in the G_0-phase [39].

Matrix homeostasis depends on the turnover of fibrillar type I and III collagens, as well as fibroblast replication and survival. Following injury, matrix and the interstitial space become a dynamic microenvironment consisting of macrophages, fibroblasts and endothelial cells, and various soluble, matrix- and cell membrane-bound molecules that operate in an orchestrated balance of reciprocal regulation between competitive stimulators and suppressors of cell behavior and matrix chemistry. AngII is both a growth stimulator and inhibitor based on its respective binding with AT_1 and AT_2 receptors [40, 41]. Expression of angiotensinogen, an aspartyl protease (e.g., renin, cathepsin D), and ACE is a differentiated function of myofibroblasts [11, 42, 43]. Local AngII also regulates multiple stimulatory and inhibitory factors involved in collagen formation and cell growth and survival. These include endothelins, aldosterone, catecholamines, and TGF-β family of polypeptides as stimulators; these also include bradykinin (BK), nitric oxide, prostaglandins and natriuretic peptides as inhibitors.

4.2. Normal heart

ACE bound to myofibroblasts residing in valve leaflet matrix demonstrates both ACE and kininase II activities. Cultured leaflet cells contain all components requisite to Ang peptide generation and AT_1 receptor-ligand binding regulates their synthesis of collagen and expression of TGF-β_1 [11, 12, 25]. Suprarenal aortic banding is associated with increased expression of type I and III collagen in tricuspid and mitral valve leaflets which suggests a role for circulating AngII and leaflet AT_1 receptor binding in response to renal ischemia [13].

Loose and dense connective tissue formation is a dynamic process during early growth and development of newborn rats. Treatment of 4-wk-old rats with enalapril in nondepressor dosage attenuates cardiac and vascular collagen accumulation in right and left ventricles, aorta, and systemic arteries compared to untreated, age-matched control rats [44]. No such study has yet been conducted with an AngII receptor antagonist. In rats with a genetic predisposition to hypertension, treatment with either quinapril or hydralazine in depressor dosage during early growth and development prevented the appearance of hypertension in adulthood, however, only quinapril attenuated expected development of vascular connective tissue seen in age-matched, untreated hypertensive controls [45].

4.3. Infarcted heart

The number of ACE transcripts found in homogenates prepared from explanted failing human heart tissue is increased, compared to non-failing donor heart tissue [46]. ACE activity has been examined in tissue obtained from the failing, infarcted and noninfarcted human heart tissue [47]. Homogenates of transmural tissue blocks obtained from tissue adjacent to visible scar tissue at the time of aneurysmectomy, revealed ACE activity. Infarct tissue ACE activity exceeded that of control tissue severalfold and the extent of activity was related to the severity of tissue damage. Rat heart tissue homogenates prepared from sites remote to a large transmural anterior MI demonstrate ACE activity the extent of which correlates with infarct size [48]. As a corollary, the presence of fibrosis at sites remote to a transmural MI is dependent on the size of the infarct [15].

ACE activity of fibrous tissue has been demonstrated in the rat heart with fibrosis of the visceral pericardium that appears 4 wks after pericardiotomy [49] and in subcutaneous pouch tissue which appeared 2 wks after instillation of chemical irritant [50]. AngII generation (from AngI substrate) found in each preparation was abrogated by lisinopril. Cultured myofibroblasts obtained from 4-wk-old infarct scar tissue possess all components requisite to generation of Ang peptides including angiotensinogen, cathepsin D, ACE, and expression of AT_1 receptors [42]. Thus, fibrous tissue with its myofibroblast composition is capable of *de novo* AngII generation, whose biologic actions include autocrine regulation of collagen turnover.

A paradigm of tissue repair has been proposed in which ACE and local AngII are integral to the orderly and sequential nature of repair that eventuates in fibrosis [51]. ACE is involved in a two-part *de novo* generation of AngII within granulation tissue that forms at sites of injury. The first component for local AngII generation is provided by activated macrophages. In an autocrine manner, macrophage-derived AngII regulates expression of TGF-β_1 that induces phenotype conversion of co-aggregating fibroblasts. VA-positive myofibroblasts next generate AngII whose autocrine induction of TGF-β_1 regulates collagen turnover at sites of fibrous tissue formation, including infarcted and noninfarcted myocardium [52].

An ongoing perivascular/interstitial fibrosis of noninfarcted myocardium is not only related to local AngII, but also to elevations in circulating effector hormones of the RAAS [53]. AngII and aldosterone each induce a proinflammatory phenotype of vasoactive segments of the arterial circulation. Inflammatory cells and myofibroblasts surround small intramyocardial coronary arteries and arterioles prior to vascular remodeling by fibrous tissue [54]. Like those found in infarct scar, these myofibroblasts express ACE independent of circulating AngII [22]. AT_1 receptor antagonism prevents this perivascular/interstitial fibrosis even when circulating levels of this peptide are suppressed by long-term aldosterone administration [55].

5. ACE inhibition and AT_1 Receptor Antagonism in the Injured Heart

Salutary clinical responses to ACE inhibition are likely to include a prevention of adverse structural remodeling of infarcted and noninfarcted myocardium by fibrous tissue. Evidence supporting a contribution of locally produced AngII in regulating myofibroblast collagen synthesis is obtained using pharmacologic probes that interfere with local AngII generation (i.e., ACE inhibition) or occupancy of its AT_1 receptor prior to circulating RAAS activation. Captopril or enalapril attenuate infarct size and expansion and attenuate the rise in hydroxyproline concentration at the infarct site in dogs following coronary artery occlusion [56-58]. The potential additional contribution of BK, an inhibitor of growth, to tissue repair and which would accompany ACE inhibition has been investigated in rats. A BK_2 receptor antagonist (HOE140 or icatibant) accentuates collagen accumulation remote to the MI site [59]. Losartan attenuates, but does not prevent infarct scar formation [60]. Moreover, the expected rise in tissue AngII concentration found at the infarct site 3 wks post coronary artery ligation is markedly attenuated by either delapril or TCV-116, an AT_1 receptor antagonist [61]. These findings raise the prospect that the number of myoFb and/or their AngII-generating activity per cell at sites of repair may be influenced by AngII.

Fibrous tissue formation at sites remote to MI is also influenced by these pharmacologic interventions. Perindopril, given 1 wk after MI, attenuates the endocardial fibrosis that appears in the non-necrotic segment of the rat left ventricle [62]. Captopril, commenced at the time of coronary artery ligation, attenuates the expected fibrosis of noninfarcted rat left and right ventricles [63, 64] and proliferation of fibroblasts and endothelial cells that appears at remote sites 1 and 2 wks following MI [63]. Under these circumstances, captopril prevents the rise in LV end diastolic pressure that appears in untreated rats and which is not the case in propranolol-treated rats. Captopril also reduces inducibility of ventricular arrhythmias in this model [64]. When initiated 3 wks post MI, well after the tissue repair process has commenced and progressed, captopril does not prevent fibrosis remote to the infarct site or the rise in ventricular stiffness [65]. Losartan prevents fibrosis at remote sites [16, 60, 66], but not the cellular proliferation that appears [16]. Others did not find an inhibition of type I and III collagen mRNA expression at remote sites [67, 68] and have suggested posttranslational modification in collagen turnover to explain why fibrosis fails to appear at remote sites [68].

These favorable tissue protective effects of ACE inhibition or AT_1 receptor antagonism are not confined to the infarcted heart. These interventions prevent the appearance of fibrosis in diverse organs with experimentally induced or naturally occurring tissue injury and where circulating RAAS is not activated. These include: pericardial fibrosis postpericardiotomy [69]; tubulointerstitial fibrosis associated with unilateral urethral obstruction [70-76], toxic nephropathy [77-79], cyclosporine [80], remnant kidney [81-84] or renal injury following irradiation [85]; cardiovascular and glomerulosclerosis that appear in stroke-prone spontaneously hypertensive rats [86-89]; interstitial pulmonary fibrosis that follows irradiation [90-92] or monocrotaline administration [69]; and subcutaneous pouch tissue in response to croton oil [50]. Attenuation of fibrous tissue formation by these interventions in diverse organs with various forms of injury supports the importance of local AngII in promoting fibrosis. A more detailed review of AngII and tissue repair involving systemic organs can be found elsewhere [24].

6. Summary

Macrophage and myofibroblast renin and ACE (recruitable) respectively regulate local concentrations of AngI and AngII involved in tissue repair. *De novo* generation of AngII modulates expression of TGF-β_1 whose autocrine/paracrine properties regulate collagen turnover in heart valve leaflets, an exteriorized portion of the normal extracellular matrix, and at sites of fibrous tissue formation that appear in response to various forms of injury involving diverse tissues. Persistent myofibroblasts and their RAS activity at the infarct site contribute to the progressive fibrosis found at and remote to sites of MI. Activation of the circulating RAAS with sustained elevations in plasma AngII and aldosterone further induce the recruitable form of ACE bound to macrophages and myofibroblasts. Locally produced AngII from this source promotes perivascular fibrosis of intramural vessels of noninfarcted myocardium. At these remote sites, such adverse structural remodeling by fibrous tissue eventuates in ischemic cardiomyopathy, a major etiologic factor involved in the appearance of chronic cardiac failure and which contributes to its progressive nature.

References

1.   Weber, K.T., Angiotensin II and connective tissue: homeostasis and reciprocal regulation. Regul Pept, 1999. 82: p. 1-17.
2.   Weber, K.T., et al., Angiotensin II and extracellular matrix homeostasis. Int J Biochem Cell Biol, 1999. 31: p. 395-403.

3.	Sun, Y., et al., Renin expression at sites of repair in the infarcted rat heart. J Mol Cell Cardiol, 2001. 33: p. 995-1003.
4.	Yamada, H., et al., Localization of angiotensin converting enzyme in rat heart. Circ Res, 1991. 68: p. 141-9.
5.	Pinto, J.E., P. Viglione, and J.M. Saavedra, Autoradiographic localization and quantification of rat heart angiotensin converting enzyme. Am J Hypertens, 1991. 4: p. 321-6.
6.	Sun, Y., A.A. Diaz-Arias, and K.T. Weber, Angiotensin-converting enzyme, bradykinin, and angiotensin II receptor binding in rat skin, tendon, and heart valves: an in vitro, quantitative autoradiographic study. J Lab Clin Med, 1994. 123: p. 372-7.
7.	Sun, Y., et al., Angiotensin II, transforming growth factor-beta1 and repair in the infarcted heart. J Mol Cell Cardiol, 1998. 30: p. 1559-69.
8.	Laurent, G.J., Dynamic state of collagen: pathways of collagen degradation in vivo and their possible role in regulation of collagen mass. Am J Physiol, 1987. 252: p. C1-9.
9.	Robinson, T.F., et al., Coiled perimysial fibers of papillary muscle in rat heart: morphology, distribution, and changes in configuration. Circ Res, 1988. 63: p. 577-92.
10.	Filip, D.A., A. Radu, and M. Simionescu, Interstitial cells of the heart valves possess characteristics similar to smooth muscle cells. Circ Res, 1986. 59: p. 310-20.
11.	Katwa, L.C., et al., Valvular interstitial cells express angiotensinogen and cathepsin D, and generate angiotensin peptides. Int J Biochem Cell Biol, 1996. 28: p. 807-21.
12.	Campbell, S.E. and L.C. Katwa, Angiotensin II stimulated expression of transforming growth factor-beta1 in cardiac fibroblasts and myofibroblasts. J Mol Cell Cardiol, 1997. 29: p. 1947-58.
13.	Willems, I.E., et al., Structural alterations in heart valves during left ventricular pressure overload in the rat. Lab Invest, 1994. 71: p. 127-33.
14.	Sun, Y., et al., Cardiac angiotensin converting enzyme and myocardial fibrosis in the rat. Cardiovasc Res, 1994. 28: p. 1423-32.
15.	van Krimpen, C., et al., Angiotensin I converting enzyme inhibitors and cardiac remodeling. Basic Res Cardiol, 1991. 86: p. 149-55.
16.	Smits, J.F., et al., Angiotensin II receptor blockade after myocardial infarction in rats: effects on hemodynamics, myocardial DNA synthesis, and interstitial collagen content. J Cardiovasc Pharmacol, 1992. 20: p. 772-8.
17.	Volders, P.G., et al., Interstitial collagen is increased in the non-infarcted human myocardium after myocardial infarction. J Mol Cell Cardiol, 1993. 25: p. 1317-23.
18.	Ratajska, A., et al., Angiotensin II associated cardiac myocyte necrosis: role of adrenal catecholamines. Cardiovasc Res, 1994. 28: p. 684-90.
19.	Benjamin, I.J., et al., Isoproterenol-induced myocardial fibrosis in relation to myocyte necrosis. Circ Res, 1989. 65: p. 657-70.
20.	Campbell, S.E., et al., Myocardial fibrosis in the rat with mineralocorticoid excess. Prevention of scarring by amiloride. Am J Hypertens, 1993. 6: p. 487-95.
21.	Darrow, D.C. and H.C. Miller, The production of carciac lesions by repeated injections of desoxycorticosterone acetate. J Clin Invest, 1942. 21: p. 601-611.
22.	Sun, Y., et al., Angiotensin-converting enzyme and myocardial fibrosis in the rat receiving angiotensin II or aldosterone. J Lab Clin Med, 1993. 122: p. 395-403.
23.	Sun, Y. and K.T. Weber, Angiotensin-converting enzyme and wound healing in diverse tissues of the rat. J Lab Clin Med, 1996. 127: p. 94-101.
24.	Weber, K.T., Fibrosis, a common pathway to organ failure: angiotensin II and tissue repair. Semin Nephrol, 1997. 17: p. 467-91.
25.	Katwa, L.C., et al., Angiotensin converting enzyme and kininase-II-like activities in cultured valvular interstitial cells of the rat heart. Cardiovasc Res, 1995. 29: p. 57-64.
26.	Messier, R.H., Jr., et al., Dual structural and functional phenotypes of the porcine aortic

valve interstitial population: characteristics of the leaflet myofibroblast. J Surg Res, 1994. 57: p. 1-21.

27. Falkenhahn, M., et al., Cellular distribution of angiotensin-converting enzyme after myocardial infarction. Hypertension, 1995. 25: p. 219-26.

28. Sun, Y. and K.T. Weber, Angiotensin converting enzyme and myofibroblasts during tissue repair in the rat heart. J Mol Cell Cardiol, 1996. 28: p. 851-8.

29. Blankesteijn, W.M., et al., A homologue of Drosophila tissue polarity gene frizzled is expressed in migrating myofibroblasts in the infarcted rat heart. Nat Med, 1997. 3: p. 541-4.

30. Willems, I.E., et al., The alpha-smooth muscle actin-positive cells in healing human myocardial scars. Am J Pathol, 1994. 145: p. 868-75.

31. Gabbiani, G., G.B. Ryan, and G. Majne, Presence of modified fibroblasts in granulation tissue and their possible role in wound contraction. Experientia, 1971. 27: p. 549-50.

32. Desmouliere, A. and G. Gabbiani, The role of the myofibroblast in wound healing and fibrocontractive diseases. In: The Molecular and Cellular Biology of Wound Repair 2nd Ed., Clark, R.A.F., ed. New York, NY: Plenum Press, 1996: p. 391-423.

33. Skalli, O., et al., Myofibroblasts from diverse pathologic settings are heterogeneous in their content of actin isoforms and intermediate filament proteins. Lab Invest, 1989. 60: p. 275-85.

34. Sun, Y. and K.T. Weber, Infarct scar: a dynamic tissue. Cardiovasc Res, 2000. 46: p. 250-6.

35. Desmouliere, A., et al., Apoptosis mediates the decrease in cellularity during the transition between granulation tissue and scar. Am J Pathol, 1995. 146: p. 56-66.

36. Sun, Y. and K.T. Weber, Cells expressing angiotensin II receptors in fibrous tissue of rat heart. Cardiovasc Res, 1996. 31: p. 518-25.

37. Weber, K.T., et al., Connective tissue: a metabolic entity? J Mol Cell Cardiol, 1995. 27: p. 107-20.

38. Pradelles, P., et al., Distribution of a negative regulator of haematopoietic stem cell proliferation (AcSDKP) and thymosin beta 4 in mouse tissues. FEBS Lett, 1991. 289: p. 171-5.

39. Azizi, M., et al., Acute angiotensin-converting enzyme inhibition increases the plasma level of the natural stem cell regulator N-acetyl-seryl-aspartyl-lysyl-proline. J Clin Invest, 1996. 97: p. 839-44.

40. Stoll, M., et al., The angiotensin AT2-receptor mediates inhibition of cell proliferation in coronary endothelial cells. J Clin Invest, 1995. 95: p. 651-7.

41. Nakajima, M., et al., The angiotensin II type 2 (AT2) receptor antagonizes the growth effects of the AT1 receptor: gain-of-function study using gene transfer. Proc Natl Acad Sci U S A, 1995. 92: p. 10663-7.

42. Katwa, L.C., et al., Cultured myofibroblasts generate angiotensin peptides de novo. J Mol Cell Cardiol, 1997. 29: p. 1375-86.

43. Katwa, L.C., et al., Pouch tissue and angiotensin peptide generation. J Mol Cell Cardiol, 1998. 30: p. 1401-13.

44. Keeley, F.W., A. Elmoselhi, and F.H. Leenen, Enalapril suppresses normal accumulation of elastin and collagen in cardiovascular tissues of growing rats. Am J Physiol, 1992. 262: p. H1013-21.

45. Albaladejo, P., et al., Angiotensin converting enzyme inhibition prevents the increase in aortic collagen in rats. Hypertension, 1994. 23: p. 74-82.

46. Studer, R., et al., Increased angiotensin-I converting enzyme gene expression in the failing human heart. Quantification by competitive RNA polymerase chain reaction. J Clin Invest, 1994. 94: p. 301-10.

47. Hokimoto, S., et al., Increased angiotensin converting enzyme activity in left ventricular

aneurysm of patients after myocardial infarction. Cardiovasc Res, 1995. 29: p. 664-9.

48. Hirsch, A.T., et al., Tissue-specific activation of cardiac angiotensin converting enzyme in experimental heart failure. Circ Res, 1991. 69: p. 475-82.

49. Ou, R., et al., In situ production of angiotensin II by fibrosed rat pericardium. J Mol Cell Cardiol, 1996. 28: p. 1319-27.

50. Sun, Y., et al., Fibrous tissue and angiotensin II. J Mol Cell Cardiol, 1997. 29: p. 2001-12.

51. Weber, K.T., Extracellular matrix remodeling in heart failure: a role for de novo angiotensin II generation. Circulation, 1997. 96: p. 4065-82.

52. Cleutjens, J.P., et al., Collagen remodeling after myocardial infarction in the rat heart. Am J Pathol, 1995. 147: p. 325-38.

53. Weber, K.T. and C.G. Brilla, Pathological hypertrophy and cardiac interstitium. Fibrosis and renin-angiotensin-aldosterone system. Circulation, 1991. 83: p. 1849-65.

54. Nicoletti, A. and J.B. Michel, Cardiac fibrosis and inflammation: interaction with hemodynamic and hormonal factors. Cardiovasc Res, 1999. 41: p. 532-43.

55. Robert, V., et al., Angiotensin AT1 receptor subtype as a cardiac target of aldosterone: role in aldosterone-salt-induced fibrosis. Hypertension, 1999. 33: p. 981-6.

56. Jugdutt, B.I., et al., Effect of enalapril on ventricular remodeling and function during healing after anterior myocardial infarction in the dog. Circulation, 1995. 91: p. 802-12.

57. Jugdutt, B.I., B.L. Schwarz-Michorowski, and M.I. Khan, Effect of long-term captopril therapy on left ventricular remodeling and function during healing of canine myocardial infarction. J Am Coll Cardiol, 1992. 19: p. 713-21.

58. Jugdutt, B.I., Effect of captopril and enalapril on left ventricular geometry, function and collagen during healing after anterior and inferior myocardial infarction in a dog model. J Am Coll Cardiol, 1995. 25: p. 1718-25.

59. Wollert, K.C., et al., Differential effects of kinins on cardiomyocyte hypertrophy and interstitial collagen matrix in the surviving myocardium after myocardial infarction in the rat. Circulation, 1997. 95: p. 1910-7.

60. De Carvalho Frimm, C., Y. Sun, and K.T. Weber, Angiotensin II receptor blockade and myocardial fibrosis of the infarcted rat heart. J Lab Clin Med, 1997. 129: p. 439-46.

61. Yamagishi, H., et al., Contribution of cardiac renin-angiotensin system to ventricular remodelling in myocardial-infarcted rats. J Mol Cell Cardiol, 1993. 25: p. 1369-80.

62. Michel, J.B., et al., Hormonal and cardiac effects of converting enzyme inhibition in rat myocardial infarction. Circ Res, 1988. 62: p. 641-50.

63. van Krimpen, C., et al., DNA synthesis in the non-infarcted cardiac interstitium after left coronary artery ligation in the rat: effects of captopril. J Mol Cell Cardiol, 1991. 23: p. 1245-53.

64. Belichard, P., et al., Markedly different effects on ventricular remodeling result in a decrease in inducibility of ventricular arrhythmias. J Am Coll Cardiol, 1994. 23: p. 505-13.

65. Litwin, S.E., et al., Contractility and stiffness of noninfarcted myocardium after coronary ligation in rats. Effects of chronic angiotensin converting enzyme inhibition. Circulation, 1991. 83: p. 1028-37.

66. Schieffer, B., et al., Comparative effects of chronic angiotensin-converting enzyme inhibition and angiotensin II type 1 receptor blockade on cardiac remodeling after myocardial infarction in the rat. Circulation, 1994. 89: p. 2273-82.

67. Hanatani, A., et al., Inhibition by angiotensin II type 1 receptor antagonist of cardiac phenotypic modulation after myocardial infarction. J Mol Cell Cardiol, 1995. 27: p. 1905-14.

68. Dixon, I.M., et al., Effect of ramipril and losartan on collagen expression in right and left heart after myocardial infarction. Mol Cell Biochem, 1996. 165: p. 31-45.

69. Molteni, A., et al., Monocrotaline-induced pulmonary fibrosis in rats: amelioration by captopril and penicillamine. Proc Soc Exp Biol Med, 1985. 180: p. 112-20.

70. Pimentel, J.L., Jr., et al., Role of angiotensin II in the expression and regulation of transforming growth factor-beta in obstructive nephropathy. Kidney Int, 1995. 48: p. 1233-46.

71. Morrissey, J.J., et al., The effect of ACE inhibitors on the expression of matrix genes and the role of p53 and p21 (WAF1) in experimental renal fibrosis. Kidney Int, 1996. 54: p. S83-7.

72. Ishidoya, S., et al., Angiotensin II receptor antagonist ameliorates renal tubulointerstitial fibrosis caused by unilateral ureteral obstruction. Kidney Int, 1995. 47: p. 1285-94.

73. Kaneto, H., et al., Enalapril reduces collagen type IV synthesis and expansion of the interstitium in the obstructed rat kidney. Kidney Int, 1994. 45: p. 1637-47.

74. Pimentel, J.L., Jr., et al., Regulation of renin-angiotensin system in unilateral ureteral obstruction. Kidney Int, 1993. 44: p. 390-400.

75. Ishidoya, S., et al., Delayed treatment with enalapril halts tubulointerstitial fibrosis in rats with obstructive nephropathy. Kidney Int, 1996. 49: p. 1110-9.

76. Yanagisawa, H., et al., Eicosanoid production by isolated glomeruli of rats with unilateral ureteral obstruction. Kidney Int, 1990. 37: p. 1528-35.

77. Diamond, J.R. and S. Anderson, Irreversible tubulointerstitial damage associated with chronic aminonucleoside nephrosis. Amelioration by angiotensin I converting enzyme inhibition. Am J Pathol, 1990. 137: p. 1323-32.

78. Lafayette, R.A., G. Mayer, and T.W. Meyer, The effects of blood pressure reduction on cyclosporine nephrotoxicity in the rat. J Am Soc Nephrol, 1993. 3: p. 1892-9.

79. Cohen, E.P., et al., Prophylaxis of experimental bone marrow transplant nephropathy. J Lab Clin Med, 1994. 124: p. 371-80.

80. Burdmann, E.A., et al., Prevention of experimental cyclosporin-induced interstitial fibrosis by losartan and enalapril. Am J Physiol, 1995. 269: p. F491-9.

81. Tanaka, R., et al., Internephron heterogeneity of growth factors and sclerosis--modulation of platelet-derived growth factor by angiotensin II. Kidney Int, 1995. 47: p. 131-9.

82. Ikoma, M., et al., Cause of variable therapeutic efficiency of angiotensin converting enzyme inhibitor on glomerular lesions. Kidney Int, 1991. 40: p. 195-202.

83. Shibouta, Y., et al., TCV-116 inhibits renal interstitial and glomerular injury in glomerulosclerotic rats. Kidney Int, 1996. 55: p. S115-8.

84. Anderson, S., H.G. Rennke, and B.M. Brenner, Therapeutic advantage of converting enzyme inhibitors in arresting progressive renal disease associated with systemic hypertension in the rat. J Clin Invest, 1986. 77: p. 1993-2000.

85. Juncos, L.I., et al., Long-term enalapril and hydrochlorothiazide in radiation nephritis. Nephron, 1993. 64: p. 249-55.

86. Nakamura, T., et al., Renal protective effects of angiotensin II receptor I antagonist CV-11974 in spontaneously hypertensive stroke-prone rats (SHR-sp). Blood Press, 1994. 5: p. 61-6.

87. Nakamura, T., et al., Involvement of angiotensin II in glomerulosclerosis of stroke-prone spontaneously hypertensive rats. Kidney Int, 1996. 55: p. S109-12.

88. Kim, S., et al., Contribution of renal angiotensin II type I receptor to gene expressions in hypertension-induced renal injury. Kidney Int, 1994. 46: p. 1346-58.

89. Kim, S., et al., Angiotensin II type I receptor antagonist inhibits the gene expression of transforming growth factor-beta 1 and extracellular matrix in cardiac and vascular tissues of hypertensive rats. J Pharmacol Exp Ther, 1995. 273: p. 509-15.

90. Ward, W.F., A. Molteni, and C.H. Ts'ao, Radiation-induced endothelial dysfunction and fibrosis in rat lung: modification by the angiotensin converting enzyme inhibitor

CL242817. Radiat Res, 1989. 117: p. 342-50.

91.   Ward, W.F., et al., Radiation pneumotoxicity in rats: modification by inhibitors of angiotensin converting enzyme. Int J Radiat Oncol Biol Phys, 1992. 22: p. 623-5.

92.   Ward, W.F., et al., Captopril reduces collagen and mast cell accumulation in irradiated rat lung. Int J Radiat Oncol Biol Phys, 1990. 19: p. 1405-9.

Chapter 9

Matrix Metalloproteinases and Myocardial Remodeling

English Chapman and Francis G. Spinale
Medical University of South Carolina, Charleston, South Carolina, U.S.A.

1. Introduction

Congestive heart failure (CHF) is a major cause of morbidity and mortality worldwide. Coupled with the development and progression of heart failure is maladaptive myocardial remodeling of the left ventricle (LV) [1-4]. LV remodeling can be defined as molecular, cellular, and interstitial changes within the myocardium that result in alterations in LV geometry and function [3-5]. Effective pharmacological treatments that form the mainstay for heart failure are often associated with favorable effects on LV remodeling, such as a reduction in, or stabilization of, LV dilation [6-8]. Therefore, specific strategies that directly target LV remodeling in heart failure may hold therapeutic value.

Myocardial remodeling may be adaptive or pathological [5]. Although initial changes in the myocardium may appear phenotypically similar for both remodeling states, the physical results of each are drastically different. Adaptive myocardial remodeling in a setting of increased workload and wall stress serves to normalize stress and maintain normal ventricular function [5, 9, 10]. For example, in the myocardium of athletes, changes in LV geometry yield increased compliance, improved filling through LV dilation, and enhanced stroke volume [9, 10]. In marked contrast, pathological remodeling and LV chamber dilation yield diminished compliance and reduced forward stroke volume [1-5]. Thus, the non-failing myocardium responds in proportion to cardiovascular stress and demands, whereas the failing myocardium undergoes disproportionate maladaptive remodeling which in turn exacerbates myocardial stress.

As pathological LV remodeling progresses beyond adaptive alterations it becomes a significant risk factor associated with myocardial disease states. For example, after myocardial infarction (MI), gross myocardial structure changes from an elliptical ventricular shape to a spherical shape, and myocardial function suffers drastically [5, 11, 12]. Because of unbridled changes in the cellular and extracellular components of the myocardium, eventually what was once an adaptive response to a myocardial insult may progress toward chronic heart failure. Furthermore, progressive LV dilation in patients with CHF is associated with a greater incidence of morbidity and mortality [7, 13]. Precisely defining the biological systems responsible for myocardial remodeling therefore holds particular relevance in the progression and treatments of heart failure.

2. Matrix Metalloproteinases

A proteolytic enzyme system known as matrix metalloproteinases (MMPs) has been demonstrated to cause tissue remodeling and is implicated in myocardial remodeling [4, 5]. The extracellular matrix (ECM)[14] remodeling process in relation to LV myocardial remodeling is the focus of other chapters in this text. The purpose of this chapter is to briefly review the biology of MMPs and the endogenous tissue inhibitors of metalloproteinases (TIMPs), as well as discuss current trends in MMP and TIMP release observed in human cardiac remodeling disease states.

MMPs are a family of over 20 species of zinc-dependent proteases that are essential in normal tissue remodeling processes such as wound healing, reproduction, and embryonic development [4, 5, 14-18]. However, a more recent research focus of MMP biology has been with respect to pathological remodeling states such as rheumatoid arthritis, cancer, and cardiovascular disease [15-17]. MMPs are responsible for turnover of the ECM, which in turn facilitates tissue remodeling [14-17]. Aside from tissue degradation, MMPs have also been demonstrated to process signaling molecules and modify extracellular signaling pathways [18].

The MMP generic structure (Figure 1) contains four well-conserved regions—a signal propeptide, propeptide, catalytic domain, and a pexin-like domain [14, 17].

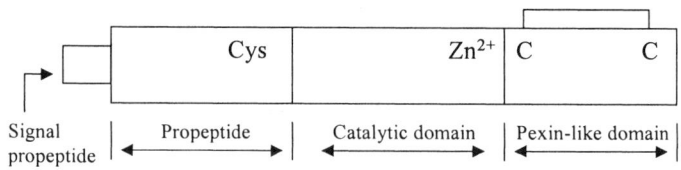

Figure 1. Generic structure of the MMP zymogen. The MMP 'signal peptide' domain directs newly synthesized MMPs for secretion. The propeptide domain maintains the enzyme in a latent state. The catalytic domain binds zinc (Zn^{2+}) and is responsible for enzymatic activity. Structural differences between classes of MMPs are conferred by the 'pexin-like' domain, which contributes to MMP substrate specificity [14, 15, 19].

The signal peptide is located at the NH_2-terminal domain of the protein and directs newly synthesized MMPs for secretion into the ECM [14]. The propeptide domain contains a highly conserved amino acid sequence and maintains the enzyme in a latent state [14, 19]. The catalytic domain, which is responsible for enzymatic activity, contains three histidine residues that bind zinc (Zn^{2+}) [14, 15, 19]. The hemopexin-like domain, located at the C-terminus of the enzyme, is present in all MMPs except matrilysin (MMP-7) [17]. The structural differences between classes of MMPs are conferred by the hemopexin-like domain, which contributes to MMP substrate specificity [14, 17]. Most MMPs are synthesized as inactive zymogens and secreted into the extracellular space in pro-enzyme form [14, 15, 17, 19]. The pro-MMP binds specific ECM proteins and remains enzymatically quiescent until the propeptide domain is cleaved, resulting in exposure of the Zn^{2+} active site of the catalytic domain, and subsequent activation. Thus, instead of a sporadic distribution of pro-MMPs throughout the ECM, there is a specific allotment of these proteolytic enzymes within the extracellular space. Moreover, a large reservoir of recruitable MMPs exists, which upon activation can result in a rapid surge of ECM proteolytic activity.

MMPs characterized thus far as having potential relevance to myocardial remodeling are shown in Table 1. Interstitial collagenase (MMP-1) and collagenase-3 (MMP-13) possess high substrate specificity for fibrillar collagens, whereas gelatinase A (MMP-2) and gelatinase B (MMP-9) demonstrate high substrate affinity for basement membrane proteins [14, 15, 17, 19]. The substrate portfolio for stromelysin (MMP-3) includes important myocardial ECM proteins such as aggrecan, fibronectin, and fibrillar collagens [14, 15, 17, 19]. Unique to the stromelysins is their participation in the MMP activational cascade [20]. Past *in vitro* studies have demonstrated MMP-3 can proteolytically process pro-MMP species [21]. For example, Murphy and colleagues reported a 12-fold increase in the conversion of pro-MMP-1 to active MMP-1 in the presence of MMP-3 [21]. In addition, other MMPs such as MMP-1 and MMP-2 can also activate pro-MMPs [15, 19]. Therefore, activation of a few select pro-MMPs can initiate a cascade of proteolytic activation.

A notable exception to the generic structure of MMPs is the class of integral MMPs, the membrane-type MMPs (MT-MMPs). In addition to the four well-conserved regions common to all MMPs, MT-MMPs also contain a transmembrane domain [14, 15, 19]. It is likely that MT-MMPs are activated intracellularly by a class of prohormone convertases, such as furin [22]. Thus, unlike the secretable MMPs, MT-MMPs are activated once positioned on the cell membrane [15].

Because MMPs degrade various components of the ECM, it is important that MMPs be tightly controlled in order to prevent abnormal tissue remodeling. Therefore, a group of endogenous proteins, the tissue inhibitors of matrix

184

Table 1. Matrix metalloproteinases identified in human myocardium

Class	Name	Number	Substrate
Collagenases	Interstitial Collagenase	MMP-1	Collagens I, II, III, gelatin, proteoglycans, glycoproteins
	Collagenase 2*	MMP-8*	Collagens I, II, III, aggrecan
	Collagenase 3	MMP-13	Collagens I, II, III, gelatin, proteoglycans, pro-MMP-1
Gelatinases	Gelatinase A	MMP-2	Gelatins, collagens I, IV, V, VII, elastin, fibronectin, laminin, proteoglycans
	Gelatinase B	MMP-9	Gelatins, collagens IV, V, XIV, elastin, proteoglycans, glycoproteins
Stromelysins	Stromelysin 1	MMP-3	Fibronectin, laminin, collagens III, IV, IX, pro-MMP-1, 7, 9
Matrilysin	Matrilysin*	MMP-7*	Fibronectin, laminin, elastin, gelatin, collagens I, IV
Membrane-type MMPs	MT1-MMP	MMP-14	Fibronectin, laminin-1, glycoproteins, proteoglycans, collagens I, II, III, pro-MMP-2, pro-MMP-13

*May exist in myocardium in states of inflammation and wound healing.

metalloproteinases (TIMPs), exists to regulate activity of the MMPs. Currently, there are four known TIMP species [23-26]. TIMPs bind to the active site of the MMPs by blocking access to extracellular matrix substrates. The MMP/TIMP complex is formed in a stoichiometric 1:1 molar ratio and forms an important endogenous system for regulating MMP activity in vivo [23-26]. The importance of inhibitory control of MMPs within the myocardium has been demonstrated through the genetic deletion of TIMP-1 expression [27]. In TIMP-1 knockout mice, LV remodeling occurred in the myocardial interstitium without a pathological stimulus [27]. Although the TIMPs are expressed in an array of cells, TIMP-4 shows a high level of expression in human myocardial tissue [24]. Several studies have suggested that TIMPs not only inhibit MMPs but also have other biological activities [20, 28]. For example, TIMPs may promote cell growth and differentiation in certain cell systems and participate in MMP activation. It has been demonstrated that TIMP-2 forms a complex with species of MT-MMPs and that the formation of this complex enhanced the activation of MMP-2 [23-26, 28].

3. MMPs and Heart Failure

There are a number of underlying factors that may contribute to the development and progression of myocardial remodeling and the symptoms of heart failure. Specifically, alterations in myocardial tissue structure occur following myocardial infarction, with the development of cardiomyopathies, and in severe myocardial hypertrophy [4, 11-13, 29, 30]. Because uncontrolled increases in MMP activity are thought to produce excessive tissue remodeling, changes in MMP expression in these three disease states has been of concern to a number of investigators [1-3, 12, 27, 30-40]. Furthermore, although each disease state may progress to the same symptoms of heart failure, all three etiologies are characterized by dissimilar patterns of remodeling and different patterns of MMP expression.

3.1. MMPs and myocardial infarction

Following myocardial infarction (MI), changes in LV myocardial collagen structure can occur, which would implicate alterations in MMP activity within the myocardial interstitium [5, 11, 13, 32-35]. Several studies have demonstrated that increased plasma levels of MMPs occur in patients in the early post-MI period [36-39]. For example, Hojo and colleagues observed an increase in plasma MMP-2 levels and activity in patients with acute MI [37]. Experimental studies have provided mechanistic evidence that increased expression and activation of MMPs contribute to the post-MI remodeling process [34, 35, 40-43]. For example, it has been demonstrated that changing the expression of certain MMP genes can alter tissue remodeling within the MI region as well as influence the degree of post-MI remodeling [40, 41]. Ducharme and colleagues observed that in MMP-9 knockout mice, following coronary ligation and MI, there was an attenuation of LV enlargement and collagen accumulation as compared to sibling wild type mice [41]. Time-dependent changes in myocardial MMP mRNA levels have been reported during the post-MI period in rats [42]. In this past study, a robust increase in MMP-13 expression occurred early in the post-MI period, which was then followed by a more gradual elevation in MMP-14 (MT1-MMP) levels [42]. Mechanical and biochemical stimuli likely contribute to the induction of MMPs in the post-MI period. The elaboration of cytokines, oxidative stress, as well as physical stimuli, which occur during the evolving MI and subsequent healing period, likely contribute to changes in myocardial MMP expression [44-48]. A cause-effect relationship between myocardial MMP activation and remodeling has been established through the use of pharmacological MMP inhibitors, and has been the subject of recent reviews [40, 49].

A loss of TIMP-mediated control has been reported in LV remodeling following MI [49, 50]. In the rat MI model, MMP mRNA levels increased early post-MI, but were not associated with a concomitant increase in TIMP mRNA levels [42]. In an *in vitro* system of ischemia and reperfusion, TIMP-1 expression was reduced in the early reperfusion period [51]. These findings suggest a loss of endogenous MMP inhibitory control occurs early in the post-MI period. [51]. Such complexities of time-dependent changes in MMPs and TIMPs throughout the post-MI myocardium must be addressed to properly develop therapeutics.

3.2. MMPs and dilated cardiomyopathy

The most extensively studied cardiac disease state with respect to myocardial MMP levels is dilated cardiomyopathy (DCM) [31, 52, 53]. The pathophysiology of DCM is characterized by an increase in LV chamber radius to wall thickness, which increases myocardial wall stress and results in further dilation [54]. While the etiologies of DCM are diverse, the general classifications are ischemic, idiopathic (non-ischemic), and infectious [54]. A model of heart failure that has reported a relationship between LV remodeling and MMP activity is the pacing-induced heart failure model [2, 3, 55]. After instituting the pacing protocol in pigs, MMP activity measured by an *in vitro* assay was increased [2, 3, 55]. Not only was an increased abundance of myocardial MMP-1, MMP-2, and MMP-3 observed, but MMP activity was also shown to coincide with the onset of LV remodeling and dilation [3]. Therefore, animal-model systems have identified important correlations between increased

MMP expression and activity, cardiac remodeling, and progression to failure.

A number of studies have examined relative MMP and TIMP expression in end-stage human DCM [50, 52, 53, 56]. For example, Tyagi and colleagues demonstrated increased *in vitro* myocardial MMP zymographic activity in DCM samples [57]. Also, it has been observed that myocardial MMP-9 was increased due to either ischemic or non-ischemic origins, and MMP-3 was increased in non-ischemic DCM myocardial extracts [52, 53]. Because MMP-3 degrades a wide portfolio of ECM substrates, increased levels may significantly contribute to pathological tissue remodeling [14-16]. Also, MMP-3 contributes to the activation of other MMPs, which in turn may amplify proteolytic activity [15, 17]. Another observation is the selective upregulation of certain MMP species in DCM. For example, the collagenase MMP-13 is expressed at low levels in normal myocardium but is significantly increased in end-stage DCM [52, 53]. Interstitial collagenase (MMP-1), in contrast, is decreased [52, 53]. In addition, because TIMP-1 binds with less affinity to MMP-13 than to MMP-1, there is a loss of endogenous control as MMP-13 levels increase [58]. Past studies have shown that MMP-13 is not only associated with direct deleterious effects on LV structure and function but also aggressive tumor metastasis [17, 18, 58]. Such findings suggest the more virulent collagenase, MMP-13, may be overall selectively expressed in pathological remodeling states [20, 52, 59]. Therefore, the emergence of MMP-13 as the predominant collagenase within the DCM myocardium may contribute to increased susceptibility of the myocardial fibrillar collagen network to degradation and subsequent maladaptive myocardial remodeling.

Myocardial TIMP levels appear to be variably expressed in end-stage DCM [53, 56]. Thomas and colleagues found an increased abundance of TIMP-1 and TIMP-2 in DCM myocardium samples [53]. In a study by Li and colleagues, TIMP-1 and TIMP-3 levels were reduced in cardiomyopathic samples whereas TIMP-2 levels were unchanged when compared to control myocardium [56]. Thus, in DCM, there is not a concordant increase of TIMPs in relation to MMP levels. Previous studies of cardiomyopathic disease have suggested that such changes occur in the stoichiometric ratio of MMPs to TIMPs [23, 52, 56]. For example, in both ischemic and non-ischemic DCM, an absolute reduction in MMP-1/TIMP-1 complex formation was observed [52]. Similar to MMPs, TIMPs are encoded by unique genes and differ in promoter regions [60, 61]. Thus, extracellular stimuli may induce differential expression of TIMPs. The quantitative reduction of MMP/TIMP complexes within the myocardium in end-stage DCM has significant relevance to the LV remodeling process. As observed post-MI, a change in the MMP/TIMP stoichiometric ratio in DCM could favor persistent MMP activation, resulting in unhindered tissue turnover and progressive LV remodeling.

3.3. MMPs and hypertrophy

Another cause of heart failure is diastolic dysfunction secondary to LV hypertrophy (LVH) [62, 63]. LVH, which is often characterized by reduced myocardial compliance and inadequate filling during diastole, occurs in response to a persistent increase in pressure or volume overload. However, the pattern of myocardial remodeling differs significantly between these two overload states [4, 5, 64-68]. Specifically, different patterns of collagen deposition and degradation result in specific changes within the fibrillar collagen network of the ECM [5, 64, 65]. An accumulation of myocardial fibrillar collagen, resulting in myocardial stiffness, is a common observation in pressure overload hypertrophy (POH) [5, 64, 65]. Changes in myocardial MMP levels and activity may cause such alterations in myocardial collagen content. Past studies have documented reduced MMP plasma levels in patients with systemic hypertension and LVH [69, 70]. For example, decreased plasma levels of MMP-1 were reported in hypertensive patients with hypertrophy, and reduced plasma levels of MMP-9 were observed in untreated hypertensive patients [69, 70]. In contrast to patterns of collagen accumulation in POH, chronic volume overload states display disruption of normal fibrillar collagens, which may be due to enhanced proteolytic activity of myocardial MMPs [67, 71].

As hypertrophy progresses to decompensation and eventual LV failure, time-dependent changes in the activation of myocardial MMPs are likely to occur. In fact, several studies have demonstrated such changes in myocardial MMP levels throughout development of POH [72, 73]. In the spontaneously hypertensive rat (SHR), the development of compensated hypertrophy is associated with an increase in TIMP-4 levels, which would imply a net reduction in MMP activity [72]. However as the SHR model progresses to decompensation and LV failure, myocardial TIMP levels decrease below normal to favor a net increase in MMP activity [72, 73]. Specifically, increased myocardial MMP-2 and MMP-9 levels occur in the SHR during the transition to LV failure [73]. More recently, Iwanaga et al. reported disparate MMP and TIMP levels in the compensation stage of LVH compared to the transition to CHF stage. While net MMP-2 and TIMP activity remained unchanged at the compensation stage, activity increased significantly, with MMP-2 activity surpassing that of the TIMPs, as LVH decompensation progressed to LV failure [74]. Net MMP activity correlated to increased wall stress and LV dilation suggests this enhanced proteolytic activity is responsible for progressive LV remodeling [74].

As demonstrated by this laboratory, time-dependent changes also occur in myocardial MMP activity following both acute and prolonged pressure overload states [66]. In these studies, acute pressure overload resulted in increased myocardial MMP-9 expression and zymographic activity. With prolonged pressure overload, MMP-9 zymographic activity began to normalize and was accompanied by changes in TIMP-1 levels [66]. More recently, it has been demonstrated in an animal model of POH that myocardial MMP and TIMP levels are altered [75]. Specifically, induction of MMP-1, MMP-2, MMP-3, and MMP-9 occurred, as well as a differential induction of TIMPs, after development of LVH [75]. Interestingly, after surgically relieving the pressure overload stimulus, alterations in MMP and TIMP levels regressed [75]. In volume overload states, such as mitral regurgitation or aorto-caval fistula, increased myocardial MMP levels and zymographic activity have been observed [66, 71, 76]. Specifically, increased myocardial MMP-2 and -9 zymographic activity was associated with changes in LV function and geometry in a rat model of volume overload [71]. In another rat model of volume overload, MMP inhibition attenuated LVH and adverse LV remodeling, with a concomitant maintenance of normal LV function [77]. Taken together, these LVH studies suggest that first, changes in myocardial MMP and TIMP levels likely contribute to the LV remodeling process given a pressure or volume overload stimulus; second, different patterns of MMP and TIMP expression occur depending on the type and duration of overload states; and third, increased levels of MMPs and TIMPs may be decreased with pharmacologic inhibition or alleviation of the overload stimulus.

4. Targeting LV Remodeling Through Modulating MMPs

The development and progression of several underlying etiologies of heart failure are accompanied by significant myocardial remodeling. The heart post-myocardial infarction (MI), in the development of dilated cardiomyopathies (DCM), and in severe hypertrophy (LVH), develops distinctly different patterns of LV remodeling. A schematic of such remodeling processes is shown in Figure 2. Not only are MMP and TIMP levels region-specific post-MI, but levels also vary along the time-course of healing. DCM is characterized by excessive matrix degradation and wall thinning, as net MMP proteolytic activity increases. In contrast, LVH yields matrix accumulation and wall thickening, concurrent with a net decrease in proteolytic activity. Recent pharmacologic studies have demonstrated that modifying MMP activity can directly influence myocardial ECM structure and function, thereby altering the course of LV remodeling and heart failure progression [1, 2, 40, 72, 77, 78]. Broad-spectrum MMP inhibition has been shown to modify the progressive development of left ventricular dysfunction while attenuating the degree of chamber dilation [1, 2, 77]. However, the clinical use of broad-spectrum MMP inhibitors in patients with tumor metastasis has been associated with undesirable side effects [79, 80], and the long-term inhibition of all MMP species will likely interfere with normal

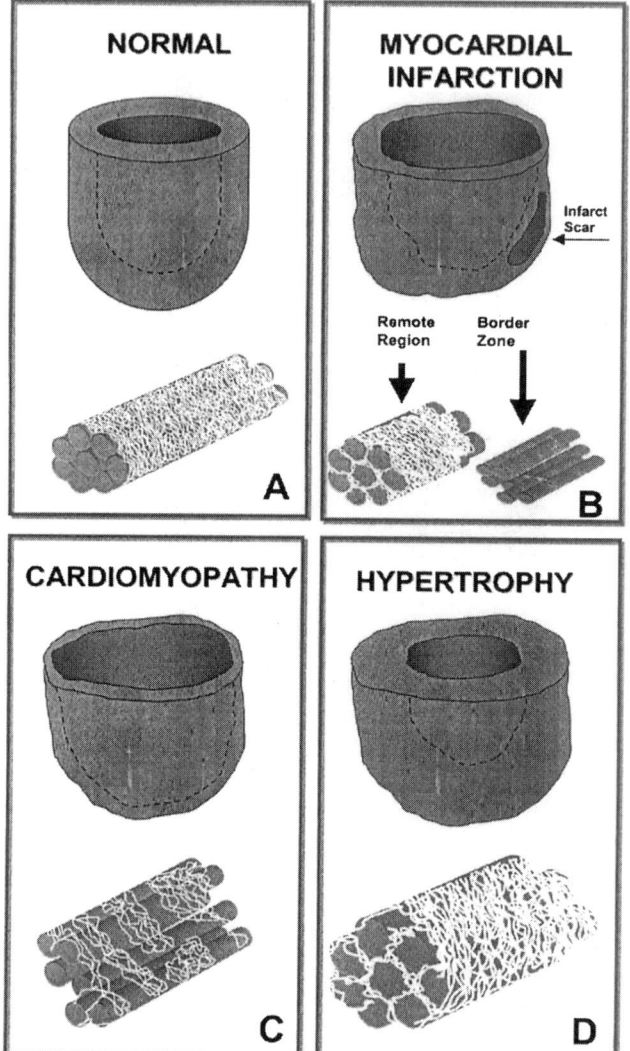

Figure 2. A simplified schematic of the different remodeling processes in three common disease states that give rise to the syndrome of heart failure. Each disease state depicts the change from normal (A) in gross LV structure as well as alignment of myofibrils and organization of ECM. (B) LV remodeling post-MI is characterized by non-uniform changes in LV myocardial wall geometry. The nonviable infarct region becomes a thinned, fibrous scar. Myocytes in the border zone are susceptible to 'infarct-expansion' and undergo realignment and slippage. Remote viable regions undergo myocyte hypertrophy and collagen accumulation. (C) Dilated cardiomyopathy remodeling is characterized by an increase in LV chamber radius to wall thickness, progressive LV dilation, focal ECM loss, and abnormalities in collagen cross-linking. (D) In LV hypertrophy, reduced compliance leads to diastolic dysfunction as individual myocytes hypertrophy and excessive matrix accumulation occurs. These three distinct forms of LV remodeling likely occur through specific patterns of MMP/TIMP expression.

tissue remodeling and ECM turnover. Hence, in a setting of disparate MMP and TIMP portfolios, each specific to a certain disease process and time-course, it follows that strategies targeted at modifying MMP and TIMP expression for each disease state should also vary. It must be recognized that while altering MMPs and TIMPs directly contributes to myocardial remodeling, the levels of MMP and TIMP expression are governed by complex biological stimuli specific to the disease process and time-course.

   Elucidating the signaling pathways of MMPs and TIMPs is an ongoing process and an exhaustive review is beyond the scope of this chapter. Several reviews regarding upstream mechanisms of MMP and TIMP expression discuss signaling pathways in detail [81-83]. Although possible mechanisms of MMP and TIMP control are multifactorial, a brief review of biological signaling factors is presented here. MMP mRNA expression can be influenced by a variety of neurohormones, corticosteroids, and cytokines [81, 83, 84]. The elaboration of cytokines such as TNF-α often occurs in states of myocardial ischemia and heart failure, and may induce MMP expression [85]. For example, as shown by Siwik et al., exposure of myocardial fibroblasts to TNF-α causes the release of MMP-9 and MMP-13 [86]. It was demonstrated more recently that in mice with cardiac-restricted over-expression of TNF-α, discordant levels of MMPs and TIMPs occurred [87]. While the mechanisms of MMP induction by these factors are complicated, a common pathway involves the formation of transcription factors and binding to the activator protein-1 (AP-1) and nuclear factor-κB promoter regions [14, 20, 81-83]. Other factors such as transforming growth factor-β (TGF-β) and glucocorticoids are thought to inhibit MMP gene expression, possibly through sequestering unbound AP-1 (DNA binding) proteins [88]. Characterization of MMP gene promoters has provided some insight into possible mechanisms that regulate MMP gene expression. For example, the AP-1 site is present in the promoter region of the MMP-1, -3 and −9 genes, while it is absent in the promoter region of MMP-2 [14, 81-83]. This may explain why MMP-2 is under less transcriptional control and therefore is considered to be constitutively expressed. Moreover, the difference in MMP-2 levels in cardiac disease states suggests differences in transcriptional and post-transcriptional regulatory mechanisms.

   Pharmacologic strategies that interfere with upstream signaling cascades involved in MMP transcription may improve our understanding of the complex myocardial remodeling process and the specific role of MMPs. For example, the use of cytokine inhibition such as that of TNF-α neutralizing proteins may prove to be useful pharmacological tools in order to identify the signaling pathways obligatory for MMP species induction[89]. Ideally, through targeting specific bioactive molecules or blocking nuclear binding sites, MMPs implicated in adverse remodeling could be inhibited while necessary basal expression of beneficial MMP levels could continue. Thus, more clinically applicable strategies specific to each disease state and time-course could be realized. In consideration of the divergent physical and biochemical pathways involved, defining the molecular triggers of MMP and TIMP expression and targeting the upstream mechanisms responsible may prove to be an important therapeutic paradigm for heart failure treatment.

References

1. Peterson, J.T., et al., Matrix metalloproteinase inhibition attenuates left ventricular remodeling and dysfunction in a rat model of progressive heart failure. Circulation, 2001. 103: p. 2303-9.
2. Spinale, F.G., et al., Matrix metalloproteinase inhibition during the development of congestive heart failure : effects on left ventricular dimensions and function. Circ Res, 1999. 85: p. 364-76.
3. Spinale, F.G., et al., Time-dependent changes in matrix metalloproteinase activity and expression during the progression of congestive heart failure: relation to ventricular and myocyte function. Circ Res, 1998. 82: p. 482-95.
4. Cohn, J.N., R. Ferrari, and N. Sharpe, Cardiac remodeling--concepts and clinical implications: a consensus paper from an international forum on cardiac remodeling. Behalf of an International Forum on Cardiac Remodeling. J Am Coll Cardiol, 2000. 35: p. 569-82.
5. Weber, K.T., et al., Remodeling and reparation of the cardiovascular system. J Am Coll Cardiol, 1992. 20: p. 3-16.
6. Hall, S.A., et al., Time course of improvement in left ventricular function, mass and geometry in patients with congestive heart failure treated with beta-adrenergic blockade. J Am Coll Cardiol, 1995. 25: p. 1154-61.
7. Greenberg, B., et al., Effects of long-term enalapril therapy on cardiac structure and function in patients with left ventricular dysfunction. Results of the SOLVD echocardiography substudy. Circulation, 1995. 91: p. 2573-81.
8. Doughty, R.N., et al., Left ventricular remodeling with carvedilol in patients with congestive heart failure due to ischemic heart disease. Australia-New Zealand Heart Failure Research Collaborative Group. J Am Coll Cardiol, 1997. 29: p. 1060-6.
9. Pelliccia, A., F.M. Di Paolo, and B.J. Maron, The athlete's heart: remodeling, electrocardiogram and preparticipation screening. Cardiol Rev, 2002. 10: p. 85-90.
10. Pelliccia, A., Athlete"s heart and hypertrophic cardiomyopathy. Curr Cardiol Rep, 2000. 2: p. 166-71.
11. Pfeffer, M.A., Left ventricular remodeling after acute myocardial infarction. Annu Rev Med, 1995. 46: p. 455-66.
12. Sutton, M.G. and N. Sharpe, Left ventricular remodeling after myocardial infarction: pathophysiology and therapy. Circulation, 2000. 101: p. 2981-8.
13. Erlebacher, J.A., et al., Early dilation of the infarcted segment in acute transmural myocardial infarction: role of infarct expansion in acute left ventricular enlargement. J Am Coll Cardiol, 1984. 4: p. 201-8.
14. Parsons, S.L., et al., Matrix metalloproteinases. Br J Surg, 1997. 84: p. 160-6.
15. Nagase, H. and J.F. Woessner, Jr., Matrix metalloproteinases. J Biol Chem, 1999. 274: p. 21491-4.
16. Dollery, C.M., J.R. McEwan, and A.M. Henney, Matrix metalloproteinases and cardiovascular disease. Circ Res, 1995. 77: p. 863-8.
17. Woessner, J.F., Jr., Matrix metalloproteinases and their inhibitors in connective tissue

remodeling. FASEB J, 1991. 5: p. 2145-54.

18. Vu, T.H. and Z. Werb, Matrix metalloproteinases: effectors of development and normal physiology. Genes Dev, 2000. 14: p. 2123-33.

19. Birkedal-Hansen, H., et al., Matrix metalloproteinases: a review. Crit Rev Oral Biol Med, 1993. 4: p. 197-250.

20. Nagase, H., Activation mechanisms of matrix metalloproteinases. Biol Chem, 1997. 378: p. 151-60.

21. Murphy, G., et al., Stromelysin is an activator of procollagenase. A study with natural and recombinant enzymes. Biochem J, 1987. 248: p. 265-8.

22. Pei, D. and S.J. Weiss, Transmembrane-deletion mutants of the membrane-type matrix metalloproteinase-1 process progelatinase A and express intrinsic matrix-degrading activity. J Biol Chem, 1996. 271: p. 9135-40.

23. Edwards, D.R., et al., The roles of tissue inhibitors of metalloproteinases in tissue remodelling and cell growth. Int J Obes Relat Metab Disord, 1996. 20: p. S9-15.

24. Green, R.M., F. Hoda, and K.L. Ward, Molecular cloning and characterization of the murine bile salt export pump. Gene, 2000. 241: p. 117-23.

25. Goldberg, G.I., et al., Human 72-kilodalton type IV collagenase forms a complex with a tissue inhibitor of metalloproteases designated TIMP-2. Proc Natl Acad Sci U S A, 1989. 86: p. 8207-11.

26. Leco, K.J., et al., Tissue inhibitor of metalloproteinases-3 (TIMP-3) is an extracellular matrix-associated protein with a distinctive pattern of expression in mouse cells and tissues. J Biol Chem, 1994. 269: p. 9352-60.

27. Roten, L., et al., Effects of gene deletion of the tissue inhibitor of the matrix metalloproteinase-type 1 (TIMP-1) on left ventricular geometry and function in mice. J Mol Cell Cardiol, 2000. 32: p. 109-20.

28. Butler, G.S., et al., Membrane-type-2 matrix metalloproteinase can initiate the processing of progelatinase A and is regulated by the tissue inhibitors of metalloproteinases. Eur J Biochem, 1997. 244: p. 653-7.

29. Francis, G.S., Pathophysiology of chronic heart failure. Am J Med, 2001. 110: p. 37S-46S.

30. Sackner-Bernstein, J.D., The myocardial matrix and the development and progression of ventricular remodeling. Curr Cardiol Rep, 2000. 2: p. 112-9.

31. Stetler-Stevenson, W.G., Dynamics of matrix turnover during pathologic remodeling of the extracellular matrix. Am J Pathol, 1996. 148: p. 1345-50.

32. Anversa, P., G. Olivetti, and J.M. Capasso, Cellular basis of ventricular remodeling after myocardial infarction. Am J Cardiol, 1991. 68: p. 7D-16D.

33. Olivetti, G., et al., Side-to-side slippage of myocytes participates in ventricular wall remodeling acutely after myocardial infarction in rats. Circ Res, 1990. 67: p. 23-34.

34. Cleutjens, J.P., et al., Collagen remodeling after myocardial infarction in the rat heart. Am J Pathol, 1995. 147: p. 325-38.

35. Zhao, M.J., et al., Profound structural alterations of the extracellular collagen matrix in postischemic dysfunctional ("stunned") but viable myocardium. J Am Coll Cardiol, 1987. 10: p. 1322-34.

36. Kai, H., et al., Peripheral blood levels of matrix metalloproteases-2 and -9 are elevated in patients with acute coronary syndromes. J Am Coll Cardiol, 1998. 32: p. 368-72.

37. Hojo, Y., et al., Expression of matrix metalloproteinases in patients with acute myocardial infarction. Jpn Circ J, 2001. 65: p. 71-5.

38. Inokubo, Y., et al., Plasma levels of matrix metalloproteinase-9 and tissue inhibitor of metalloproteinase-1 are increased in the coronary circulation in patients with acute coronary syndrome. Am Heart J, 2001. 141: p. 211-7.

39. Hirohata, S., et al., Time dependent alterations of serum matrix metalloproteinase-1 and

metalloproteinase-1 tissue inhibitor after successful reperfusion of acute myocardial infarction. Heart, 1997. 78: p. 278-84.

40.   Rohde, L.E., et al., Matrix metalloproteinase inhibition attenuates early left ventricular enlargement after experimental myocardial infarction in mice. Circulation, 1999. 99: p. 3063-70.

41.   Ducharme, A., et al., Targeted deletion of matrix metalloproteinase-9 attenuates left ventricular enlargement and collagen accumulation after experimental myocardial infarction. J Clin Invest, 2000. 106: p. 55-62.

42.   Peterson, J.T., et al., Evolution of matrix metalloprotease and tissue inhibitor expression during heart failure progression in the infarcted rat. Cardiovasc Res, 2000. 46: p. 307-15.

43.   Etoh, T., et al., Myocardial and interstitial matrix metalloproteinase activity after acute myocardial infarction in pigs. Am J Physiol, 2001. 281: p. H987-94.

44.   Brenner, D.A., et al., Prolonged activation of jun and collagenase genes by tumour necrosis factor-alpha. Nature, 1989. 337: p. 661-3.

45.   MacNaul, K.L., et al., Discoordinate expression of stromelysin, collagenase, and tissue inhibitor of metalloproteinases-1 in rheumatoid human synovial fibroblasts. Synergistic effects of interleukin-1 and tumor necrosis factor-alpha on stromelysin expression. J Biol Chem, 1990. 265: p. 17238-45.

46.   Bond, M., et al., Synergistic upregulation of metalloproteinase-9 by growth factors and inflammatory cytokines: an absolute requirement for transcription factor NF-kappa B. FEBS Lett, 1998. 435: p. 29-34.

47.   Kurrelmeyer, K.M., et al., Endogenous tumor necrosis factor protects the adult cardiac myocyte against ischemic-induced apoptosis in a murine model of acute myocardial infarction. Proc Natl Acad Sci U S A, 2000. 97: p. 5456-61.

48.   Siwik, D.A., P.J. Pagano, and W.S. Colucci, Oxidative stress regulates collagen synthesis and matrix metalloproteinase activity in cardiac fibroblasts. Am J Physiol, 2001. 280: p. C53-60.

49.   Creemers, E.E., et al., Matrix metalloproteinase inhibition after myocardial infarction: a new approach to prevent heart failure? Circ Res, 2001. 89: p. 201-10.

50.   Li, Y.Y., et al., Downregulation of matrix metalloproteinases and reduction in collagen damage in the failing human heart after support with left ventricular assist devices. Circulation, 2001. 104: p. 1147-52.

51.   Baghelai, K., et al., Decreased expression of tissue inhibitor of metalloproteinase 1 in stunned myocardium. J Surg Res, 1998. 77: p. 35-9.

52.   Spinale, F.G., et al., A matrix metalloproteinase induction/activation system exists in the human left ventricular myocardium and is upregulated in heart failure. Circulation, 2000. 102: p. 1944-9.

53.   Thomas, C.V., et al., Increased matrix metalloproteinase activity and selective upregulation in LV myocardium from patients with end-stage dilated cardiomyopathy. Circulation, 1998. 97: p. 1708-15.

54.   Wynne, J. and E. Braunwald, The cardiomyopathies and myocarditis. Heart Disease, 1997: p. 1404-1463.

55.   Coker, M.L., et al., Myocardial matrix metalloproteinase activity and abundance with congestive heart failure. Am J Physiol, 1998. 274: p. H1516-23.

56.   Li, Y.Y., et al., Differential expression of tissue inhibitors of metalloproteinases in the failing human heart. Circulation, 1998. 98: p. 1728-34.

57.   Tyagi, S.C., et al., Matrix metalloproteinase activity expression in infarcted, noninfarcted and dilated cardiomyopathic human hearts. Mol Cell Biochem, 1996. 155: p. 13-21.

58.   Knauper, V., et al., Biochemical characterization of human collagenase-3. J Biol Chem,

1996. 271: p. 1544-50.

59. Kim, H.E., et al., Disruption of the myocardial extracellular matrix leads to cardiac dysfunction. J Clin Invest, 2000. 106: p. 857-66.

60. De Clerck, Y.A., et al., Characterization of the promoter of the gene encoding human tissue inhibitor of metalloproteinases-2 (TIMP-2). Gene, 1994. 139: p. 185-91.

61. Wick, M., et al., Structure of the human TIMP-3 gene and its cell cycle-regulated promoter. Biochem J, 1995. 311: p. 549-54.

62. Little, W.C. and R.J. Applegate, Congestive heart failure: systolic and diastolic function. J Cardiothorac Vasc Anesth, 1993. 7: p. 2-5.

63. Smith, V.-E. and Zile,M., Relaxation and diastolic properties of the heart. In: The Heart and Cardiovascular System: Scientific Foundations, Fozzard, H.A., et al., New York, NY: Raven Press, 1992: p. 1353-1367.

64. Weber, K.T., Cardiac interstitium in health and disease: the fibrillar collagen network. J Am Coll Cardiol, 1989. 13: p. 1637-52.

65. Weber, K.T. and C.G. Brilla, Pathological hypertrophy and cardiac interstitium. Fibrosis and renin-angiotensin-aldosterone system. Circulation, 1991. 83: p. 1849-65.

66. Nagatomo, Y., et al., Differential effects of pressure or volume overload on myocardial MMP levels and inhibitory control. Am J Physiol, 2000. 278: p. H151-61.

67. Spinale, F.G., et al., Structural basis for changes in left ventricular function and geometry because of chronic mitral regurgitation and after correction of volume overload. J Thorac Cardiovasc Surg, 1993. 106: p. 1147-57.

68. Grossman, W., D. Jones, and L.P. McLaurin, Wall stress and patterns of hypertrophy in the human left ventricle. J Clin Invest, 1975. 56: p. 56-64.

69. Li-Saw-Hee, F.L., et al., Matrix metalloproteinase-9 and tissue inhibitor metalloproteinase-1 levels in essential hypertension. Relationship to left ventricular mass and anti-hypertensive therapy. Int J Cardiol, 2000. 75: p. 43-7.

70. Laviades, C., et al., Abnormalities of the extracellular degradation of collagen type I in essential hypertension. Circulation, 1998. 98: p. 535-40.

71. Brower, G.L. and J.S. Janicki, Contribution of ventricular remodeling to pathogenesis of heart failure in rats. Am J Physiol, 2001. 280: p. H674-83.

72. Li, H., et al., MMP/TIMP expression in spontaneously hypertensive heart failure rats: the effect of ACE- and MMP-inhibition. Cardiovasc Res, 2000. 46: p. 298-306.

73. Mujumdar, V.S. and S.C. Tyagi, Temporal regulation of extracellular matrix components in transition from compensatory hypertrophy to decompensatory heart failure. J Hypertens, 1999. 17: p. 261-70.

74. Iwanaga, Y., et al., Excessive activation of matrix metalloproteinases coincides with left ventricular remodeling during transition from hypertrophy to heart failure in hypertensive rats. J Am Coll Cardiol, 2002. 39: p. 1384-91.

75. Walther, T., et al., Regression of left ventricular hypertrophy after surgical therapy for aortic stenosis is associated with changes in extracellular matrix gene expression. Circulation, 2001. 104: p. I54-8.

76. Dolgilevich, S.M., et al., Changes in collagenase and collagen gene expression after induction of aortocaval fistula in rats. Am J Physiol, 2001. 281: p. H207-14.

77. Chancey, A.L., et al., Effects of matrix metalloproteinase inhibition on ventricular remodeling due to volume overload. Circulation, 2002. 105: p. 1983-8.

78. Lindsey, M.L., et al., Selective matrix metalloproteinase inhibition reduces left ventricular remodeling but does not inhibit angiogenesis after myocardial infarction. Circulation, 2002. 105: p. 753-8.

79. Rasmussen, H.S. and P.P. McCann, Matrix metalloproteinase inhibition as a novel anticancer strategy: a review with special focus on batimastat and marimastat. Pharmacol Ther, 1997. 75: p. 69-75.

80. Hidalgo, M. and S.G. Eckhardt, Development of matrix metalloproteinase inhibitors in cancer therapy. J Natl Cancer Inst, 2001. 93: p. 178-93.

81. Crawford, H.C. and L.M. Matrisian, Mechanisms controlling the transcription of matrix metalloproteinase genes in normal and neoplastic cells. Enzyme Protein, 1996. 49: p. 20-37.

82. Benbow, U. and C.E. Brinckerhoff, The AP-1 site and MMP gene regulation: what is all the fuss about? Matrix Biol, 1997. 15: p. 519-26.

83. Fini, M., et al., Regulation of matrix metalloproteinase gene expression. In: Matrix Metalloproteinases, Parks, W.C. and Mecham, R.P., eds. San Diego, CA: Academic, 1998: p. 299-356.

84. Ries, C. and P.E. Petrides, Cytokine regulation of matrix metalloproteinase activity and its regulatory dysfunction in disease. Biol Chem Hoppe Seyler, 1995. 376: p. 345-55.

85. Bradham, W.S., et al., Tumor necrosis factor-alpha and myocardial remodeling in progression of heart failure: a current perspective. Cardiovasc Res, 2002. 53: p. 822-30.

86. Siwik, D.A., D.L. Chang, and W.S. Colucci, Interleukin-1beta and tumor necrosis factor-alpha decrease collagen synthesis and increase matrix metalloproteinase activity in cardiac fibroblasts in vitro. Circ Res, 2000. 86: p. 1259-65.

87. Sivasubramanian, N., et al., Left ventricular remodeling in transgenic mice with cardiac restricted overexpression of tumor necrosis factor. Circulation, 2001. 104: p. 826-31.

88. Schroen, D.J. and C.E. Brinckerhoff, Nuclear hormone receptors inhibit matrix metalloproteinase (MMP) gene expression through diverse mechanisms. Gene Expr, 1996. 6: p. 197-207.

89. Deswal, A., et al., Safety and efficacy of a soluble P75 tumor necrosis factor receptor (Enbrel, etanercept) in patients with advanced heart failure. Circulation, 1999. 99: p. 3224-6.

Chapter 10

Cardiac Mast Cells as Mediators of Ventricular Remodeling

Joseph S. Janicki, Gregory L. Brower, Amanda L.Chancey, Mary F. Forman, Lynetta J. Jobe
University of Auburn, Auburn, Alabama U.S.A.

1. Introduction

The increased workload imposed on the heart by chronic ventricular volume or pressure overload induces compensatory structural remodeling of the muscular, vascular, and extracellular matrix components of the myocardium. However, progressive ventricular hypertrophy and dilatation ultimately have a detrimental effect on ventricular function, resulting in heart failure [1, 2]. A suitable explanation for this pathologic remodeling has not been established, although degradation of fibrillar myocardial collagen represents a common pathway that could produce these adverse structural and architectural alterations. Fibrillar collagen provides the framework which interconnects cardiomyocytes and blood vessels, thereby maintaining ventricular size and shape and contributing to tissue stiffness [3, 4]. This being the case, then these myocardial collagen fibers must be disrupted for ventricular dilatation, sphericalization and wall thinning to occur. The presence of a latent matrix metalloproteinase (MMP) system which is colocalized with the interstitial myocardial collagen matrix has been documented [5], indicating the potential for collagen degradation to exceed synthesis should there be significant activation of this latent MMP system. Mast cells are known to store and release a variety of biologically active mediators, including tumor necrosis factor-α (TNF-α) and proteases such as tryptase, chymase and stromelysin, which are capable of MMP activation [6-11]. Increased numbers of mast cells have been reported in explanted human hearts with dilated cardiomyopathy [12, 13], and in animal models of experimentally-induced hypertension [14-16], myocardial infarction [17], and chronic volume overload secondary to aortocaval fistula [18, 19] and mitral regurgitation [20, 21]. Accordingly, mast cells have been thought to have a major role in the pathophysiology of these cardiovascular disorders. Thus, the potential role of cardiac mast cells and in particular mast cell-derived tryptase, chymase and TNF-α in activating MMPs and causing myocardial fibrillar collagen degradation and adverse remodeling will be the focus of this chapter.

2. Historical Background

Mast cells were first described in 1878 by Paul Ehrlich, who referred to this relatively large cell as MASTZELLEN or the "well-fed cell" because its cytoplasm was stuffed with prominent granules. However, it took an additional 80 to 100 years for reports addressing cardiac mast cells to appear. These studies, however, were focused on observations of increased numbers of cardiac mast cells associated with endomyocardial fibrosis and eosinophilic myocarditis [22, 23]. Since then, several other articles which document significant increases in cardiac mast cells and/or their functional role in cardiac diseases have been published. For example, clear evidence of cardiac mast cell degranulation and massive interstitial edema in endomyocardial biopsies from two cardiac patients was reported in 1986 [24]. In 1992, Li and his coworkers concluded from serial endomyocardial biopsies taken from transplanted human hearts that cardiac mast cells are associated with interstitial and perimyocytic fibrosis [25]. In 1995, the role of increased mast cells in atherosclerotic plaques and the erosion or rupture of coronary atheromas was clearly delineated in a review article by Petri T. Kovanen [26]. Also in 1995, Engels et al. [17] reported their finding of a marked accumulation of cardiac mast cells in the subepicardial layer of the infarcted region following experimental myocardial infarction in rats. In 1996, our laboratory reported a marked increase in cardiac mast cell density in the first week after creation of an infrarenal aortocaval fistula in rats which was correlated with significant MMP activation and subsequent degradation of the extracellular collagen matrix [19]. During the following year, Patella et al. [12] described a significantly higher than normal mast cell density in explanted hearts from patients with dilated cardiomyopathy, and Dell'Italia and colleagues [20] reported an elevation in the number of mast cells in dog hearts four months after the onset of experimental mitral regurgitation. From this brief overview, it is obvious that cardiac mast cells are significantly elevated when subjected to the increased stress of pressure or volume overload. Also, an understanding of their role as mediators of ventricular remodeling is beginning to emerge. This role is detailed in the next subsection.

3. Mast Cell Characteristics and Their Role in Ventricular Remodeling

There are at least two distinct mast cell phenotypes that have been best characterized in the mucosa, skin and lungs [8, 27]. The traditional classification of human mast cells has been based on their neutral protease content, with the MC_T subset having granules which contain only tryptase, while MC_{TC} cells contain chymase, cathepsin G and carboxypeptidase, in addition to tryptase. The few studies characterizing human cardiac mast cells have described them as being consistent with the MC_{TC} subtype [28, 29]. Mature cardiac mast cells are relatively large and when stained with toluidine blue are easily discernible with light microscopy (Figure 1). In vitro studies have verified that mast cell proteases are capable of activating collagenase (MMP-1) and stromelysin (MMP-3) [6, 11]. Also, stromelysin was found to be present in normal murine skin and lung mast cells [30], but whether it is also stored in cardiac mast cell granules remains to be determined. Recently, cardiac mast cells have also been shown to contain preformed TNF-α [10].

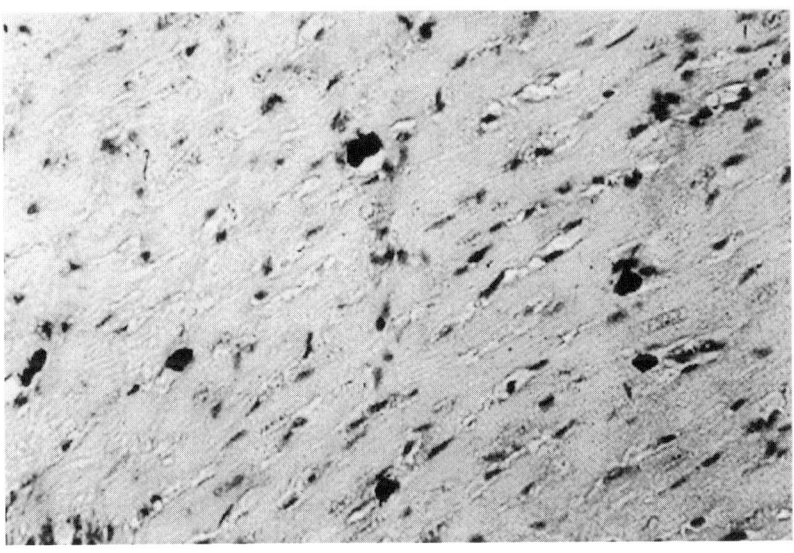

Figure 1. Toluidine blue stained histologic section demonstrating the prominent appearance of cardiac mast cells which are shown as cells with a large, dark appearance.

While the above represents only a partial listing of the enzymes and cytokines contained within mast cells, they are specifically mentioned because of their potential roles in activating MMPs and initiating a ventricular remodeling process. In fact, recent results from our laboratory clearly demonstrate that the relatively low density of mast cells in the normal heart (i.e., 2.1 ± 0.5 cells/mm^2) is sufficient to produce significant activation of cardiac MMPs [31]. Using a blood-perfused isolated rat heart, compound 48/80-induced degranulation of cardiac mast cells produced a substantial activation of MMP-2 (i.e., 126% increase relative to control) together with a nearly 50% decrease in myocardial collagen within a thirty minute period. A tendency for the left ventricle (LV) to dilate was also evident despite a significant increase in myocardial edema (i.e., water content of $80.1\pm3.4\%$ vs. $77.4\pm1.1\%$). It is of interest to note that MacKenna et al. [32] obtained similar results in isolated rat hearts perfused with bacterial collagenase. Specifically, they noted a similar rightward parallel shift in the LV pressure-volume relationship (i.e., dilatation) together with a 36% decrease in myocardial collagen following one hour of exposure to bacterial collagenase.

The above findings using compound 48/80 led us to conclude that mast cell density in the normal rat heart is sufficient to produce rapid activation of MMPs, degradation of the extracellular matrix, and ventricular dilatation. Such mast cell-mediated events are likely to contribute to the adverse ventricular remodeling associated with the development of congestive heart failure. For example, explanted human hearts from patients with idiopathic dilated cardiomyopathy have been reported to have increased mast cell numbers [12], as well as marked increases in MMP activity [33, 34]. Also, we and others have reported increases in MMPs that were correlated with ventricular dilatation in the cardiomyopathic Syrian hamster [35, 36] and the rapid pacing model of heart failure [37]. Unfortunately, studies such as these which utilize tissue obtained from hearts at autopsy, endomyocardial biopsies, explanted human hearts or hearts subjected to a prolonged experimentally induced pathologic conditions do not provide insight into a cause and effect relationship. Instead they merely indicate the status of interstitial collagen and MMP activity after the ventricle has remodeled sufficiently for heart failure to have occurred. Obviously a cause and effect relation between MMP activation, collagen degradation and ventricular remodeling can only be obtained from a temporal study that precedes the onset of heart failure.

Recently, we investigated the temporal response of myocardial mast cell density, MMP activity and collagen concentration in rats sacrificed at 0.5, 1, 2, 3, 5, 14, 21, 35, and 56 days following the creation of an aortocaval fistula [18]. Marked increases in MMP activity and mast cell density occurred rapidly and remained significantly elevated for the first week before returning toward normal by 14 days. At 35 and 56 days, MMP activity was again significantly elevated but to a lesser extent than that seen during the first 5 days. At days 14 and 21, mast cell densities were statistically similar to control values, but mast cell density increased again on days 35 and 56. The temporal response of mast cell density in the right ventricle was similar to that in the left ventricle in this biventricular chronic volume overload model. The marked increase in MMP activity was associated with a subsequent degradation of the extracellular matrix, with the myocardial collagen volume fraction being significantly decreased by 48% and 61% at 3 and 5 days post-fistula, respectively. Collagen volume fraction then rebounded and actually exceeded control values by 55% and 43% on days 14 and 35, respectively. At all other times investigated, it was statistically similar to the control values. Thus, there was a close association (correlation coefficient of r = 0.75) between mast cell density and MMP activity and, as a result of the initial increase in MMP activity, collagen degradation was noted to occur. A similar relation between mast cell density and MMP activity (correlation coefficient of r = 0.67) was obtained in a dog model of experimental mitral regurgitation [21]. The following results were obtained in this dog study: 1) mast cell density was significantly elevated by four- and three-fold after 2 and 4 weeks of mitral regurgitation, respectively, but had returned to normal by 24 weeks: 2) MMP activity was increased by 60 to 100% above control values throughout; and 3) collagen volume fraction was reduced by 30 to 50% from control levels at all three study periods.

As recently reported, degradation of the extracellular matrix in the rat aortocaval fistula model of volume overload-induced heart failure is clearly associated with alterations in LV size and stiffness [38]. Within the first week post-fistula there is a 65μl parallel shift of the LV end-diastolic relationship to the right (i.e., a 30% increase in unstressed ventricular volume indicative of LV dilatation), which in all likelihood was the result of the initial decrease in interstitial myocardial collagen. During the subsequent seven weeks, there was a progressive increase in LV size together with decreased myocardial stiffness and depressed contractility. Beyond 8 weeks, no additional changes were noted in compensated hearts, while in failing hearts there was a further marked ventricular dilatation, increased compliance and decreased contractility [39]. Thus, with the exception of the first week of remodeling, the relationships between the subsequent responses in mast cell density, MMP activity and collagen volume fraction and the progressive LV enlargement and increasing compliance are not straightforward. Clearly, however, the up and down fluctuations in mast cell density, collagen volume fraction, and MMP activity beyond the first week of volume overload reflect a dynamic imbalance between collagen synthesis and degradation as LV remodeling continues [18, 21, 38, 39]. Moreover, the characteristics of the remodeled extracellular matrix would also be expected to have an impact on ventricular size and distensibility. For example, newly synthesized collagen may have a greater proportion of the more compliant type III collagen and be less extensively cross-linked, with both of these alterations resulting in a more compliant ventricle. While additional factors, such as decreased integrin-mediated cardiomyocyte adhesion to the extracellular matrix could also lead to ventricular dilatation and increased compliance, it remains to be determined whether mast-cell release of TNF-α or related substances are capable of decreasing cardiomyocyte adhesion to extracellular matrix components.

4. Mast Cell Stabilizing Compounds Prevent Cardiac Remodeling

As discussed above, there is a rapid increase in mast cell density induced by ventricular volume overload which is associated with a concomitant increase in MMP activity and extracellular matrix degradation. Since myocardial fibrillar collagen maintains ventricular chamber stiffness and preserves ventricular size and shape, a significant disruption and degradation of collagen fibers might be expected to initiate a ventricular remodeling process resulting in altered myocardial compliance and dilatation. Thus, if this myocardial remodeling is indeed initiated by mast cell secretory products, then it should not occur when mast cell degranulation is pharmacologically prevented. In a recent study [18], cromolyn sodium, a mast cell membrane stabilizing compound commonly used for the treatment of allergies and asthma, was administered to rats 7 days prior to creating an aortocaval fistula and continued until the rats were terminally studied on days 1, 2 and 3 post-fistula. It was found that cromolyn was not only able to prevent the expected activation of MMP, but also the expected acute increase in mast cell density. In addition to implicating mast cell secretory products as the cause of MMP activation, these results suggest that mast cell degranulation may be responsible for the increase in mast cell density.

In another study [40], the long term implications of preventing mast cell degranulation on ventricular remodeling were tested utilizing rats with chronic ventricular volume overload. Instead of cromolyn, however, a related mast cell stabilizing compound, nedocromil sodium, was continuously administered beginning seven days prior to creating the fistula and thereafter for eight weeks. At the end of the study period, LV end diastolic pressure-volume (EDP-EDV) curves were obtained and compared to curves obtained from untreated fistula rats. The average LV EDP-EDV relationship for the fistula group receiving nedocromil was found to be essentially unchanged from that of a sham operated control group while EDP-EDV curves obtained from the untreated fistula group were significantly shifted to the right in a non-parallel fashion (dilatation with increased compliance). In addition to preventing adverse LV remodeling, nedocromil preserved contractility, which was significantly reduced (58%) in the untreated fistula group. The morbidity/mortality typically seen during the study period was also significantly reduced. That is, only 17% of the nedocromil treated fistula rats died versus 54% in the untreated fistula group [40]. Thus, the ability of a mast cell membrane stabilizing drug to prevent cardiac remodeling for periods of up to 8 weeks of chronic LV volume overload is strong evidence that mast cells are centrally involved in the regulation of ventricular remodeling. Furthermore, given the ability of cromolyn to prevent both the fistula-induced increase in mast cell density and MMP activity [18], it is highly likely that mast cell mediated MMP activation is directly responsible for initiating the ventricular remodeling process in this model.

A recent study by Hara et al. [14] using the mast cell membrane stabilizing drug, tranilast, indicates that mast cell secretory products are also responsible for the adverse remodeling that occurs as the pressure overloaded ventricle decompensates and fails. Tranilast significantly limited the development of LV dilatation and hypertrophy as well as the decrease in fractional shortening that occurred in the untreated pressure overloaded mice. This study also evaluated the response to abdominal aortic constriction in mast cell deficient mice and found LV performance to be preserved throughout the 15 week study period.

5. Source of Increased Mast Cell Density

As stated earlier, increases in mast cell density have been noted by several groups of investigators. In humans with end-stage dilated cardiomyopathy, a nearly four-fold increase from a normal value of 5.3 ± 0.7 cells/mm^2 was measured in explanted hearts [13]. In rats, normal and increased values were as follows: 1) hypertension - 0.87 ± 0.05 and 1.36 ± 0.05 cells/mm^2 [16]; volume overload - 2.1 ± 0.5 and 3.6 ± 1.1 cells/mm^2 [18]; and 3) myocardial infarction - 1.8 ± 0.3 and 26.3 ± 7.4 cells/mm^2 [17]. The mast cell density of 14 cells/mm^2 reported in dogs with mitral valve regurgitation was approximately three times greater than that in control dogs [20]. However, the source of the increased number of cardiac mast cells in these various conditions was not addressed.

In their studies of ischemia/reperfusion, Frangogiannis et al. [10] noted an increased number of mast cells in the reperfused region of the heart. While they did not find evidence of mast cell proliferation, they did observe intravascular cells expressing the mast cell-specific protease, tryptase, in both the ischemic and reperfused areas of myocardium. This led them to hypothesize that chemotaxis of circulating mast cell precursors may be responsible for the mast cell accumulation in the healing myocardium. Another source potentially responsible for the increased mast cell density that was not considered by this group is the rapid maturation of resident immature mast cells. The maturation and differentiation of these resident cardiac mast cells is associated with progressive sulfation of heparin, the formation of mast cell chymase and histamine, and the loss of mitotic activity [41, 42]. Using differential staining criteria, four stages of mast cell maturation have been identified using alcian and safranin stains as follows: Stage I or immature mast cells are those that are completely blue; stage II mast cells are those that stain predominantly blue (i.e., >60%), with lesser amounts of red (i.e., <40%); stage III mast cells are those that stain predominantly red (i.e., >60%), with lesser amounts of blue (i.e., <40%); and stage IV or totally differentiated, mature mast cells which appear completely brick-red. In addition to staining differences, Yong et al. [42, 43] found the mean diameter of mast cells from adult rat peritoneal washings and cardiac tissue to be smallest in stage I and largest in stage IV. Similarly, using thymidine-H^3 uptake, stage I and II mast cells were identified as being capable of mitosis, while stages III and IV were mitotically inactive [42, 43]. We recently found similar evidence of maturation/differentiation in cardiac mast cells isolated from rats with an aortocaval fistula. Using these staining and cell size criteria, we demonstrated a significant decrease in stage II, coupled with a significant increase in stage III and no change in the number of stage IV cardiac mast cells relative to sham-operated rats within 24 hours of inducing biventricular volume overload. Thus, the acute rise in cardiac mast cells following volume overload appears to be due to the maturation/differentiation of resident immature cardiac mast cells.

The initiation of this maturation process in all likelihood was triggered by the degranulation of mature mast cells, as mast cell degranulation products have previously been shown to induce mast cell hyperplasia [44]. Given that circulating atrial natriuretic peptide (ANP) levels are known to increase rapidly in chronic volume overload [45], and ANP induces degranulation of peritoneal mast cells [46, 47], it makes sense that ANP might potentially be responsible for inducing cardiac mast cell degranulation. While this remains to be proven, based on these observations one can propose the following sequence of events depicted in Figure 2: an aortocaval fistula causes atrial distension, thereby producing ANP secretion, cardiac mast cell degranulation, and the stimulation of rapid maturation in stage I and II mast cells.

6. Contribution of Mast Cell-Derived TNF-α to Ventricular Remodeling

Strong evidence that TNF-α secreted by cardiac mast cells contributes to the ventricular remodeling secondary to chronic volume overload is rapidly emerging. Recently, Bozkurt et al. [48] reported that a continuous infusion of pathophysiologic levels of TNF-α in normal rats led to a time dependent depression of LV function, cardiac myocyte fractional shortening, and LV dilation. Preliminary results from our laboratory indicate that a similar two week administration of TNF-α to normal rats resulted in a significant rightward shift of the LV EDP-EDV relationship that was identical to that obtained following three weeks of chronic LV volume overload. Another interesting observation was that the level of TNF-α in the myocardium was elevated to the same extent in rats with a fistula as in those receiving an infusion of TNF-α. Moreover, a two week administration of TNF-α superimposed on a fistula did not produce additional LV dilatation beyond that seen in fistula rats that were not treated with TNF-α. These observations indicating TNF-α mediated myocardial remodeling occurs during the compensated phase of chronic volume overload led to a subsequent study testing the ability of a p75 TNF-α receptor fusion protein, etanercept (TNFR:Fc), to prevent adverse myocardial remodeling. The administration of etanercept was initiated in rats three weeks after creation of a fistula, continued for three weeks and its effect on LV remodeling assessed at eight weeks post-fistula. TNF-α antagonism using etanercept essentially prevented any further myocardial remodeling, while those rats that were untreated post-fistula progressed on to develop heart failure. Bradham et al. [49] reported similar efficacious results using etanercept in a rapid pacing model of heart failure.

In another preliminary study, the response in the heart during the first week of a TNF-α infusion was assessed. Like the response observed in the hearts with volume overload, TNF-α infusion resulted in increased MMP activity and a decrease in myocardial collagen concentration. These results and the findings of Bradham et al. [49], which demonstrated that etanercept prevented the increase in MMP activity in their model of heart failure, indicate that mast cell-derived TNF-α or one of the downstream cytokines it induces is responsible for the early MMP activation occurring in these experimental models. Thus, it would appear that, as a consequence of ventricular volume overload, TNF-α from the increased number of mast cells plays an important role in ventricular remodeling.

7. Summary

Mast cells are known to store and release a variety of biologically active mediators including TNF-α, and proteases such as tryptase and chymase. With cardiac chamber distension there is a release of atrial natriuretic peptide, which is known to cause mast cell degranulation (Figure 2). Secreted TNF-α, tryptase and chymase are all capable of activating matrix metalloproteinases, which in turn are responsible for fibrillar collagen degradation. Since one of the roles of the extracellular collagen matrix is to maintain ventricular size and shape, its disruption results in adverse remodeling. Also secreted from the mast cell is a yet to be identified substance that stimulates the maturation of resident immature mast cells. Proof of mast cell involvement in these processes is provided by the use of mast cell membrane stabilizing compounds such as cromolyn sodium, which prevent mast cell degranulation. These drugs prevent the activation of MMPs, degradation of collagen, the increase in mast cell density, adverse ventricular remodeling, and the decrease in contractility as well as attenuate the morbidity/mortality associated with chronic volume overload. They are similarly efficacious in preventing the onset of heart failure in hearts subjected to chronic pressure overload. Finally, evidence is rapidly emerging which identifies mast cell-derived TNF-α and/or the downstream cytokine cascade it induces as a major contributor to adverse ventricular remodeling and associated contractile dysfunction.

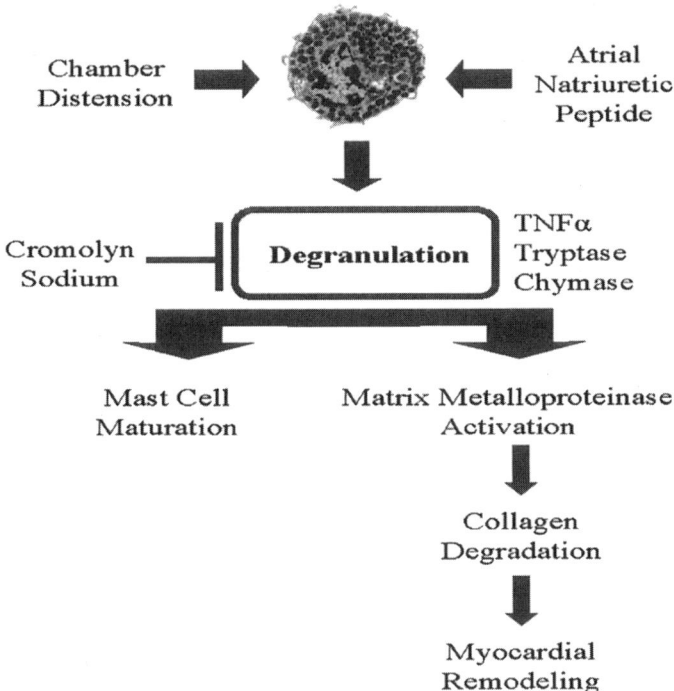

Figure 2. Schematic depicting cardiac mast cell mediated myocardial remodeling.

Acknowledgements

This work was supported in part by grants from NHLBI (Dr. Janicki, RO1-HL-59981 and HL-62228; Dr. Jobe, HL-59981-02S) and from American Heart Association Southern Research Consortium (Dr. Brower, 0051505B & Dr. Forman 00020383B).

References

1. Grossman, W., D. Jones, and L.P. McLaurin, Wall stress and patterns of hypertrophy in the human left ventricle. J Clin Invest, 1975. 56: p. 56-64.
2. Pfeffer, J.M., et al., Progressive ventricular remodeling in rat with myocardial infarction. Am J Physiol, 1991. 260: p. H1406-14.
3. Borg, T.K. and J.B. Caulfield, The collagen matrix of the heart. Fed Proc, 1981. 40: p. 2037-41.
4. Robinson, T.F., et al., Structure and function of connective tissue in cardiac muscle: collagen types I and III in endomysial struts and pericellular fibers. Scanning Microsc, 1988. 2: p. 1005-15.
5. Montfort, I. and R. Perez-Tamayo, The distribution of collagenase in normal rat tissues. J Histochem Cytochem, 1975. 23: p. 910-20.
6. Lees, M., D.J. Taylor, and D.E. Woolley, Mast cell proteinases activate precursor forms of collagenase and stromelysin, but not of gelatinases A and B. Eur J Biochem, 1994. 223: p. 171-7.
7. Marone, G., et al., Immunological modulation of human cardiac mast cells. Neurochem Res, 1999. 24: p. 1195-202.
8. Metcalfe, D.D., D. Baram, and Y.A. Mekori, Mast cells. Physiol Rev, 1997. 77: p. 1033-79.
9. Nagase, H., Activation mechanisms of matrix metalloproteinases. Biol Chem, 1997. 378: p. 151-60.
10. Frangogiannis, N.G., et al., Resident cardiac mast cells degranulate and release preformed TNF-alpha, initiating the cytokine cascade in experimental canine myocardial ischemia/reperfusion. Circulation, 1998. 98: p. 699-710.
11. Suzuki, K., et al., Activation of precursors for matrix metalloproteinases 1 (interstitial collagenase) and 3 (stromelysin) by rat mast-cell proteinases I and II. Biochem J, 1995. 305: p. 301-6.
12. Patella, V., et al., Increased cardiac mast cell density and mediator release in patients with dilated cardiomyopathy. Inflamm Res, 1997. 46: p. S31-2.
13. Patella, V., et al., Stem cell factor in mast cells and increased mast cell density in idiopathic and ischemic cardiomyopathy. Circulation, 1998. 97: p. 971-8.
14. Hara, M., et al., Evidence for a role of mast cells in the evolution to congestive heart failure. J Exp Med, 2002. 195: p. 375-81.
15. Olivetti, G., et al., Long-term pressure-induced cardiac hypertrophy: capillary and mast cell proliferation. Am J Physiol, 1989. 257: p. H1766-72.
16. Panizo, A., et al., Are mast cells involved in hypertensive heart disease? J Hypertens, 1995. 13: p. 1201-8.
17. Engels, W., et al., Transmural changes in mast cell density in rat heart after infarct induction in vivo. J Pathol, 1995. 177: p. 423-9.

208

18. Brower, G.L., et al., Cause and effect relationship between myocardial mast cell number and matrix metalloproteinase activity. Am J Physiol, 2002. 283: p. H518-25.

19. Henegar, J.R. et al., Myocardial Mast Cell Response to Chronic Ventricular Volume Overload. J Mol Cell Cardiol, 1996. 28: p. A202.

20. Dell'Italia, L.J., et al., Volume-overload cardiac hypertrophy is unaffected by ACE inhibitor treatment in dogs. Am J Physiol, 1997. 273: p. H961-70.

21. Stewart, J.A., et al., Cardiac mast cell- and chymase-mediated matrix metalloproteinase activity and left ventricular remodeling in mitral regurgitation in the dog. J Mol Cell Cardiol, 2003. 35: p. 311-9.

22. Estensen, R.D., Eosinophilic myocarditis: a role for mast cells? Arch Pathol Lab Med, 1984. 108: p. 358-9.

23. Fernex, M., The Mast-Cell System: Its Relationship to Atherosclerosis, Fibrosis and Eosinophils. Baltimore: The Williams & Wilkins Company, 1968: p. 93-95.

24. Dvorak, A.M., Mast-cell degranulation in human hearts. N Engl J Med, 1986. 315: p. 969-70.

25. Li, Q.Y., et al., The relationship of mast cells and their secreted products to the volume of fibrosis in posttransplant hearts. Transplantation, 1992. 53: p. 1047-51.

26. Kovanen, P.T., Role of mast cells in atherosclerosis. Chem Immunol, 1995. 62: p. 132-70.

27. Galli, S.J., The Paul Kallos Memorial Lecture. The mast cell: a versatile effector cell for a challenging world. Int Arch Allergy Immunol, 1997. 113: p. 14-22.

28. Patella, V., et al., Human heart mast cells: a definitive case of mast cell heterogeneity. Int Arch Allergy Immunol, 1995. 106: p. 386-93.

29. Patella, V., et al., Human heart mast cells. Isolation, purification, ultrastructure, and immunologic characterization. J Immunol, 1995. 154: p. 2855-65.

30. Brownell, E., et al., Immunolocalization of stromelysin-related protein in murine mast cell granules. Int Arch Allergy Immunol, 1995. 107: p. 333-5.

31. Chancey, A.L., G.L. Brower, and J.S. Janicki, Cardiac mast cell-mediated activation of gelatinase and alteration of ventricular diastolic function. Am J Physiol, 2002. 282: p. H2152-8.

32. MacKenna, D.A., et al., Contribution of collagen matrix to passive left ventricular mechanics in isolated rat hearts. Am J Physiol, 1994. 266: p. H1007-18.

33. Gunja-Smith, Z., et al., Remodeling of human myocardial collagen in idiopathic dilated cardiomyopathy. Role of metalloproteinases and pyridinoline cross-links. Am J Pathol, 1996. 148: p. 1639-48.

34. Thomas, C.V., et al., Increased matrix metalloproteinase activity and selective upregulation in LV myocardium from patients with end-stage dilated cardiomyopathy. Circulation, 1998. 97: p. 1708-15.

35. Dixon, I.M., et al., Cardiac collagen remodeling in the cardiomyopathic Syrian hamster and the effect of losartan. J Mol Cell Cardiol, 1997. 29: p. 1837-50.

36. Janicki, J.S. el al., Interstitial Collagen Remodeling in Chronic Heart Failure. Basic Applied Myology, 1995. 5: p. 339-48.

37. Spinale, F.G., et al., Cellular and extracellular remodeling with the development and recovery from tachycardia-induced cardiomyopathy: changes in fibrillar collagen, myocyte adhesion capacity and proteoglycans. J Mol Cell Cardiol, 1996. 28: p. 1591-608.

38. Brower, G.L., J.R. Henegar, and J.S. Janicki, Temporal evaluation of left ventricular remodeling and function in rats with chronic volume overload. Am J Physiol, 1996. 271: p. H2071-8.

39. Brower, G.L. and J.S. Janicki, Contribution of ventricular remodeling to pathogenesis of heart failure in rats. Am J Physiol, 2001. 280: p. H674-83.

40. Brower, G.L., W.D. Berry, and J. Janicki, Pharmacologic Inhibition of Mast Cell Degranulation Prevents Left Ventricular Remodeling Induced by Chronic Volume Overload in Rats. Circulation, 1997. 96:p. I-519.

41. Combs, J.W., D. Lagunoff, and E.P. Benditt, Differentiation and proliferation of embryonic mast cells of the rat. J Cell Biol, 1965. 25: p. 577-92.

42. Yong, L.C., S. Watkins, and D.L. Wilhelm, The mast cell: distribution and maturation in the peritoneal cavity of the adult rat. Pathology, 1975. 7: p. 307-18.

43. Yong, L.C., S.G. Watkins, and J.E. Boland, The mast cell: III. Distribution and maturation in various organs of the young rat. Pathology, 1979. 11: p. 427-45.

44. Marshall, J.S., et al., The role of mast cell degranulation products in mast cell hyperplasia. I. Mechanism of action of nerve growth factor. J Immunol, 1990. 144: p. 1886-92.

45. Huang, M., R.L. Hester, and A.C. Guyton, Hemodynamic changes in rats after opening an arteriovenous fistula. Am J Physiol, 1992. 262: p. H846-51.

46. Opgenorth, T.J., et al., Atrial peptides induce mast cell histamine release. Peptides, 1990. 11: p. 1003-7.

47. Yoshida, H., et al., Histamine release induced by human natriuretic peptide from rat peritoneal mast cells. Regul Pept, 1996. 61: p. 45-9.

48. Bozkurt, B., et al., Pathophysiologically relevant concentrations of tumor necrosis factor-alpha promote progressive left ventricular dysfunction and remodeling in rats. Circulation, 1998. 97: p. 1382-91.

49. Bradham, W.S., et al., TNF-alpha and myocardial matrix metalloproteinases in heart failure: relationship to LV remodeling. Am J Physiol, 2002. 282: p. H1288-95.

IV. *MATRIX METALLOPROTEINASES IN CARDIAC REMODELING*

Chapter 11

Turmoil in the Cardiac Myocyte: Acute Intracellular Activation of Matrix Metalloproteinases

Manoj M. Lalu, Hernando Leon, Richard Schulz
University of Alberta, Edmonton, Alberta, Canada.

1. Introduction

Matrix metalloproteinases (MMPs) are a family of zinc-dependent endopeptidases. From their first description in the process of tadpole tail resorption [1] it was recognized that they play an active role in regulating the extracellular matrix. Thus, over the course of the last forty years most research has focused on the extracellular role of MMPs in long term (days to months) physiological and pathological processes such as wound healing, embryogenesis, metastasis, atherosclerosis, and heart failure. Recently, however, it has been recognized that MMPs may also act: a) intracellularly, b) on non-extracellular matrix substrates, and c) over an acute time frame (i.e., seconds-minutes). This chapter will review these recent developments with a special focus on the intracellular role MMPs play in acute ischemia-reperfusion injury in the heart.

2. MMPs: Structure and Activation

MMPs can be classified according to their *in vitro* substrate specificity: there are collagenases (MMPs -1, -8, and -13), stromelysins (e.g., MMP-3), membrane-type MMPs (MT-MMPs 1 through 8), gelatinases (MMP-2 and -9), and a host of other MMPs (e.g., matrilysin). Many of these MMPs have been identified in a variety of cardiovascular cell types as detailed in Table 1.

Regardless of *in vitro* substrate preference, all the MMPs have quite similar domain structures [2]. Starting at the N-terminus, most MMPs have a signal peptide which allows for secretion into the endoplasmic reticulum and eventual transport out of the cell. Next to the signal peptide, most MMPs have a hydrophobic propeptide domain which shields the catalytic domain next to it. Finally, at the C-terminus most MMPs also have a hemopexin domain which confers some substrate specificity and allows docking with other proteins. The catalytic domain of all MMPs is known as the 'matrixin fold' and consists of five-stranded beta-sheets and three alpha-helices. This fold forms substrate binding pockets, coordinates with the catalytic Zn^{2+} ion, and also binds two Ca^{2+} ions [3]. In its zymogen form, the catalytic Zn^{2+} is coordinated to a cysteinyl sulphydryl group on the propeptide domain and is rendered inactive.

Since MMPs are initially synthesized with the propeptide domain shielding the matrixin fold, they must be activated to expose the catalytic Zn^{2+} ion. To date, four different mechanisms have been described: a) stepwise activation in the extracellular space, b) activation at the cell surface by MT-MMPs, c) intracellular activation, and d) activation by oxidative stress. Although most of these mechanisms have not been specifically elucidated in any type of cardiac cells, it is generally believed that MMP activation occurs by similar means regardless of the cell type.

In the extracellular stepwise activation process, another proteinase (such as plasmin, trypsin, elastase, or an MMP) cleaves at a susceptible loop region (which acts as 'bait') in the propeptide domain of the MMP. Upon cleavage, the prodomain structure breaks down and its shielding of the catalytic cleft is withdrawn. Water is then allowed to enter and hydrolyze the coordination of the cysteine to the Zn^{2+} ion. Ultimately, this renders the MMP prone to autocatalytic activity which cleaves off the propeptide domain and produces a lower molecular weight active enzyme [3, 4].

Alternatively, MMPs may also be activated at the cell surface by MT-MMPs. In order for MT-MMP1 to process MMP-2, MT-MMP1 forms a complex with tissue inhibitor of matrix metallproteinase-2 (TIMP-2), which serves as a receptor for MMP-2. When MMP-2 docks with this complex, proteolytic activation occurs at the cell surface, and an active MMP is released from the cell [5]. This method of activation would allow for increased MMP activity within the immediate pericellular space and, as noted in MT-MMP1 knockout mice, is important for MMP-2 activation [6]. Specifically, fibroblasts derived from MT-MMP1 knockout mice were unable to show MMP-2 activity in response to stimulation with collagen [7].

In contrast to extracellular or pericellular activation, a number of MMPs (e.g., MMP-11, MT-MMPs) are activated intracellularly by furin-like proprotein convertases [8-10]. After intracellular activation has occurred, the active MMP is shuttled either to the cell membrane for insertion (in the case of MT-MMPs) or secretion (e.g., MMP-11). Thus, in these circumstances, intracellular activation does not necessarily lead to intracellular activity of MMPs.

A final pathway of MMP activation, which can occur both extracellularly and intracellularly, is oxidative stress (Figure 1). In this latter mechanism, a variety of endogenous oxidants (e.g., superoxide anion, hydrogen peroxide, peroxynitrite) oxidize the sulphydryl bond of the cysteinyl group which binds to the catalytic Zn^{2+}. As with other mechanisms, disruption of this

Table 1. Synthesis of MMPs in cells relevant to the cardiovascular system.

Cell Type	MMP	Reference #
Cardiomyocyte	MMP-2	27
	MMP-3	11
	MMP-9	11
	TIMP-1	6
	TIMP-2	6
	TIMP-3	6
	TIMP-4	6,22
Cardiac fibroblasts	MMP-1	12,113
	MMP-2	28,112,114-116
	MMP-3	14
	MMP-9	28,115
	MMP-13	28
	MT-1 MMP (MMP-14)	114
	TIMP-4	23,29
Endocardial cells	MMP-2	117
	TIMP-2	118
	TIMP-3	118
Smooth muscle cells	MMP-1	19,120
	MMP-2	31, 121-123
	MMP-3	120
	MMP-9	31,119-121,123
	MMP-12	124
	MT- MMP1 (MMP-14)	125
	MT- MMP3 (MMP-16)	125,126
	TIMP-1	123,127
	TIMP-2	123
	TIMP-3	123
Endothelial cells	MMP-1	128-130
	MMP-2	30,115,128,130
	MMP-3	128
	MMP-9	30,115
	MT- MMP1 (MMP-14)	129-131
	TIMP-1	128,132
	TIMP-2	128,132
Platelets	MMP-1	46
	MMP-2	45,133,134
	MMP-9	134
	MT- MMP1 (MMP-14)	133
	TIMP-1	133
	TIMP-2	133
	TIMP-4	135

inhibitory bond allows hydration of the catalytic site. However, unlike other mechanisms, oxidative activation produces an 'activated proenzyme' in which the propeptide domain has not been removed [11, 12].

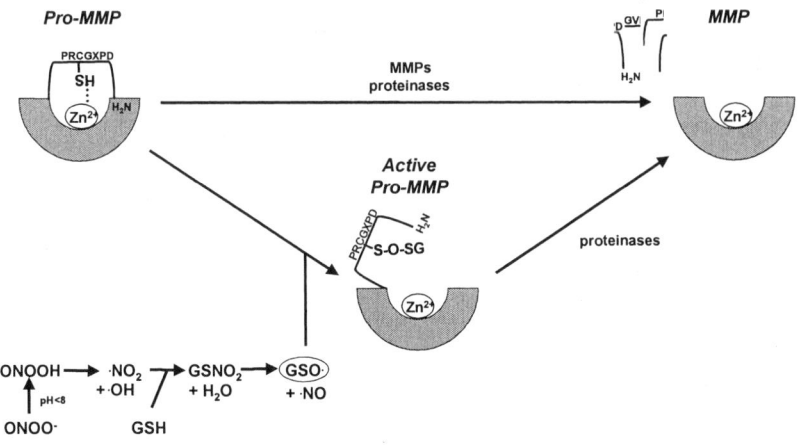

Figure 1. Activation of MMPs by oxidative stress. ProMMPs are synthesized with a propeptide domain shielding the catalytic Zn^{2+}. ProMMPs can be activated by classic proteolytic cleavage of its propeptide domain by MMPs or other proteinases (top arrow). Alternatively, MMPs can be activated by oxidants like peroxynitrite which interfere with the propeptide domain's ability to coordinate with the catalytic Zn^{2+} ion. Peroxynitrite (ONOO⁻), in an aqueous environment, forms the species ONOOH which spontaneously releases nitrogen dioxide radical (·NO₂) and hydroxyl radical (·OH). Glutathione (GSH, a tripeptide sulfhydryl containing compound) reacts with nitrogen dioxide radical and hydroxyl radical to form S-nitroglutathione (GSNO₂) and water. A concerted rearrangement of the S-nitroglutathione then occurs [GSNO₂ ➔ (GS⁻, ⁻NO₂) ➔ GSONO ➔ GSO⁻ + NO⁻] and GSH sulfinyl radical (GSO·) is produced along with nitric oxide (NO·). GSH sulfinyl radical can S-glutathiolate the cysteine (thiol) containing PRCGVPD sequence of the propeptide domain to form a glutathione S-oxide (GS(O)SR). S-glutathiolation of the propeptide inhibits its ability to coordinate and shield the catalytic Zn^{2+} ion, thus, an active proMMP is formed. This active proMMP may be further processed by other proteinases which can cleave the propeptide domain. (Figure based on work from References 12 and 13).

Mechanistically, it has been shown that as little as 1-20 μM peroxynitrite, one of the most powerful endogenous oxidants formed by the diffusion limited reaction between nitric oxide and superoxide, causes S-glutathiolation of a sequence within the propoeptide [13]. This S-glutathiolation takes place via disulfide S-oxide formation and produces a modification in size which is too small to detect by regular SDS-PAGE but can be detected using mass spectroscopy. Such a process, if it occurs intracellularly, could lead to inappropriate activation and proteolysis of proteins by MMPs within the cell. MMP-1, -2 -8, and -9, have been shown to be activated by peroxynitrite in this way and [11-13], although it has not been specifically tested whether other

MMPs can also be activated in this manner. Generally, the activation of MMPs by peroxynitrite (or other oxidants) occurs without the loss of the propeptide domain. This suggests that commonly used nomenclature which labels an MMP as being a 'proMMP' only by virtue of its higher molecular weight is both inaccurate and misleading. Such nomenclature does not take into account the potential for the higher molecular weight form of MMPs to be proteolytically active during oxidative stress, a condition common to several cardiovascular pathologies.

3. Inhibition of MMPs: TIMPs and Chemical Inhibitors

The tissue inhibitors of matrix metalloproteinases (TIMPs) provide another level of proteinase regulation by complexing with these MMPs and inhibiting their activity. Four TIMPs have been identified thus far and each binds to MMPs in a 1:1 stoichiometric ratio [14]. Although there is some binding preference of TIMP-2 with MMP-2, and TIMP-1 with MMP-9, the TIMPs in general do not show a high degree of specificity for any one MMP [14]. Indeed, there is no one study which has thoroughly examined the comparative inhibitory activity of all known TIMPs to a particular MMP. Structurally, TIMPs are two-domain molecules having an N-terminal MMP inhibiting domain and a smaller C-terminal domain. Three disulfide bonds stabilize each of these domains [15]. All four TIMPs have been observed in cardiac myocytes [16].

TIMP-1 and TIMP-2 are the best characterized TIMPs and they inhibit all known MMPs. TIMP-2 is constitutively expressed in a variety of cells of the heart, while TIMP-1 mRNA expression can be increased in response to signals such as cytokines in myocytes [16]. Both TIMP-1 and -2 also have diverse actions which are unrelated to inhibition of MMP activity, such as growth stimulating effects on erythroid precursor cells [17, 18].

TIMP-3, unlike the other TIMPs, binds tightly to the extracellular matrix and is usually difficult to extract from tissues [14, 19]. High levels of TIMP-3 transcripts have been detected in rat kidney, lungs, and heart [20]. Recently, TIMP-3 has been found to play an important role in inhibiting angiogenesis by attenuating the binding of vascular endothelial growth factor to its receptor [21]. Surprisingly, this effect was independent of its MMP inhibitory activity, since other TIMPs and synthetic MMP inhibitors could not inhibit binding of the growth factor to its receptor. Future studies will undoubtedly uncover novel biological actions for other TIMPs which are independent of their MMP inhibitory action.

TIMP-4 appears to be the most abundant of the TIMPs in the myocardium – and in fact the heart is the only organ where abundant transcripts for TIMP-4 have been found [22]. With this pronounced expression in the cardiovascular system, it has been suggested that TIMP-4 protects against cardiomyopathy, tumor development, and metastasis [23]. It may also protect against vascular injury and remodeling, since TIMP-4 was found to be acutely upregulated in rat carotid arteries following experimental vascular damage [24]. Our own investigation of TIMP-4 in the rat heart has localized this TIMP to the sarcomeres of rat ventricular myocytes, thus it may also play an important regulatory role in cardiomyocyte homeostasis [25].

Aside from the natural endogenous inhibitors of MMPs, a number of synthetic inhibitors of MMPs have been developed. These inhibitors include o-phenanthroline, tetracycline-class antibiotics (this is an additional effect independent of their antibiotic actions [26]) and many proprietary compounds developed by drug companies primarily as potential anti-arthritic and anti-cancer drugs. They share the common characteristic in having high specificity in binding to and inactivating the active site of MMPs. Use of these MMP inhibitors has proven very effective in discovering novel roles of MMPs in both physiological and pathological processes.

4. MMP-2, MMP-9, and TIMPs in Chronic Cardiovascular Pathology

A variety of MMPs have been implicated in long term cardiovascular remodeling and their roles have been detailed elsewhere in this book. For the purposes of this chapter, focus will be placed on the gelatinases since they are abundant in a variety of cardiac cells and a number of novel and acute functions have been elucidated for these MMPs. The gelatinases, MMP-2 (72 kDa and 62 kDa) and MMP-9 (92 kDa and 84 kDa), have been described in cardiac myocytes [27], cardiac fibroblasts [28], endocardial cells [29], as well as endothelial [30] and vascular smooth muscle cells [31] (Table 1). MMP-2 is usually regarded as a constitutive MMP which is ubiquitously expressed throughout the body [32] whereas MMP-9 is often viewed as a cytokine inducible MMP [33].

Peterson and colleagues found that MMP-2 activity is increased in the myocardium of spontaneously hypertensive heart failure rats compared to that of normotensive control rats. Interestingly, this increase corresponded with ventricular dilation and dysfunction as the animals aged, and the inhibition of MMP-2 activity by a four month treatment with a broad spectrum inhibitor could ameliorate remodeling and dysfunction [34]. Lee and colleagues demonstrated that targeted deletion of MMP-9 attenuated left ventricular remodeling after experimental myocardial infarction in mice [35]. In this study, the importance of MMP-9 in ventricular remodeling was highlighted by the fact that less dilation occurred in these knockout mice even though other MMPs were upregulated. Other evidence for the importance of MMP-2 and -9 in the heart was provided by a demonstration that selective MMP inhibitors (i.e., an MMP-1 sparing inhibitor) could ameliorate ventricular dysfunction in a model of heart failure in pigs caused by rapid ventricular pacing [36].

MMP-2 and MMP-9 also play crucial roles in human cardiovascular disease. It has been found that the activities of both these MMPs were increased in the ventricles of patients suffering from dilated cardiomyopathy [37]. Specifically, MMP-2 activity was increased in the hearts of patients with non-ischemic dilated cardiomyopathy while MMP-9 activity was increased in both ischemic and non-ischemic dilated cardiomyopathy. In a recent study the concentration of plasma MMP-9 was identified as a novel predictor of adverse cardiovascular events [38]. Patients who experienced a fatal cardiovascular event were found to have higher concentrations of circulating MMP-9. Moreover, a particular polymorphism of MMP-9 (R279Q) was found to be highly associated with patients suffering from stable angina. Thus, several lines of evidence underscore the importance of MMP-2 and -9 in long term cardiovascular pathology.

A number of studies have also demonstrated that dysregulation of TIMPs also contributes to long term cardiovascular pathologies. The deletion of TIMP-1 in mice was recently found to potentiate adverse remodeling following experimental myocardial infarction [39]. In these mice, ventricular weight and cross-sectional area of left ventricular myocytes were significantly increased, indicative of a pronounced hypertrophic response. In addition, fibrillar collagen content was reduced and myocardial infarct length was increased, which led the authors to conclude that TIMP-1 is important in myocardial structural remodeling. Thus, dysregulation of TIMPs likely plays a role in cardiovascular pathology.

Studies in humans with ischemic and non-ischemic dilated cardiomyopathy have produced conflicting data in regards to TIMP protein levels. In one study, myocardial TIMP-1 and -2 protein content increased five times in dilated cardiomyopathy compared to control patients [40] while another study found no differences in TIMP-1 content [37]. In contrast, Li et al. found a decrease in TIMP-1 and -3 protein content and no change in TIMP-2 and -4 protein content, [41] while Rouet-Benzinab et al. found both TIMP-1 and -2 were decreased in patients with dilated cardiomyopathy [42]. In patients with end stage congestive heart failure, TIMP-1, -2, and -4 protein content was no different from control patients, but TIMP-3 content was significantly decreased [43].

From these conflicting results it is evident that further research is necessary. However, one promising result was found in the fact that chronic unloading of the ventricle through the use of ventricular assist devices was associated with an increase in TIMP-1 and -3 content and a reduction in chamber dilation [44].

5. Novel Roles for MMP-2

Given the abundant evidence linking MMPs with chronic cardiovascular disease, most researchers have focused on the long term proteolytic effects of MMPs on extracellular matrix substrates (ie. collagen breakdown). Nonetheless, a number of novel acute effects for MMPs (on a seconds to minutes timescale) have been uncovered in the past few years. For instance, in a seminal investigation Sawicki et al. demonstrated that MMP-2 is released by activated platelets and promotes platelet aggregation [45]. Moreover, exogenously added MMP-2 could stimulate aggregation while recombinant TIMP-2, as well as neutralizing antibodies against MMP-2, could prevent it. Further work has shown that MMP-1 also contributes to platelet aggregation through 'outside-in' signaling pathways [46]. In other investigations, MMPs were found to be involved in cell-to-cell signaling through the cleavage of chemokines. Overall and colleagues demonstrated that MMP-2 could cleave monocyte chemoattractant protein-3 to an inactive peptide which acts as a chemokine receptor antagonist to dampen inflammation [47].

MMPs also appear to acutely regulate vascular tone since MMP-2 can cleave big endothelin (ET) to yield the novel vasoconstrictor ET-1[48]. Further investigations found that MMP-9 could also produce this potent vasoconstrictor, and that ET-1 could promote neutrophil adhesion to endothelial cells [49]. MMP-2 was also shown to cleave and inactivate the vasodilator calcitonin gene related peptide [50]. Thus, in conditions where MMP-2 activity is upregulated, increased vascular tone may result from the combined activation of a vasoconstrictor propeptide and the inactivation of a vasodilator peptide.

These studies suggest that MMPs have as-yet-unknown regulatory actions in both normal and pathological conditions. Moreover, the word "matrix" in MMPs does not accurately reflect the full spectrum of their biological activities. In seeking a role for MMPs in regulating cardiac function, our laboratory has investigated whether MMP-2 contributes to acute cardiac mechanical dysfunction in ischemia-reperfusion injury.

6. Myocardial Ischemia-Reperfusion Injury

The first description of myocardial ischemia-reperfusion (I/R) injury was made in 1975 when Heyndrickx et al. observed persistent regional myocardial mechanical dysfunction after brief coronary ligation and reperfusion in dogs [51]. Braunwald and Kloner [52] later named this phenomenon myocardial 'stunning' injury. Clinically, I/R and stunning injuries are seen with the use of thrombolytics following infarction [53] and also as a result of surgery involving cardiopulmonary bypass [52]. The pathogenesis of I/R injury in the heart includes several mechanisms that have been intensively studied in recent years [54]. These include damage caused by: a) alterations in cardiac metabolism [55], b) the production of reactive oxygen species [56], c) alterations in calcium handling [57], and, most recently, d) the intracellular activation of myocardial MMPs. In the following sections of this chapter this latter mechanism of MMP induced myocardial I/R injury will be described and discussed.

7. Role of MMP-2 in Acute Myocardial I/R Injury

The acute activation of MMP-2 in myocardial I/R injury was first demonstrated by our group in isolated rat hearts [58]. Hearts were excised, aerobically perfused, and then subjected to 15 to 25 min of global, no-flow ischemia followed by 30 min of aerobic reperfusion. During periods of aerobic perfusion cardiac mechanical function was noted and samples of coronary effluent were collected. In this model, 15 min of ischemia produces a cardiac mechanical dysfunction which is fully reversible within the 30 min of reperfusion, whereas 20 min of ischemia produces dysfunction which is not fully reversible within 30 min of reperfusion.

It was found that 20 min of ischemia significantly increased the release of MMP-2 activity into the coronary effluent upon reperfusion, and this increase actually peaked within the first 5 min of reperfusion. This increase in coronary effluent MMP-2 activity was coupled with a decrease in its activity in the ventricles. Thus, in the setting of I/R, the most abundant gelatinase in the rat myocardium was activated and released from the heart in a time frame of seconds to minutes.

The acute activation of MMP-2 was demonstrated to be functionally significant by several experiments. Increasing the ischemic time was found to increase MMP-2 activation and release, and this increase in activity correlated inversely with the resulting cardiac dysfunction. In other words, increasing the ischemic insult resulted in greater MMP-2 activation, and greater MMP-2 activation was related with poorer cardiac function upon reperfusion. Infusion of a preparation of MMP-2 into the heart subjected only to a 15 min period of ischemia (insufficient in itself to cause stunning upon reperfusion) diminished the recovery of function during reperfusion. The recovery of cardiac function upon reperfusion was enhanced in a concentration-dependent manner by a neutralizing antibody against MMP-2, or with the use of MMP inhibitors such as doxycycline or o-phenanthroline.

A recent study confirmed the damaging role of MMPs in acute I/R injury. Prasan et al. used isolated rabbit hearts to demonstrate that MMP-2 is activated and released during reperfusion following prolonged (60 min) ischemia

[59]. These results suggested that MMP-2 may contribute to myocardial dysfunction following prolonged ischemia in rabbit hearts.

We also examined the role of MMP-2 in I/R injury and ischemic preconditioning in isolated rat hearts [60]. Ischemic preconditioning is a well described adaptive response in which brief exposure to ischemia markedly enhances the ability of the heart to withstand a subsequent I/R injury [61]. In our study hearts were subjected to a preconditioning protocol followed by 30 min ischemia and 5 min reperfusion. This preconditioning protocol not only protected against I/R cardiac dysfunction, [62] it also prevented the activation and release of MMP-2 [60]. Thus, myocardial protection against I/R was associated with decreased MMP-2 activation and release.

Other studies have confirmed a role for MMPs in I/R injury *in vivo*. Romanic et al. used an *in vivo* mouse model in which the lower anterior descending coronary artery was occluded for 30 min followed by 4 to 24 h reperfusion [63]. Following I/R injury myocardial MMP-2 activity was significantly increased and MMP-9 activity (associated with neutrophil infiltration) was also found. Selectively deleting MMP-9 decreased both the infarct size and neutrophil infiltration following I/R, with homozygous (-/-) mice showing greater protection than heterozygous (+/-) mice.

Mehta and colleagues have used an *in vivo* rat model of I/R in which the left anterior descending coronary artery was occluded for 60 min and then the heart was reperfused for 60 min [64, 65]. Associated with the cardiac dysfunction upon reperfusion was a significant upregulation of MMP-1 protein within the myocardium. Pretreating these animals with either transforming growth factor-β_1 [64] or an antibody against lectin-like oxidized low-density lipoprotein receptor (LOX-1) [65] not only protected the myocardium, but also inhibited the upregulation of MMP-1. Moreover, pretreatment of adult rat myocytes with a broad spectrum MMP inhibitor (PD-166793) attenuated MMP-1 mediated inury [64].

In summary, evidence is accumulating that MMP activation during acute I/R injury contributes to impaired cardiac function. However, several limiting factors exist in the studies done to date. First is the piecemeal approach, not strongly rationalized in many cases, in measuring a limited number of MMPs. As well, some studies measure protein content, others enzyme activity, but rarely are both analyzed together. This is likely due to the fact that no one simple technique that can readily measure all MMPs and their activities. Furthermore, our studies highlighted the fact that the *release* of MMPs from the heart into the coronary circulation acts as a marker of their activation and the injury process, [58, 66] thus measurement of myocardial MMP activity alone may not paint a complete picture (Figure 2). Moreover, the time course of MMP activation/release in the immediate phase of reperfusion (ie. minutes timescale) is also lacking in many studies. Finally, in the future it will also be necessary to perform follow-up studies of myocardial I/R injury with the use of more specific MMP inhibitors, if and when they become available, as well as studies with transgenic animals.

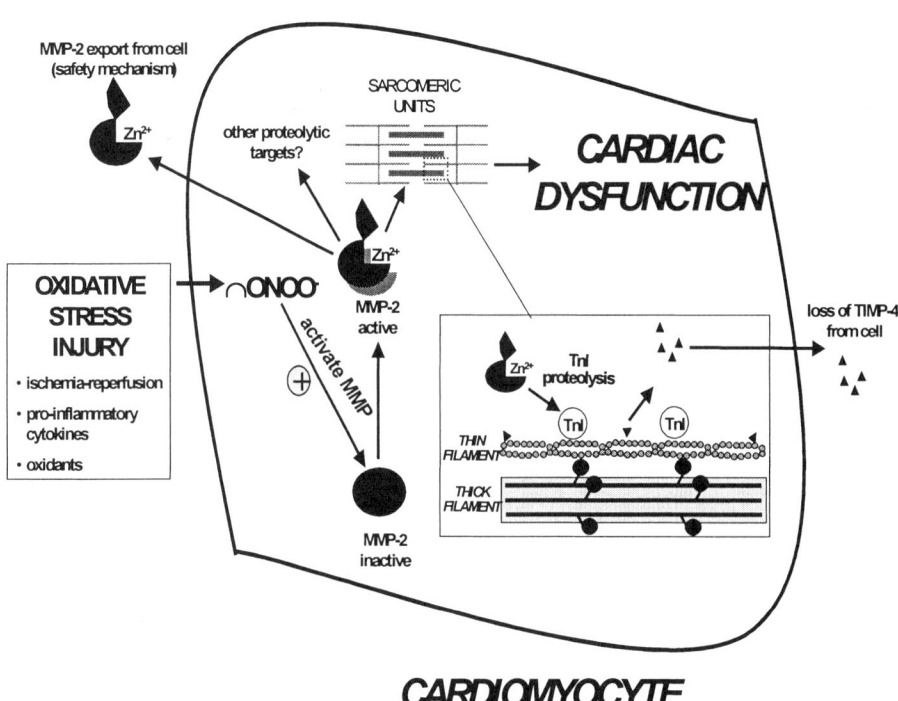

Figure 2. Paradigm of MMP-2/TIMP-4 in oxidatively stressed cardiomyocytes: intracellular localization, inappropriate activation, and proteolysis of troponin I (TnI). Peroxynitrite (ONOO⁻) production is increased during oxidative stress injuries (such as ischemia-reperfusion or insult by pro-inflammatory cytokines). Peroxynitrite activates MMP-2, forming an active enzyme with the prodomain still intact. Both MMP-2 and TIMP-4 are localized with thin myofilaments located in the sarcomere. The activation of MMP-2 coupled with the loss of TIMP-4 (black triangles) from the thin myofilaments produces a localized area of increased proteolysis within the sarcomere, and leads to troponin I degradation. The release of MMP-2 by the cell may be a means to limit the proteolytic stress. The net result of this inappropriate intracellular activation of MMP-2 is cardiac dysfunction.

8. TIMPs and I/R Injury

Although evidence of MMP involvement in I/R injury is rapidly accumulating, little work has been done to examine TIMPs in this pathology.

Considering that a dysregulation of TIMPs has been implicated in long term cardiovascular pathology, a similar dysregulation may contribute to acute cardiac injury as well.

We examined the role of TIMPs in I/R using isolated rat hearts. As described above, TIMP-4 is highly expressed in the cardiovascular system and is likely an important inhibitory regulator of MMPs in the heart [22]. In rat hearts subjected to 20 min of global, no-flow ischemia followed by 30 min of reperfusion there was a rapid release of TIMP-4 from the heart during the first minutes of reperfusion [25]. This release could be detected by both western blot and reverse zymography of the perfusate during reperfusion. Importantly, no other TIMP activity was measurable in the coronary effluent apart from TIMP-4. Immunogold electron microscopy of myocardial tissue revealed the association of TIMP-4 with the sarcomeres as well as a loss of TIMP-4 following I/R. Mirroring the release of MMP-2 described above, there was a negative correlation between the recovery of cardiac mechanical function and the release of TIMP-4 during reperfusion in hearts subjected to longer durations of ischemia. The loss of TIMP-4 resulted in a net increase in myocardial MMP activity as demonstrated using *in situ* zymography. Thus, although we see a loss of myocardial MMP-2 as a consequence of I/R, it appears that there may be a greater loss of inhibitory TIMP-4 activity and this results in a tip in the balance towards enhanced intracellular proteolytic activity (Figure 2).

Using isolated rabbit hearts subjected to 20 min of ischemia followed by 120 min reperfusion Baghelai et al. [67] found that the level of TIMP-1 mRNA was significantly depressed following reperfusion, although no functional data were reported in this paper. Thus, in I/R a decrease in TIMP-1 may also favor enhanced MMP activity and a balance in favor of proteolysis.

9. Consequences of MMP Activation in I/R Injury

Given the evidence for a net increase in MMP activity following I/R, proteolysis of susceptible target proteins is a potential mechanism by which MMPs cause diminshed contractile function. Collagen and other extracellular matrix proteins are the most obvious targets. However, in the isolated rat heart model of I/R there is no evidence for collagen degradation in the time scale of 20 min of ischemia followed by reperfusion (Reference 68 and unpublished observations, Schulz lab). Indeed others have shown that protective actions of MMP inhibitors on myocardial contractile function following infarction were independent of changes in collagen content [68]. As extracellular matrix degradation does not fully explain the observed mechanical dysfunction other proteolytic targets for MMP-2 likely exist.

Proteolysis of myocardial sarcomeric and cytoskeletal proteins has been proposed as a mechanism of the cardiac dysfunction seen following I/R injury.

Proteins which have been found to be degraded following acute myocardial injury include troponin I (TnI) [69-71], troponin C [69], α-actinin [70, 72], myosin heavy and light chains [70], actin [73], desmin [72, 74], and spectrin [72], with TnI being the most studied molecule to date. In 1997, Marban and colleagues first proposed that TnI is degraded in an isolated rat heart model of brief ischemia (20 min) followed by 30 min reperfusion [71]. In the following years many investigators have confirmed this observation and the evidence for the involvement of TnI degradation (as well as its post-translational modification) in I/R injury is accumulating [75]. Van Eyk and colleagues identified one of the TnI fragments as TnI_{1-193} in isolated rat hearts subjected to a 60 min duration of ischemia [69]. In a subsequent study this fragment was overexpressed in mice, and the resulting transgenic mouse displayed depressed cardiac contractility which resembled features of I/R injury [76]. Moreover, in myocardial biopsy samples taken from patients undergoing cardiopulmonary bypass surgery, a scenario mimicking acute I/R injury, the proteolysis of TnI was also demonstrated [69, 76].

The enzyme(s) responsible for the proteolysis of TnI is not clear. Calpain, a calcium-activated protease, may contribute to the degradation of TnI in severe, prolonged ischemia (ie. greater than 60 min) [77]. However, there is no evidence to show that calpain activity is increased in acute myocardial I/R injury associated with stunning injury (ie. brief periods of ischemia) [78]. We therefore speculated that MMP-2 may contribute to the degradation of TnI during I/R. Recombinant human MMP-2 was found to rapidly degrade recombinant TnI and TnC (but not TnT) *in vitro*, and also TnI alone when assembled in the intact troponin complex (consisting of TnI, TnC and TnT) [66]. This degradation was inhibited by TIMP-2 and other synthetic MMP inhibitors. When isolated perfused rat hearts were subjected to 20 min global ischemia followed by 30 min reperfusion, troponin I was found to be significantly degraded in the myocardium. Treatment of hearts with o-phenanthroline or doxycycline significantly improved the recovery of mechanical function and also protected against TnI degradation.

These results not only indicate that TnI may be a proteolytic target for MMP-2, but also suggest a novel intracellular locus of action for this MMP within the cardiac myocyte. Until this time there had been no clear evidence suggesting any biological roles for any MMP inside the cell. Thus it was not a simple task to convince the research community that MMP-2 could possibly have an intracellular action.

Using immunogold electron microscopy we showed that MMP-2 is found in close association within the sarcomeres of the cardiac myocyte (Figure 3). Immunofluorescent confocal microscopy confirmed this result, showing that MMP-2 colocalizes with troponin I [66]. Interestingly, when TnI was immunoprecipitated from myocardial homogenates MMP-2 activity was present in the precipitated complex. As well, MMP-2 activity could be found in a purified preparation of thin myofilaments (consisting of actin, tropomyosin, and TnI) prepared from ischemic-reperfused rat hearts, and the level of MMP-2 protein was enhanced after I/R injury. Thus, several lines of evidence suggest MMP-2 has an intracellular locus of action on TnI in mediating I/R injury [66].

Moreover, as described above, TIMP-4 was also localized to the sarcomeres by electron microscopy, and less TIMP-4 is present within the sarcomeres following I/R injury [25]. Based on the findings of these studies, it is likely that MMP-2 was inappropriately activated during I/R injury and this, coupled with the loss of TIMP-4 from the sarcomere, produces a localized area of increased proteolysis within the cardiomyocyte (Figure 2). The activated MMP-2 cleaves TnI and the former is then rapidly released from the cardiomyocyte. In acute myocardial I/R injury, the rapid release of MMP-2 may act as a safety mechanism to protect the heart from further proteolytic stress.

Other immunohistochemical evidence exists which demonstrates an intracellular locus of MMPs in the cardiomyocyte. Using serial section confocal microscopy Rouet-Benzinab et al. demonstrated that MMP-2 and MMP-9 are associated with the sarcomeres of cardiomyocytes in biopsies taken from patients with dilated cardiomyopathy [42]. Immunohistochemical evidence has also shown dense intracellular staining for MMP-2 [79] and MMP-9 [44, 79] within human cardiomyocytes. Confocal microscopy demonstrated that MT-MMP1 and MMP-2 are associated with a sarcomeric banding pattern in isolated porcine left ventricular myocytes [27, 80]. Finally, in a study of isolated human left ventricular myocytes using immunofluorescent staining, MT1-MMP was colocalized to the sarcomeric protein α-actinin [37]. Colocalization with this latter protein suggests that MMPs may also be responsible for α-actinin degradation seen following I/R [72].

10. Activation of MMPs by Peroxynitrite

Oxidative activation of MMPs is the most likely mechanism by which inappropriate intracellular activation of MMPs occurs. Despite the absence of oxygen supply to the myocardium during ischemia, the production of reactive oxygen species such as superoxide anion, hydrogen peroxide, and hydroxyl radical is enhanced [81]. Upon reperfusion, the reintroduction of oxygen after ischemia leads to an even higher generation of reactive oxygen species as well as nitric oxide, leading to cardiotoxic levels of peroxynitrite (the reaction product of nitric oxide and superoxide anion) within the first minute [56, 82]. In addition, direct administration of peroxynitrite decreases cardiac mechanical function in isolated rat hearts [83, 84].

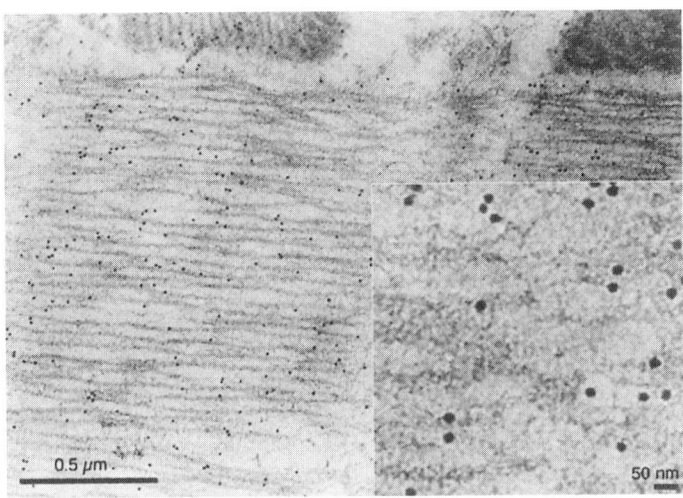

Figure 3. Intracellular localization of MMP-2 in association with thin myofilaments. Representative transmission electron micrographs from rat hearts subjected to 20 min ischemia and 30 min reperfusion. Black dots indicate immunogold labeling of MMP-2 with anti-MMP-2 antibody. Magnification x56 000 (insert x100 000). (Reproduced from *Circulation*, Wang et al. Intracellular action of matrix metalloproteinase-2 accounts for acute myocardial ischemia and reperfusion injury. 2002;106:1543-1549 by copyright permission).

In order to investigate whether oxidative stress plays a role in the activation of myocardial MMPs, we infused authentic peroxynitrite into isolated perfused rat hearts and measured MMP activity [84]. Infusion of 80 μM peroxynitrite for 15 min caused the release of MMP-2 into the perfusate, which was rapidly followed by a significant depression in cardiac mechanical function. It was found that both the release of MMP-2 and the loss of function were both blocked with either the peroxynitrite scavenger glutathione or a synthetic MMP inhibitor (PD-166793). These results suggested that the acute cardiac toxicity induced by exogenous peroxynitrite was mediated by MMP-2.

Peroxynitrite can also be generated within the heart following exposure to pro-inflammatory cytokines. These cytokines produce a rapidly developing acute heart failure caused by the concurrent upregulation of both nitric oxide and superoxide levels [85]. Accordingly, this model provided yet another method to investigate oxidative activation of MMPs in a setting of acute myocardial dysfunction. In accordance with the oxidative stress hypothesis, MMP-2 was found to be activated in cytokine treated hearts and this was followed by a significant decline in cardiac function and degradation of TnI [86]. These effects could be ameliorated with the use of either neutralizing MMP-2 antibody or MMP inhibitors [86]. Thus, this investigation demonstrated that MMPs are not only activated by oxidative stress but are also mediators of acute cardiac dysfunction.

Interestingly, it has been observed that the inhibitory activity of TIMP-1 can be reduced upon exposure to peroxynitrite *in vitro* [87]. Thus, in conditions of oxidative stress, particularly those caused by peroxynitrite, it could be predicted that TIMP inhibitory activity would decrease while MMP activity would increase. Whether other TIMPs are susceptible to inactivation by peroxynitrite is unknown.

11. How Does MMP Release Equate to Activation?

This is an important question which is yet to be fully answered. However, our working hypothesis is that peroxynitrite generated within the heart by ischemia-reperfusion [56, 82], pro-inflammatory cytokines [85], or infusion of authentic peroxynitrite into the heart results in the activation of MMP-2, without proteolytic removal of its autoinhibitory propeptide domain [58, 66, 84, 86]. This MMP associates with and degrades TnI and possibly other intracellular structural or contractile proteins, which contributes to the acute mechanical changes to the heart [66, 86]. As a means to protect itself, the oxidatively activated MMP is released from the cardiomyocyte, in order to reduce the proteolytic stress. In contrast, TIMP-4 is simply lost from the sarcomeres to the extracellular space as a result of I/R injury. Moreover, oxidative stress may also impair the ability of TIMPs to block MMP activity.

The exact mechanism of MMP-2 and TIMP-4 release from the myocyte remains obscure at this time, however we do know that it is a very early event of peroxynitrite mediated toxicity. Several other potential targets of peroxynitrite in causing cellular toxicity exist, including direct actions on lipids, DNA, and many other proteins [88]. However, our notion is that MMPs are particularly sensitive to peroxynitrite and that they represent some of the earliest targets and mediators of the detrimental actions of oxidative stress to the heart. Thus, by inhibiting MMP activity it may be possible to stop the ensuing cascade of changes that can result in further, and possibly irreversible, damage to the heart.

12. Species Specific Differences in Acute Activation of Myocardial MMPs

Interestingly, some species specific differences in the acute activation of MMPs have been noted. Two studies have used *in vivo* pig models in which the left anterior descending coronary artery was temporarily occluded for 90 min [89] or 6 h [90] to produce a regional I/R injury. In these models MMP-1, -2, and -9 activities were observed in nonischemic regions of the heart, but only MMP-1 and –9 activities were increased in ischemic-reperfused regions after 90 min [89] or 3 h [90] reperfusion. This increase in MMP activity was accompanied by profound cardiac dysfunction [89] even though collagen ultrastructure remained undamaged [89, 91]. Surprisingly, the increases in MMP activity were found to be functionally insignificant since treatment with a broad spectrum MMP inhibitor (GM-2487) did not improve cardiac dysfunction following I/R [89]. One explanation for the lack of functional significance of MMP activity may lie in the fact that TnI proteolysis has not been observed in porcine models of I/R [92, 93]. Instead, alternate hypotheses have been forwarded to explain I/R induced myocardial dysfunction in pigs [93, 94]. Similar results have also been noted in canine models of I/R, in which I/R is associated with neither an increase in myocardial MMP-2 activity [95] nor TnI proteolysis [96]. Thus, it appears that in models of I/R in which MMP-2 activity is not increased the degradation of TnI also does not occur.

13. Clinical I/R injury and MMPs/TIMPs

In order to address whether myocardial MMPs are activated with I/R in humans, patients undergoing coronary artery bypass grafting with cardiopulmonary bypass have been studied. In this surgery, the heart is subjected to a type of I/R injury which results in a well-characterized reversible cardiac dysfunction [97-99].

Mayers et al. analyzed MMP activity in plasma and myocardial samples from patients prior to the ischemic insult (i.e., placement of the aortic cross clamp) and immediately following reperfusion (i.e., release of the aortic cross clamp) [79]. It was found that both MMP-2 and -9 activities were significantly increased at the termination of bypass in the reperfused myocardium, while only MMP-9 activity was increased in the plasma at this time. Biopsy samples were immunostained following reperfusion and MMP-2 and -9 were diffusely expressed in the cytoplasm of cardiac myocytes. These results were supported by findings from our laboratory which demonstrated that MMP-2 and -9 activities were increased in myocardium reperfused for less than 10 min [100]. Moreover, both of these activities were found to correlate directly with ischemic (aortic cross clamp) time and inversely with cardiac function upon reperfusion. In other words, longer ischemic times were correlated with larger MMP-2 and -9 myocardial activities upon reperfusion, and larger MMP activities were correlated with poorer cardiac function upon reperfusion.

TIMP protein content was also analyzed in the biopsy samples taken before and after I/R. TIMP-1 protein content was significantly decreased in the immediate post-reperfusion period and was found to correlate directly with cardiac function and indirectly with ischemic time. On the other hand, TIMP-2 and -4 protein content did not change upon reperfusion. Thus, in the first ten minutes of reperfusion when MMP activities are increased in the heart, the level of their primary inhibitors are either decreased or unchanged, resulting in an imbalance between MMPs and TIMPs and increased proteolysis. Moreover, the inhibitory activity of TIMP-1 following I/R may be further diminished as peroxynitrite has also been shown to inactivate it [87].

In order to determine whether MMPs were being released during reperfusion (as MMP-2 is in the rat heart), we determined across the heart differences in MMP activity by sampling blood simultaneously from the radial artery and the coronary sinus [100]. Unlike rat hearts, it was found that the human myocardium does not release MMPs in the early minutes of reperfusion. However, another study has demonstrated that MMP-2 (which may be of myocardial origin) is only increased in the plasma 6 h after coronary artery bypass grafting with cardiopulmonary bypass [101]. Thus, a detectable release of activated gelatinase activity does not occur as rapidly in human hearts as it does in isolated perfused rat hearts. This MMP activity may not be readily detected due to its binding to plasma or tissue components, its rapid inactivation, or its rapid removal.

The exact targets for acutely increased MMP activity in the human myocardium remain to be determined. TnI degradation has been proposed as a possible mechanism for I/R injury in humans, and the proteolysis of TnI has been demonstrated in patients subjected to coronary artery bypass grafting with cardiopulmonary bypass [76, 102]. Since MMPs have been localized intracellulary in myocytes [44, 79], and specifically to the sarcomeres of human ventricular myocytes [37, 42], it is possible that MMPs are involved in the intracellular proteolysis of contractile proteins.

14. Future Directions

The acute activation of MMPs during myocardial I/R injury may be a target for future clinical trials of MMP inhibitors. However, given the very limited success in clinical trials of broad spectrum MMP inhibitors as anti-cancer drugs due to limiting side-effects [103], it will be necessary to 'fine-tune' the approach to inhibit MMPs. First, the full spectrum and extent of MMP activation will need to be established, possibly through 'degradomic' [104] or *in vivo* imaging techniques [105]. Since a number of polymorphisms for MMP-2 and MMP-9 have been identified, future studies will also need to examine whether these polymorphisms affect MMP expression and activity in clinical settings of ischemia-reperfusion. Next, specific inhibitors of select MMPs (or perhaps only MMP-2) need to be developed in order to avoid the theoretical pitfalls and practical side effects associated with broad spectrum MMP inhibition (e.g., paradoxical activation of MMPs and muscular pain, respectively) [106]. Finally, other novel intracellular targets for MMPs need to be examined in order to better understand the pathology underlying myocardial I/R injury .

Retrospective epidemiological studies may also shed light on the clinical utility of MMP inhibition. Interestingly, in surveying antibiotic usage and the risk of first time acute myocardial infarct in more than 16,000 patients, a statistically significant risk reduction was seen only in those who had taken tetracycline class, but not in any other class of antibiotics [107, 108]. Tetracyclines inhibit MMPs through a mechanism which is independent of their antimicrobial effect [26]. Thus, it has been suggested that the cardiovascular benefits associated with tetracycline usage may be connected to the suppression of pathological MMP activity [58], and stabilization of atheromatous coronary artery plaques [109]. Studies to assess the potential cardioprotective properties of a tetracycline-class antibiotic in the setting of cardiopulmonary bypass are underway in our group.

In summary, a great deal of evidence points to acute, intracellular actions of MMPs on a number of non-extracellular matrix substrates. An acute reduction in TIMP inhibitory activity through oxidative damage and cellular loss may also contribute to dysfunction at a cellular level. Based on studies in myocardial I/R injury, and localization of MMPs intracellularly, we urge other investigators to begin to look inside the cardiomyocyte when investigating MMPs.

Acknowledgements

M. Lalu is a graduate trainee of the Alberta Heritage Foundation for Medical Research and the Canadian Institutes of Health Research (CIHR). H. Leon is a graduate trainee of the CIHR-TORCH strategic training program. R. Schulz is a Senior Scholar of the Alberta Heritage Foundation for Medical Research. Studies in the Schulz lab mentioned in this review are supported by grants from the CIHR (MT-14741, MT-11563) as well as the Heart and Stroke Foundation of Alberta, NWT and Nunavut.

References

1.   Gross, J. and C. Lapiere, Collagenolytic activity in amphibian tissues: a tissue culture assay. Proc Natl Acad Sci USA, 1962. 54: p. 1197-1204.
2.   Woessner, J., The matrix metalloproteinase family. In: Matrix Metalloproteinases Parks, W., Mecham, R., eds. San Diego, CA: Academic Press., 1998: p. 1-14.

3.   Morgunova, E., et al., Structure of human pro-matrix metalloproteinase-2: activation mechanism revealed. Science, 1999. 284: p. 1667-70.

4.   Nagase, H., Activation mechanisms of matrix metalloproteinases. Biol Chem, 1997. 378: p. 151-60.

5.   Strongin, A.Y., et al., Mechanism of cell surface activation of 72-kDa type IV collagenase. Isolation of the activated form of the membrane metalloprotease. J Biol Chem, 1995. 270: p. 5331-8.

6.   Holmbeck, K., et al., MT1-MMP-deficient mice develop dwarfism, osteopenia, arthritis, and connective tissue disease due to inadequate collagen turnover. Cell, 1999. 99: p. 81-92.

7.   Ruangpanit, N., et al., Gelatinase A (MMP-2) activation by skin fibroblasts: dependence on MT1-MMP expression and fibrillar collagen form. Matrix Biol, 2001. 20: p. 193-203.

8.   Kang, T., H. Nagase, and D. Pei, Activation of membrane-type matrix metalloproteinase 3 zymogen by the proprotein convertase furin in the trans-Golgi network. Cancer Res, 2002. 62: p. 675-81.

9.   Sato, H., et al., Activation of a recombinant membrane type 1-matrix metalloproteinase (MT1-MMP) by furin and its interaction with tissue inhibitor of metalloproteinases (TIMP)-2. FEBS Lett, 1996. 393: p. 101-4.

10.   Pei, D. and S.J. Weiss, Furin-dependent intracellular activation of the human stromelysin-3 zymogen. Nature, 1995. 375: p. 244-7.

11.   Rajagopalan, S., et al., Reactive oxygen species produced by macrophage-derived foam cells regulate the activity of vascular matrix metalloproteinases in vitro. Implications for atherosclerotic plaque stability. J Clin Invest, 1996. 98: p. 2572-9.

12.   Okamoto, T., et al., Activation of human neutrophil procollagenase by nitrogen dioxide and peroxynitrite: a novel mechanism for procollagenase activation involving nitric oxide. Arch Biochem Biophys, 1997. 342: p. 261-74.

13.   Okamoto, T., et al., Activation of matrix metalloproteinases by peroxynitrite-induced protein S-glutathiolation via disulfide S-oxide formation. J Biol Chem, 2001. 276: p. 29596-602.

14.   Brew, K., D. Dinakarpandian, and H. Nagase, Tissue inhibitors of metalloproteinases: evolution, structure and function. Biochim Biophys Acta, 2000. 1477: p. 267-83.

15.   Williamson, R.A., et al., Disulphide bond assignment in human tissue inhibitor of metalloproteinases (TIMP). Biochem J, 1990. 268: p. 267-74.

16.   Li, Y.Y., C.F. McTiernan, and A.M. Feldman, Proinflammatory cytokines regulate tissue inhibitors of metalloproteinases and disintegrin metalloproteinase in cardiac cells. Cardiovasc Res, 1999. 42: p. 162-72.

17.   Stetler-Stevenson, W.G., N. Bersch, and D.W. Golde, Tissue inhibitor of metalloproteinase-2 (TIMP-2) has erythroid-potentiating activity. FEBS Lett, 1992. 296: p. 231-4.

18.   Hayakawa, T., et al., Growth-promoting activity of tissue inhibitor of metalloproteinases-1 (TIMP-1) for a wide range of cells. A possible new growth factor in serum. FEBS Lett, 1992. 298: p. 29-32.

19.   Pavloff, N., et al., A new inhibitor of metalloproteinases from chicken: ChIMP-3. A third member of the TIMP family. J Biol Chem, 1992. 267: p. 17321-6.

20.   Wu, I. and M.A. Moses, Cloning and expression of the cDNA encoding rat tissue inhibitor of metalloproteinase 3 (TIMP-3). Gene, 1996. 168: p. 243-6.

21.   Qi, J.H., et al., A novel function for tissue inhibitor of metalloproteinases-3 (TIMP3): inhibition of angiogenesis by blockage of VEGF binding to VEGF receptor-2. Nat Med, 2003. 9: p. 407-15.

22.   Greene, J., et al., Molecular cloning and characterization of human tissue inhibitor of

metalloproteinase 4. J Biol Chem, 1996. 271: p. 30375-80.

23. Tummalapalli, C.M., B.J. Heath, and S.C. Tyagi, Tissue inhibitor of metalloproteinase-4 instigates apoptosis in transformed cardiac fibroblasts. J Cell Biochem, 2001. 80: p. 512-21.

24. Dollery, C.M., et al., TIMP-4 is regulated by vascular injury in rats. Circ Res, 1999. 84: p. 498-504.

25. Schulze, C., et al., Imbalance between tissue inhibitor of metalloproteinase-4 and matrix metalloproteinases during acute myocardial ischemia-reperfusion injury. Circulation, 2003. 107: p. 2487-92.

26. Golub, L.M., et al., Tetracyclines inhibit connective tissue breakdown by multiple non-antimicrobial mechanisms. Adv Dent Res, 1998. 12: p. 12-26.

27. Coker, M.L., et al., Matrix metalloproteinase synthesis and expression in isolated LV myocyte preparations. Am J Physiol, 1999. 277: p. H777-87.

28. Siwik, D.A., D.L. Chang, and W.S. Colucci, Interleukin-1beta and tumor necrosis factor-alpha decrease collagen synthesis and increase matrix metalloproteinase activity in cardiac fibroblasts in vitro. Circ Res, 2000. 86: p. 1259-65.

29. Tyagi, S.C., S. Kumar, and G. Glover, Induction of tissue inhibitor and matrix metalloproteinase by serum in human heart-derived fibroblast and endomyocardial endothelial cells. J Cell Biochem, 1995. 58: p. 360-71.

30. May, A.E., et al., Engagement of glycoprotein IIb/IIIa (alpha(IIb)beta3) on platelets upregulates CD40L and triggers CD40L-dependent matrix degradation by endothelial cells. Circulation, 2002. 106: p. 2111-7.

31. Galis, Z.S., et al., Cytokine-stimulated human vascular smooth muscle cells synthesize a complement of enzymes required for extracellular matrix digestion. Circ Res, 1994. 75: p. 181-9.

32. Yu, A., A. Murphy, and W. Stetler-Stevenson, 72-kDa gelatinase (gelatinase A): structure activation, regulation and substrate specificity. In: Matrix Metalloproteinases. Parkds, W., Mecham, R., eds San Diego, CA: Academic Press., 1998: p. 85-114.

33. Vu, T. and Z. Werb, Gelatinase B: Structure, Regulation and Function, In: Matrix Metalloproteinases. Parks, WC., Mecham, R., eds. San Diego, CA: Academic Press., 1998: p. 115-148.

34. Peterson, J.T., et al., Matrix metalloproteinase inhibition attenuates left ventricular remodeling and dysfunction in a rat model of progressive heart failure. Circulation, 2001. 103: p. 2303-9.

35. Ducharme, A., et al., Targeted deletion of matrix metalloproteinase-9 attenuates left ventricular enlargement and collagen accumulation after experimental myocardial infarction. J Clin Invest, 2000. 106: p. 55-62.

36. King, M.K., et al., Selective matrix metalloproteinase inhibition with developing heart failure: effects on left ventricular function and structure. Circ Res, 2003. 92: p. 177-85.

37. Spinale, F.G., et al., A matrix metalloproteinase induction/activation system exists in the human left ventricular myocardium and is upregulated in heart failure. Circulation, 2000. 102: p. 1944-9.

38. Blankenberg, S., et al., Plasma concentrations and genetic variation of matrix metalloproteinase 9 and prognosis of patients with cardiovascular disease. Circulation, 2003. 107: p. 1579-85.

39. Creemers, E.E., et al., Deficiency of TIMP-1 exacerbates LV remodeling after myocardial infarction in mice. Am J Physiol, 2003. 284: p. H364-71.

40. Thomas, C.V., et al., Increased matrix metalloproteinase activity and selective upregulation in LV myocardium from patients with end-stage dilated cardiomyopathy. Circulation, 1998. 97: p. 1708-15.

41. Li, Y.Y., et al., Differential expression of tissue inhibitors of metalloproteinases in the

failing human heart. Circulation, 1998. 98: p. 1728-34.

42.   Rouet-Benzineb, P., et al., Altered balance between matrix gelatinases (MMP-2 and MMP-9) and their tissue inhibitors in human dilated cardiomyopathy: potential role of MMP-9 in myosin-heavy chain degradation. Eur J Heart Fail, 1999. 1: p. 337-52.

43.   Fedak, P.W., et al., Matrix remodeling in experimental and human heart failure: a possible regulatory role for TIMP-3. Am J Physiol, 2003. 284: p. H626-34.

44.   Li, Y.Y., et al., Downregulation of matrix metalloproteinases and reduction in collagen damage in the failing human heart after support with left ventricular assist devices. Circulation, 2001. 104: p. 1147-52.

45.   Sawicki, G., et al., Release of gelatinase A during platelet activation mediates aggregation. Nature, 1997. 386: p. 616-9.

46.   Galt, S.W., et al., Outside-in signals delivered by matrix metalloproteinase-1 regulate platelet function. Circ Res, 2002. 90: p. 1093-9.

47.   McQuibban, G.A., et al., Inflammation dampened by gelatinase A cleavage of monocyte chemoattractant protein-3. Science, 2000. 289: p. 1202-6.

48.   Fernandez-Patron, C., M.W. Radomski, and S.T. Davidge, Vascular matrix metalloproteinase-2 cleaves big endothelin-1 yielding a novel vasoconstrictor. Circ Res, 1999. 85: p. 906-11.

49.   Fernandez-Patron, C., et al., Matrix metalloproteinases regulate neutrophil-endothelial cell adhesion through generation of endothelin-1[1-32]. FASEB J, 2001. 15: p. 2230-40.

50.   Fernandez-Patron, C., et al., Vascular matrix metalloproteinase-2-dependent cleavage of calcitonin gene-related peptide promotes vasoconstriction. Circ Res, 2000. 87: p. 670-6.

51.   Heyndrickx, G.R., et al., Regional myocardial functional and electrophysiological alterations after brief coronary artery occlusion in conscious dogs. J Clin Invest, 1975. 56: p. 978-85.

52.   Braunwald, E. and R.A. Kloner, The stunned myocardium: prolonged, postischemic ventricular dysfunction. Circulation, 1982. 66: p. 1146-9.

53.   Markis, J.E., et al., Myocardial salvage after intracoronary thrombolysis with streptokinase in acute myocardial infarction. N Engl J Med, 1981. 305: p. 777-82.

54.   Bolli, R. and E. Marban, Molecular and cellular mechanisms of myocardial stunning. Physiol Rev, 1999. 79: p. 609-34.

55.   Lopaschuk, G.D., Treating ischemic heart disease by pharmacologically improving cardiac energy metabolism. Am J Cardiol, 1998. 82(5A): p. 14K-17K.

56.   Yasmin, W., K.D. Strynadka, and R. Schulz, Generation of peroxynitrite contributes to ischemia-reperfusion injury in isolated rat hearts. Cardiovasc Res, 1997. 33: p. 422-32.

57.   Gao, W.D., et al., Intrinsic myofilament alterations underlying the decreased contractility of stunned myocardium. A consequence of $Ca2+$-dependent proteolysis? Circ Res, 1996. 78: p. 455-65.

58.   Cheung, P.Y., et al., Matrix metalloproteinase-2 contributes to ischemia-reperfusion injury in the heart. Circulation, 2000. 101: p. 1833-9.

59.   Prasan, A.M., et al., Duration of ischaemia determines matrix metalloproteinase-2 activation in the reperfused rabbit heart. Proteomics, 2002. 2: p. 1204-10.

60.   Lalu, M.M., et al., Preconditioning decreases ischemia/reperfusion-induced release and activation of matrix metalloproteinase-2. Biochem Biophys Res Commun, 2002. 296: p. 937-41.

61.   Ferdinandy, P. and R. Schulz, Nitric oxide, superoxide, and peroxynitrite in myocardial ischaemia-reperfusion injury and preconditioning. Br J Pharmacol, 2003. 138: p. 532-43.

62.   Csonka, C., et al., Preconditioning decreases ischemia/reperfusion-induced peroxynitrite formation. Biochem Biophys Res Commun, 2001. 285: p. 1217-9.

63.   Romanic, A.M., et al., Myocardial protection from ischemia/reperfusion injury by targeted deletion of matrix metalloproteinase-9. Cardiovasc Res, 2002. 54: p. 549-58.

64.   Chen, H., et al., TGF-beta 1 attenuates myocardial ischemia-reperfusion injury via inhibition of upregulation of MMP-1. Am J Physiol, 2003. 284: p. H1612-7.

65.   Li, D., et al., LOX-1 inhibition in myocardial ischemia-reperfusion injury: modulation of MMP-1 and inflammation. Am J Physiol, 2002. 283: p. H1795-801.

66.   Wang, W., et al., Intracellular action of matrix metalloproteinase-2 accounts for acute myocardial ischemia and reperfusion injury. Circulation, 2002. 106: p. 1543-9.

67.   Baghelai, K., et al., Decreased expression of tissue inhibitor of metalloproteinase 1 in stunned myocardium. J Surg Res, 1998. 77: p. 35-9.

68.   Rohde, L.E., et al., Matrix metalloproteinase inhibition attenuates early left ventricular enlargement after experimental myocardial infarction in mice. Circulation, 1999. 99: p. 3063-70.

69.   McDonough, J.L., D.K. Arrell, and J.E. Van Eyk, Troponin I degradation and covalent complex formation accompanies myocardial ischemia/reperfusion injury. Circ Res, 1999. 84: p. 9-20.

70.   Van Eyk, J.E., et al., Breakdown and release of myofilament proteins during ischemia and ischemia/reperfusion in rat hearts: identification of degradation products and effects on the pCa-force relation. Circ Res, 1998. 82: p. 261-71.

71.   Gao, W.D., et al., Role of troponin I proteolysis in the pathogenesis of stunned myocardium. Circ Res, 1997. 80: p. 393-9.

72.   Matsumura, Y., et al., Inhomogeneous disappearance of myofilament-related cytoskeletal proteins in stunned myocardium of guinea pig. Circ Res, 1996. 79: p. 447-54.

73.   Eberhardt, F., et al., Structural myocardial changes after coronary artery surgery. Eur J Clin Invest, 2000. 30: p. 938-46.

74.   Papp, Z., J. van der Velden, and G.J. Stienen, Calpain-I induced alterations in the cytoskeletal structure and impaired mechanical properties of single myocytes of rat heart. Cardiovasc Res, 2000. 45: p. 981-93.

75.   Van Eyk, J.E. and A.M. Murphy, The role of troponin abnormalities as a cause for stunned myocardium. Coron Artery Dis, 2001. 12: p. 343-7.

76.   Murphy, A.M., et al., Transgenic mouse model of stunned myocardium. Science, 2000. 287: p. 488-91.

77.   Toyo-Oka, T., Phosphorylation with cyclic adenosine 3':5' monophosphate-dependent protein kinase renders bovine cardiac troponin sensitive to the degradation by calcium-activated neutral protease. Biochem Biophys Res Commun, 1982. 107: p. 44-50.

78.   Sorimachi, Y., et al., Downregulation of calpastatin in rat heart after brief ischemia and reperfusion. J Biochem (Tokyo), 1997. 122: p. 743-8.

79.   Mayers, I., et al., Cardiac surgery increases the activity of matrix metalloproteinases and nitric oxide synthase in human hearts. J Thorac Cardiovasc Surg, 2001. 122: p. 746-52.

80.   Coker, M.L., et al., Matrix metalloproteinase expression and activity in isolated myocytes after neurohormonal stimulation. Am J Physiol, 2001. 281: p. H543-51.

81.   Ferrari, R., et al., Occurrence of oxidative stress during reperfusion of the human heart. Circulation, 1990. 81: p. 201-11.

82.   Wang, P. and J.L. Zweier, Measurement of nitric oxide and peroxynitrite generation in the postischemic heart. Evidence for peroxynitrite-mediated reperfusion injury. J Biol Chem, 1996. 271: p. 29223-30.

83.   Schulz, R., et al., Peroxynitrite impairs cardiac contractile function by decreasing cardiac efficiency. Am J Physiol, 1997. 272: p. H1212-9.

84.   Wang, W., G. Sawicki, and R. Schulz, Peroxynitrite-induced myocardial injury is mediated through matrix metalloproteinase-2. Cardiovasc Res, 2002. 53: p. 165-74.

85.   Ferdinandy, P., et al., Peroxynitrite is a major contributor to cytokine-induced myocardial contractile failure. Circ Res, 2000. 87: p. 241-7.

86.   Gao, C.Q., et al., Matrix metalloproteinase-2 mediates cytokine-induced myocardial contractile dysfunction. Cardiovasc Res, 2003. 57: p. 426-33.

87.   Frears, E.R., et al., Inactivation of tissue inhibitor of metalloproteinase-1 by peroxynitrite. FEBS Lett, 1996. 381: p. 21-4.

88.   Beckman, J.S. and W.H. Koppenol, Nitric oxide, superoxide, and peroxynitrite: the good, the bad, and ugly. Am J Physiol, 1996. 271: p. C1424-37.

89.   Lu, L., et al., Matrix metalloproteinases and collagen ultrastructure in moderate myocardial ischemia and reperfusion in vivo. Am J Physiol, 2000. 279: p. H601-9.

90.   Danielsen, C.C., H. Wiggers, and H.R. Andersen, Increased amounts of collagenase and gelatinase in porcine myocardium following ischemia and reperfusion. J Mol Cell Cardiol, 1998. 30: p. 1431-42.

91.   Wiggers, H., et al., Ischemia and reperfusion of the porcine myocardium: effect on collagen. J Mol Cell Cardiol, 1997. 29: p. 289-99.

92.   Thomas, S.A., et al., Absence of troponin I degradation or altered sarcoplasmic reticulum uptake protein expression after reversible ischemia in swine. Circ Res, 1999. 85: p. 446-56.

93.   Kim, S.J., et al., A novel mechanism for myocardial stunning involving impaired Ca(2+) handling. Circ Res, 2001. 89: p. 831-7.

94.   Kudej, R.K., et al., Brief increase in carbohydrate oxidation after reperfusion reverses myocardial stunning in conscious pigs. Circulation, 2002. 106: p. 2836-41.

95.   Lindsey, M., et al., Matrix-dependent mechanism of neutrophil-mediated release and activation of matrix metalloproteinase 9 in myocardial ischemia/reperfusion. Circulation, 2001. 103: p. 2181-7.

96.   Sherman, A.J., et al., Myofibrillar disruption in hypocontractile myocardium showing perfusion-contraction matches and mismatches. Am J Physiol, 2000. 278: p. H1320-34.

97.   Kloner, R.A., K. Przyklenk, and G.L. Kay, Clinical evidence for stunned myocardium after coronary artery bypass surgery. J Card Surg, 1994. 9: p. S397-402.

98.   Gray, R., et al., Scintigraphic and hemodynamic demonstration of transient left ventricular dysfunction immediately after uncomplicated coronary artery bypass grafting. J Thorac Cardiovasc Surg, 1979. 77: p. 504-10.

99.   Breisblatt, W.M., et al., Acute myocardial dysfunction and recovery: a common occurrence after coronary bypass surgery. J Am Coll Cardiol, 1990. 15: p. 1261-9.

100.   Pasini, E., et al., Cardiac matrix metalloproteinase activation in patients undergoing coronary artery bypass grafting. J Mol Cell Cardiol, 2002. 34: p. A49.

101.   Joffs, C., et al., Cardiopulmonary bypass induces the synthesis and release of matrix metalloproteinases. Ann Thorac Surg, 2001. 71: p. 1518-23.

102.   McDonough, J.L., et al., Cardiac troponin I is modified in the myocardium of bypass patients. Circulation, 2001. 103: p. 58-64.

103.   Overall, C.M. and C. Lopez-Otin, Strategies for MMP inhibition in cancer: innovations for the post-trial era. Nat Rev Cancer, 2002. 2: p. 657-72.

104.   Lopez-Otin, C. and C.M. Overall, Protease degradomics: a new challenge for proteomics. Nat Rev Mol Cell Biol, 2002. 3: p. 509-19.

105.   Bremer, C., C.H. Tung, and R. Weissleder, In vivo molecular target assessment of matrix metalloproteinase inhibition. Nat Med, 2001. 7: p. 743-8.

106.   Bernardo, M.M., et al., Design, synthesis, and characterization of potent, slow-binding inhibitors that are selective for gelatinases. J Biol Chem, 2002. 277: p. 11201-7.

107.   Glenn, L., Antibiotic use and risk of myocardial infarction. JAMA, 1999. 282: p. 1997.

108.   Meier, C.R., et al., Antibiotics and risk of subsequent first-time acute myocardial infarction. JAMA, 1999. 281: p. 427-31.

109.   Golub, L.M., R.A. Greenwald, and R.W. Thompson, Antibiotic use and risk of myocardial infarction. JAMA, 1999. 282: p. 1997-8.

Chapter 12

Regulation of Cardiac Extracellular Matrix Remodeling Following Myocardial Infarction.

Jack P.M. Cleutjens
University of Maastricht, Maastricht, The Netherlands

1. Introduction

A three-dimensional structural network of the interstitial, types I and III, collagens forms the structural backbone of the cardiac extracellular matrix (ECM). Other matrix components, including collagens types V and VI, proteoglycans, growth factors and cell-matrix receptors (integrins), can attach to this backbone [1]. The main physiological functions of the ECM are to retain tissue integrity and cardiac pump function, but it is also a dynamic entity which interacts with cells and regulates cell phenotype [2]. Collagen deposition is controlled and can be modulated by hormones, growth factors, cytokines, regulatory proteins and/or hemodynamic forces. Some of these components can be attached to the ECM network [3]. In order to prevent dilatation of the infarcted area, collagen deposition is increased and properly aligned. Excessive accumulation of collagen can lead to diastolic and systolic dysfunction, disturbances in conduction, and can contribute to the development of heart failure.

For normal morphogenesis and maintenance of tissue architecture a balance between extracellular matrix synthesis and degradation is required. An imbalance in the extracellular matrix turnover either by decreased matrix synthesis and/or increased degradation can yield decreased myocardial extracellular matrix content which can lead to cardiac dilatation or even rupture [4]. Extracellular matrix-degrading enzymes expressed after myocardial infarction belong to the families of serine protease and matrix metalloproteinases (MMPs) and are in general secreted as latent proenzymes which need to be activated. The MMPs are the driving force behind myocardial matrix degradation. It is essential to keep the activity of these enzymes under tight control by either influencing the synthesis, activation and/or inhibition by tissue inhibitors of MMPs (TIMPs) or α2-macroglobulin.

Recent studies demonstrate that preventing the breakdown of the myocardial extracellular matrix with pharmacological broad spectrum MMP inhibitors in animal models of cardiomyopathy and myocardial infarction yields favorable effects on left ventricular (LV) remodeling. This led to the proposal that MMP inhibitors could potentially be used as therapy for patients at risk for the development of heart failure.

The plasminogen-plasmin system plays a central role in the activation of MMPs. Invasion of inflammatory cells and subsequently the next phases of wound healing are inhibited in plasminogen- or uPA-deficient mice, most likely by the inhibition of MMP activity. Thus regulation of extracellular matrix remodeling either by influencing extracellular matrix synthesis or degradation might be one of the possible prevention mechanisms for cardiac remodeling in the near future.

2. Extracellular Matrix Synthesis

New interstitial collagen synthesis and deposition starts in the infarcted area within the first 3-4 days post-infarction and reaches its maximum at approximately 14 days after infarction. The non-infarcted myocardium is also affected after infarction and interstitial collagen is deposited around the cardiomyocytes and surrounding the coronary vasculature [2]. Over time the increased collagen deposition in the non-infarcted myocardium might further decrease the already impaired cardiac function.

Collagen deposition has a dual effect on cardiac structure and function. Increased collagen deposition is a prerequisite for preventing dilatation of the infarcted area. However, excessive accumulation of collagen in the infarcted and non-infarcted myocardium leads to increased tissue stiffness, increases the incidence of arrhythmias, and adversely affects myocardial viscoelasticity, which leads to ventricular diastolic and systolic dysfunction and might ultimately contribute to heart failure. Not only the amount of collagen but also collagen cross-linking, localization, and direction and alignment of the collagen fibers in the tissue will determine the contribution to myocardial viscoelasticity.

Collagen is produced and deposited by mesenchymal fibroblast-like cells. In the non-infarcted area collagen is predominantly deposited by fibroblasts whereas in the infarcted area, besides fibroblasts, myofibroblasts contribute to the synthesis and deposition of collagens. The origin of these myofibroblasts is most likely due to differentiation of fibroblasts or pericytes, by gaining smooth muscle-like appearances. The myofibroblasts are not only able to deposit collagen in the infarcted area, but they are also involved in the contraction of the fibrotic area, prevention of dilatation by cell-cell and cell-matrix interactions, and they play a role in the architectural control of scar tissue formation [5-7]. In contrast to dermal wound healing where all myofibroblasts disappear by apoptosis 3-4 weeks after wound healing, myofibroblasts in the infarct remain, although reduced in numbers, for years in the cardiac scar tissue [8]. This might suggest that the myofibroblasts are actively involved in collagen turnover and fine tuning of scar contraction.

Collagen deposition is controlled by hormonal factors, growth factors, cytokines, regulatory proteins and/or hemodynamic factors [9]. The renin-angiotensin-aldosterone system is one of the main hormonal regulators of fibrosis. Elevation of angiotensin II or aldosterone will increase collagen deposition in both right and left ventricle. Prevention of excessive fibrosis can be achieved by inhibition of these hormones (e.g. angiotensin converting enzyme inhibitors) or inhibition of receptor-ligand binding (e.g. angiotensin type I receptor antagonists) [10]. Besides inhibition, also regression of fibrosis and cardiac stiffness can also be achieved by inhibition of the receptor-ligand binding of angiotensin receptors [11].

A major source of cytokine and growth factor production in the area of wound healing post infarction are the non-cardiomyocytes, e.g., myofibroblasts and endothelial cells. These factors such as tumor necrosis factor-alpha (TNF-α), interleukin (IL)-1β, IL-6, transforming growth factors (TGF)-β1 and TGF-β3 can act via autocrine and paracrine pathways to stimulate collagen synthesis and deposition [12]. Besides growth factors and cytokines other proteins can also be involved in the regulation of collagen synthesis and deposition. Heat-shock protein (Hsp47) is such a protein, which stimulates intracellular procollagen synthesis by interaction with procollagen during its folding, assembly and transport through the endoplasmic reticulum (ER) [13, 14]. Inhibition of cytokines, growth factors or other proteins which stimulate collagen synthesis or deposition might in the future be of interest for clinical application. The use of therapeutics for these factors is not yet well established. The involvement of these interventions with other processes in the body are unknown and the various treatments, either receptor blockers, soluble decoy receptors, or gene transfer have so far only been tested in experimental animal settings. Also models using transgenic or gene-deficient animals are only experimental but can give a good insight into the role of these factors in the remodeling and wound healing process.

Hemodynamic factors alone can regulate cardiac myocyte work and induce cardiomyocyte hypertrophy, but can also stimulate extracellular matrix synthesis. Hypertension leads to thickening of the collagen network (reactive fibrosis), increased stiffness, and impaired cardiac function. The choice of a pharmacological intervention for treatment of heart failure must take into account that not only the extracellular matrix but also cardiomyocytes and non-cardiomyocytes are involved in the remodeling process during heart failure.

3. Extracellular Matrix (ECM) Remodeling

An appropriate regulated balance of ECM synthesis and degradation is required for normal morphogenesis and maintenance of tissue architecture (Figure 1). ECM molecules and their receptors, as well as proteinases and their inhibitors, are all involved in matrix remodeling. An imbalance in ECM turnover either by decreased synthesis and/or increased degradation leads to a decrease of ECM in the myocardium which may lead to cardiac dilatation or rupture.

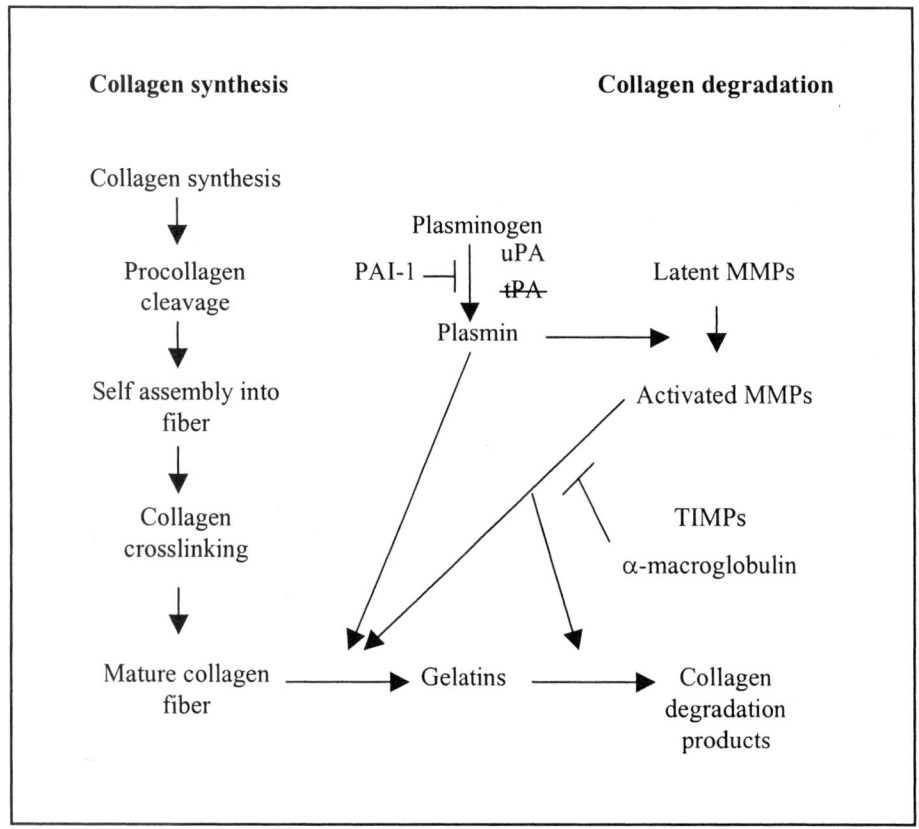

Figure 1. Graphical depiction of collagen synthesis and degradation (i.e. turnover).

4. Extracellular Matrix Degradation

4.1. Matrix metalloproteinases (MMPs)

Extracellular matrix-degrading enzymes which are activated after myocardial infarction belong to the families of serine proteases (plasmin, uPA, tPA, thrombin, elastase, cathepsin G) and matrix metalloproteinases. Matrix metalloproteinases (MMPs) are a family of zinc-containing endoproteinases that share structural domains, but differ in substrate specificity, cellular sources, and inducibility. MMPs share several functions: they a) are able to degrade ECM components, b) are in general secreted in a latent proform and have to be activated before they can perform proteolytic activity, c) contain Zn^{2+} at their active site, d) function at neutral pH, and e) can be inhibited by specific tissue inhibitors of metalloproteinases (TIMPs).

The MMP family can be subdivided into the four groups based on their substrate specificity and primary structure: 1) the collagenases, which are able to cleave fibrillar type I, II and III collagens, which are known to be resistant to cleavage by most other proteinases because they are tightly apposed and consist of highly cross-linked fibrils, 2) the gelatinases, which are able to degrade gelatins and basement membrane type IV collagen, 3) the stromelysins, which are able to degrade a broad spectrum of ECM components, 4) the membrane-type MMPs, which are anchored to the cell membrane by a transmembrane domain and concentrate the proteolytic activity at the cell surface and can activate other MMPs. However, the substrate profiles of the enzymes are more gradual than absolute and besides matrix degradation, MMPs can also act on a variety of non-matrix proteins as well, e.g., activation of growth factors and chemokines [15-20].

4.2. Activation of MMPs

MMPs act at physiologic pH and have certain substrate specificities [2]. The majority of proMMPs are stored extracellularly, bound to different extracellular matrix components. This latent proenzyme pool can be rapidly activated and mobilized upon stimulation [21, 22]. Activated MMPs are able to degrade the complete extracellular matrix. Therefore it is crucial to keep the activity of these enzymes under tight control. This is regulated at several levels. First of all the MMPs are regulated by transcription, which will ultimately generate a latent enzyme pool. Transcription is controlled by several cytokines, growth factors, corticosteroids and other inducers like Extracellular Matrix MetalloPRoteinase INducer (EMMPRIN) [22, 23] or matrix fragments such as matrikines [22, 24].

Besides transcriptional regulation of MMP mRNA's, activation of the latent proenzymes is the most crucial step in the regulation of the proteolytic cleavage. The following three different activation mechanisms have been described: stepwise activation, cell-surface-activation by MT-MMPs and intracellular activation [25]. During stepwise activation of MMPs proteinases such as plasmin, trypsin, chymase, elastase or kallikrein are involved. These proteinases are able to attack the proteinase-susceptible region in the "propeptide domain" of the MMP, which induces conformational changes in the propeptide and allows the activation site to be cleaved by a second proteinase, usually another MMP [25] or other inducers like EMMPRIN [22, 23].

Cell surface activation of MMPs is considered to be important for pericellular degradation of the ECM during cell migration. The plasminogen system but also the recently cloned membrane-types of MMP (MT-MMPs) are capable of activating MMPs at the plasma membrane [25-28]. Intracellular activation is the third possible activation mechanism. Stromelysin-3 (MMP-11) can be activated by the Golgi-associated subtilisin-like proteinase, furin [29]. Also MT-MMPs are likely to be activated intracellularly [30].

4.3. Endogenous MMP inhibitors

The tissue inhibitors of matrix metalloproteinases (TIMPs) are able to fully inhibit activated MMPs. TIMPs are expressed by a variety of cell types and are present in most tissues and body fluids. The TIMP family consists of four structurally related members, TIMP-1, -2, -3, and -4. TIMPs bind non-covalently to active MMPs in a 1: 1 molar ratio [2]. TIMPs interact with the zinc-binding site within the catalytic domain of active MMPs. TIMP-1 potently inhibits the activity of most MMPs, with the exception of MMP-2 and MT1-MMP, whereas TIMP-2 is a potent inhibitor of most MMPs, except MMP-9. Besides the role of TIMPs in the prevention of matrix degradation by inhibition of MMP activity, TIMPs are also involved in other biological actions. TIMP-1 and TIMP-2 exhibit growth factor-like activity and can inhibit angiogenesis [31-33], whereas TIMP-3 is involved in the inhibition of apoptosis [34].

Adenoviral human TIMP-1 overexpression in mice led to delayed wound healing, characterized by reduced leukocyte influx, reduced neovascularization, reduced collagen content, larger necrotic areas and a decreased incidence of cardiac rupture [35]. These data suggest that inhibition of MMP activity might be beneficial for cardiac remodeling. However, administration of TIMPs have not been found suitable for pharmacological applications due to their short half-life in vivo .

Besides TIMPs other naturally occurring inhibitors are also able to inhibit MMPs and other proteases. Alpha 2-macroglobulin is the most prominent occurring circulating protease inhibitor but due to its large size its effectiveness as an inhibitor in areas of wound healing after infarction may be limited.

4.4. MMPs post-myocardial infarction

Collagenase (MMP-1) was demonstrated in the normal myocardium, where it could be located in the interstitium, in the neighborhood of its substrate, fibrillar collagen [36]. Myocardial MMPs are produced by fibroblast-like cells inflammatory cells as well as by cardiomyocytes, and are predominantly present in their latent form. MMPs are increasingly expressed and activated in several pathological conditions of the heart [37-41]. Increased MMP-1, -2, -3 and -9 expression and activity was demonstrated in infarcted rat and porcine hearts [21, 42-45]. Differences in the time course of post-infarction myocardial MMP activity have been demonstrated but most authors agree that MMP activation starts early, within one day post-infarction. This is most probably due to the influx of inflammatory cells, e.g. granulocytes, which need active MMPs (e.g., MMP-9) to invade the infarcted ventricle. TIMPs are normally in delicate balance with the MMP activity, but loss of TIMP-mediated inhibitory control has also been reported to occur in several cardiac pathologies [39, 40, 46].

4.5. Pharmacological inhibition

During wound healing, including wound healing post-myocardial infarction, extracellular matrix turnover is one of the essential processes which occurs. Proteolytic degradation of the extracellular matrix takes place at different levels: 1) degradation of pre-existing extracellular matrix components, 2) promoting cell migration during inflammation, angiogenesis and granulation tissue formation, 3) remodeling of synthesized extracellular matrix in for example, scar tissue and 4) regulation of growth factors (e.g. TNF-α, TGF-β and IL-1β) by proteolytic release and activation [20]. Activation of MMPs after infarction might result in collagen degradation and progressive ventricular dilatation [47]. Chronic treatment with MMP inhibitors after infarction might have a beneficial effect on long-term left ventricular remodeling by increasing collagen deposition, decreased LV dilatation and increased cardiac function. The synthetic MMP inhibitors used belong to the hydroxamate class or are chemically modified tetracyclines. One of the greatest problems of the hydroxamic acid-based MMP inhibitors is the poor oral bioavailability and toxicity [48]. In a porcine study of 3 weeks rapid pacing, MMP inhibition limits LV remodeling and reduces wall stress [49].

In a mouse myocardial infarction study a broad range MMP inhibitor attenuates left ventricular dilatation 4 days after infarction [50]. A selective MMP inhibitor that does not inhibit MMP-1 was used in a rabbit model of myocardial infarction. At 4 weeks after MI, there were no differences between untreated and treated animals in infarct size or collagen deposition but the MMP inhibitor attenuated ventricular dilation, reduced infarct wall thinning and increased neovascularization in the subendocardium [51]. Treatment of mouse infarcts for 1 or 2 weeks with the MMP inhibitor GM6001 (Galardin, Ilomastat) delayed wound healing. After 1 week of treatment attenuation of left ventricular dilatation and thinning and reduction of collagen deposition in the infarct was observed [52]. The thicker infarcts might be explained by preservation of the pre-existing extracellular matrix, which inhibits myocyte slippage. The larger necrotic areas observed after 1 week of MMP inhibition, the reduction of cell numbers and the reduced collagen deposition may seem a paradox to MMP inhibition, but inflammatory cells including granulocytes and macrophages responsible for the removal of the necrotic debris and the influx of myofibroblasts responsible for collagen production need MMPs to migrate towards the area of infarction. By the use of MMP inhibitors both inflammatory cell and myofibroblast influx, and therefore debris removal and collagen deposition are transiently impeded. MMP inhibition might also interfere with the release and activation of growth factors and cytokines, like TNF-α, TGF-β and IL-1,which will in turn also influence the remodeling process after infarction [20, 53]. Decreased activity of TGF-β might reduce the synthesis of new collagen fibers [54].

5. Genetically Modified Animal Studies

The plasminogen-plasmin system is not only involved in fibrinolysis but is also a key regulatory system for MMP activity in the heart. Plasmin, a serine protease, is the active enzyme of the plasminogen (Plg) system and degrades a variety of ECM components [55]. The generation of plasmin is primarily controlled by the balance between the plasminogen activators (tPA and uPA) and their physiological inhibitors, the plasminogen activator inhibitors (PAIs). Inhibition of wound healing after infarction even 5 weeks after myocardial infarction was observed in urokinase plasminogen activator uPA-deficient [35] and plasminogen Plg-deficient [56] mice but not in tissue type plasminogen activator tPA-deficient mice. Others also described a plasmin-independent activation pathway of MMP activation [57-59]. These findings suggest a central role for the uPA-mediated plasminogen-plasmin system in cardiac wound healing after myocardial infarction in mice. Three distinct observations strongly suggest that the effects of plasminogen and uPA deficiency are mediated by reduced activation of MMPs. First, uPA was coexpressed with MMP-9 in infiltrating leukocytes [35]. Second, MMP activity was reduced in both uPA-/- and Plg-/- infarcts. Third, MMP inhibition by pharmacological tools has comparable, although less pronounced, effects on infarct healing and cardiac rupture as uPA/plasminogen deficiency.

Cardiac rupture was also inhibited in the uPA-deficient and MMP-9 deficient mice and in mice treated after infarction with adenovirus mediated TIMP-1 or plasminogen activator inhibitor-1 (PAI-1) gene transfer but not in MMP-2 or MMP-12 deficient mice. These data suggest that uPA and MMP-9 activity predispose to cardiac rupture, whereas increased levels of PAI-1 and TIMP-1 may be protective [35]. Prevention of cardiac rupture by local TIMP-1 or PAI-1 overexpression may thus, be developed as a non-surgical treatment.

As already demonstrated in the study of Heymans et al. [35], MMP-9 plays a crucial role in the influx of inflammatory cells in the early phase of the wound healing. The significance of MMP-9 activity in early infarct healing and rupture was emphasized by the observation that MMP-9 was predominantly found in leukocytes and macrophages and that its activity peaked around day 2, the period in which most of the ruptures occur. This indicates that by degrading matrix molecules, MMP-9 allows inflammatory cells to infiltrate the infarct and to disrupt the collagen network, a prerequisite for cardiac rupture [60]. Furthermore, targeted deletion of MMP-9 attenuated LV dilatation as well, at 15 days post-MI [61]. Limited LV dilatation was accompanied with a reduced inflammatory response and a decrease in collagen deposition in the infarct of MMP-9 deficient mice [62]. Interestingly, these MMP-9 null mice had increased expression of MMP-3 and MMP-13 in ventricular tissue compared to the wild-types. This indicates that compensatory upregulation of other MMPs with possible overlapping MMP substrates should be taken into consideration when interpreting MMP deletion experiments [63].

6. Future Directions

It is still a matter of debate whether a stiffer scar is more beneficial than a more compliant scar. Therefore, the timing of treatment should be a topic of future research in order to determine whether interfering with extracellular matrix remodeling should be started either immediately after infarction or later on. In some of the described animal studies MMP inhibitor administration started before the onset of infarction. Therefore the effect might be smaller when the inhibitors are given after infarction. Inhibition of MMP activity might be a new therapeutic concept to retard development of heart failure and cardiac dysfunction but long term effects have not yet been established.

The positive effects of MMP inhibition on LV dilatation in animal models led to the proposal to use MMP inhibitors as a potential therapy for patients at risk for the development of heart failure after MI. Although the promising results in animal studies encourage the design of clinical trails with MMP inhibitors, several issues have to be studied more extensively. First, the precise effect of MMP inhibitor treatment on cardiac function has to be studied more extensively. Second, the timing of MMP inhibitor administration after infarction has to be resolved and third, the choice between narrow versus broad range MMP inhibitors has to be made.

Another important issue is which type of MMP inhibitors should be used, broad range MMP inhibitors or inhibitors with a selective specificity? Broad range MMP inhibitors could be maximally effective, but could also induce negative side effects and might effect normal tissue as well. The use of selective MMP inhibitors could be favorable in some processes after infarction but the wound healing after infarction is orchestrated by numerous MMPs, each of them having specific tasks, which might be taken over by other less specific MMPs when they are specifically inhibited. When a selection of MMP inhibitors is to be made, inhibition of MMP-9 may be a suitable candidate, since the targeted deletion of MMP-9 attenuates left ventricular enlargement and collagen content after infarction [61] and decreases the influx of leukocytes and the incidence of cardiac rupture [35].

Regulating the balance of extracellular matrix remodeling either by extracellular matrix synthesis or degradation is a possible mechanism to prevent heart failure after infarction. MMP inhibitors might serve in the future as new therapeutic strategies for heart failure.

Besides regulation of extracellular matrix remodeling, restoring the contractile behavior of the infarcted area could beneficial for the treatment of heart failure. Recent studies described that infused bone marrow derived angioblasts or implanted adult stem cells in the infarcted heart reduce remodeling and improve cardiac function by promoting vasculogenesis/angiogenesis and regeneration of ischemic cardiac muscle [64-66]. On the other hand stimulation of differentiation to myofibroblasts and prevention of myofibroblast loss might be another new therapeutic strategy to generate a more functional scar tissue post-infarction. Although these new techniques have to deal with a lot of pitfalls, such as the access and selection of these cells and silencing the immune system, they represent potential promising new heart failure treatments.

Acknowledgements

Supported by grants of the Netherlands Heart Foundation (NHS 94.012 and NHS 99.054) and the Netherlands Organization for Scientific Research (NWO 902-16-098).

References

1. Ross, R.S. and T.K. Borg, Integrins and the myocardium. Circ Res, 2001. 88: p. 1112-9.
2. Cleutjens, J.P., The role of matrix metalloproteinases in heart disease. Cardiovasc Res, 1996. 32: p. 816-21.
3. Park, P.W., O. Reizes, and M. Bernfield, Cell surface heparan sulfate proteoglycans: selective regulators of ligand-receptor encounters. J Biol Chem, 2000. 275: p. 29923-6.
4. Kim, H.E., et al., Disruption of the myocardial extracellular matrix leads to cardiac dysfunction. J Clin Invest, 2000. 106: p. 857-66.
5. Cleutjens, J.P., et al., Collagen remodeling after myocardial infarction in the rat heart. Am J Pathol, 1995. 147: p. 325-38.
6. Cleutjens, J.P., et al., The infarcted myocardium: simply dead tissue, or a lively target for therapeutic interventions. Cardiovasc Res, 1999. 44: p. 232-41.

7.   Blankesteijn, W.M., et al., A homologue of Drosophila tissue polarity gene frizzled is expressed in migrating myofibroblasts in the infarcted rat heart. Nat Med, 1997. 3: p. 541-4.

8.   Willems, I.E., et al., The alpha-smooth muscle actin-positive cells in healing human myocardial scars. Am J Pathol, 1994. 145: p. 868-75.

9.   Burlew, B.S. and K.T. Weber, Connective tissue and the heart. Functional significance and regulatory mechanisms. Cardiol Clin, 2000. 18: p. 435-42.

10.   Smits, J.F., et al., Angiotensin II receptor blockade after myocardial infarction in rats: effects on hemodynamics, myocardial DNA synthesis, and interstitial collagen content. J Cardiovasc Pharmacol, 1992. 20: p. 772-8.

11.   Brilla, C.G., R.C. Funck, and H. Rupp, Lisinopril-mediated regression of myocardial fibrosis in patients with hypertensive heart disease. Circulation, 2000. 102: p. 1388-93.

12.   Yue, P., et al., Cytokine expression increases in nonmyocytes from rats with postinfarction heart failure. Am J Physiol, 1998. 275: p. H250-8.

13.   Tasab, M., M.R. Batten, and N.J. Bulleid, Hsp47: a molecular chaperone that interacts with and stabilizes correctly-folded procollagen. Embo J, 2000. 19: p. 2204-11.

14.   Takeda, K., et al., Greater than normal expression of the collagen-binding stress protein heat-shock protein-47 in the infarct zone in rats after experimentally-induced myocardial infarction. Coron Artery Dis, 2000. 11: p. 57-68.

15.   McQuibban, G.A., et al., Inflammation dampened by gelatinase A cleavage of monocyte chemoattractant protein-3. Science, 2000. 289: p. 1202-6.

16.   McGeehan, G.M., et al., Regulation of tumour necrosis factor-alpha processing by a metalloproteinase inhibitor. Nature, 1994. 370: p. 558-61.

17.   Suzuki, M., et al., Matrix metalloproteinase-3 releases active heparin-binding EGF-like growth factor by cleavage at a specific juxtamembrane site. J Biol Chem, 1997. 272: p. 31730-7.

18.   Yu, Q. and I. Stamenkovic, Cell surface-localized matrix metalloproteinase-9 proteolytically activates TGF-beta and promotes tumor invasion and angiogenesis. Genes Dev, 2000. 14: p. 163-76.

19.   Levi, E., et al., Matrix metalloproteinase 2 releases active soluble ectodomain of fibroblast growth factor receptor 1. Proc Natl Acad Sci U S A, 1996. 93: p. 7069-74.

20.   Gearing, A.J., et al., Processing of tumour necrosis factor-alpha precursor by metalloproteinases. Nature, 1994. 370: p. 555-7.

21.   Cleutjens, J.P., et al., Regulation of collagen degradation in the rat myocardium after infarction. J Mol Cell Cardiol, 1995. 27: p. 1281-92.

22.   Li, Y.Y., C.F. McTiernan, and A.M. Feldman, Interplay of matrix metalloproteinases, tissue inhibitors of metalloproteinases and their regulators in cardiac matrix remodeling. Cardiovasc Res, 2000. 46: p. 214-24.

23.   Spinale, F.G., et al., A matrix metalloproteinase induction/activation system exists in the human left ventricular myocardium and is upregulated in heart failure. Circulation, 2000. 102: p. 1944-9.

24.   Simeon, A., et al., Expression and activation of matrix metalloproteinases in wounds: modulation by the tripeptide-copper complex glycyl-L-histidyl-L-lysine-Cu2+. J Invest Dermatol, 1999. 112: p. 957-64.

25.   Nagase, H., Activation mechanisms of matrix metalloproteinases. Biol Chem, 1997. 378: p. 151-60.

26.   Knauper, V., et al., The role of the C-terminal domain of human collagenase-3 (MMP-13) in the activation of procollagenase-3, substrate specificity, and tissue inhibitor of metalloproteinase interaction. J Biol Chem, 1997. 272: p. 7608-16.

27.   Sato, H., et al., A matrix metalloproteinase expressed on the surface of invasive tumour cells. Nature, 1994. 370: p. 61-5.

28.   Sato, H., et al., Cell surface binding and activation of gelatinase A induced by expression of membrane-type-1-matrix metalloproteinase (MT1-MMP). FEBS Lett, 1996. 385: p. 238-40.

29.   Pei, D. and S.J. Weiss, Furin-dependent intracellular activation of the human stromelysin-3 zymogen. Nature, 1995. 375: p. 244-7.

30.   Sato, H., et al., Activation of a recombinant membrane type 1-matrix metalloproteinase (MT1-MMP) by furin and its interaction with tissue inhibitor of metalloproteinases (TIMP)-2. FEBS Lett, 1996. 393: p. 101-4.

31.   Hayakawa, T., et al., Cell growth-promoting activity of tissue inhibitor of metalloproteinases-2 (TIMP-2). J Cell Sci, 1994. 107: p. 2373-9.

32.   Hayakawa, T., et al., Growth-promoting activity of tissue inhibitor of metalloproteinases-1 (TIMP-1) for a wide range of cells. A possible new growth factor in serum. FEBS Lett, 1992. 298: p. 29-32.

33.   Thorgeirsson, U.P., et al., Breast cancer; tumor neovasculature and the effect of tissue inhibitor of metalloproteinases-1 (TIMP-1) on angiogenesis. In Vivo, 1996. 10: p. 137-44.

34.   Fata, J.E., et al., Accelerated apoptosis in the Timp-3-deficient mammary gland. J Clin Invest, 2001. 108: p. 831-41.

35.   Heymans, S., et al., Inhibition of plasminogen activators or matrix metalloproteinases prevents cardiac rupture but impairs therapeutic angiogenesis and causes cardiac failure. Nat Med, 1999. 5: p. 1135-42.

36.   Montfort, I. and R. Perez-Tamayo, The distribution of collagenase in normal rat tissues. J Histochem Cytochem, 1975. 23: p. 910-20.

37.   Tyagi, S.C., L. Matsubara, and K.T. Weber, Direct extraction and estimation of collagenase(s) activity by zymography in microquantities of rat myocardium and uterus. Clin Biochem, 1993. 26: p. 191-8.

38.   Spinale, F.G., et al., Time-dependent changes in matrix metalloproteinase activity and expression during the progression of congestive heart failure: relation to ventricular and myocyte function. Circ Res, 1998. 82: p. 482-95.

39.   Thomas, C.V., et al., Increased matrix metalloproteinase activity and selective upregulation in LV myocardium from patients with end-stage dilated cardiomyopathy. Circulation, 1998. 97: p. 1708-15.

40.   Li, Y.Y., et al., Differential expression of tissue inhibitors of metalloproteinases in the failing human heart. Circulation, 1998. 98: p. 1728-34.

41.   Coker, M.L., et al., Myocardial matrix metalloproteinase activity and abundance with congestive heart failure. Am J Physiol, 1998. 274: p. H1516-23.

42.   Tyagi, S.C., et al., Post-transcriptional regulation of extracellular matrix metalloproteinase in human heart end-stage failure secondary to ischemic cardiomyopathy. J Mol Cell Cardiol, 1996. 28: p. 1415-28.

43.   Carlyle, W.C., et al., Delayed reperfusion alters matrix metalloproteinase activity and fibronectin mRNA expression in the infarct zone of the ligated rat heart. J Mol Cell Cardiol, 1997. 29: p. 2451-63.

44.   Danielsen, C.C., H. Wiggers, and H.R. Andersen, Increased amounts of collagenase and gelatinase in porcine myocardium following ischemia and reperfusion. J Mol Cell Cardiol, 1998. 30: p. 1431-42.

45.   Sato, S., et al., Connective tissue changes in early ischemia of porcine myocardium: an ultrastructural study. J Mol Cell Cardiol, 1983. 15: p. 261-75.

46.   Peterson, J.T., et al., Evolution of matrix metalloprotease and tissue inhibitor expression during heart failure progression in the infarcted rat. Cardiovasc Res, 2000. 46: p. 307-15.

47.   Creemers, E.E., et al., Matrix metalloproteinase inhibition after myocardial infarction:

a new approach to prevent heart failure? Circ Res, 2001. 89: p. 201-10.

48. Hodgson, J., Remodeling MMPIs. Biotechnology (N Y), 1995. 13: p. 554-7.

49. Spinale, F.G., et al., Matrix metalloproteinase inhibition during the development of congestive heart failure : effects on left ventricular dimensions and function. Circ Res, 1999. 85: p. 364-76.

50. Rohde, L.E., et al., Matrix metalloproteinase inhibition attenuates early left ventricular enlargement after experimental myocardial infarction in mice. Circulation, 1999. 99: p. 3063-70.

51. Lindsey, M.L., et al., Selective matrix metalloproteinase inhibition reduces left ventricular remodeling but does not inhibit angiogenesis after myocardial infarction. Circulation, 2002. 105: p. 753-8.

52. Creemers, E., et al., Inhibition of matrix metalloproteinase (MMP) activity in mice reduces LV remodeling and depresses cardiac function after myocardial infarction (Abstract). Circulation, 1999. 100: p. I-250.

53. Schonbeck, U., F. Mach, and P. Libby, Generation of biologically active IL-1 beta by matrix metalloproteinases: a novel caspase-1-independent pathway of IL-1 beta processing. J Immunol, 1998. 161: p. 3340-6.

54. Narayanan, A.S., R.C. Page, and J. Swanson, Collagen synthesis by human fibroblasts. Regulation by transforming growth factor-beta in the presence of other inflammatory mediators. Biochem J, 1989. 260: p. 463-9.

55. Carmeliet, P., et al., Urokinase-generated plasmin activates matrix metalloproteinases during aneurysm formation. Nat Genet, 1997. 17: p. 439-44.

56. Creemers, E., et al., Disruption of the plasminogen gene in mice abolishes wound healing after myocardial infarction. Am J Pathol, 2000. 156: p. 1865-73.

57. Lim, Y.T., et al., Independent regulation of matrix metalloproteinases and plasminogen activators in human fibrosarcoma cells. J Cell Physiol, 1996. 167: p. 333-40.

58. Lijnen, H.R., et al., Stromelysin-1 (MMP-3)-independent gelatinase expression and activation in mice. Blood, 1998. 91: p. 2045-53.

59. Andreasen, P.A., R. Egelund, and H.H. Petersen, The plasminogen activation system in tumor growth, invasion, and metastasis. Cell Mol Life Sci, 2000. 57: p. 25-40.

60. Przyklenk, K., et al., Effect of myocyte necrosis on strength, strain, and stiffness of isolated myocardial strips. Am Heart J, 1987. 114: p. 1349-59.

61. Ducharme, A., et al., Targeted deletion of matrix metalloproteinase-9 attenuates left ventricular enlargement and collagen accumulation after experimental myocardial infarction. J Clin Invest, 2000. 106: p. 55-62.

62. Romanic, A.M., et al., Myocardial protection from ischemia/reperfusion injury by targeted deletion of matrix metalloproteinase-9. Cardiovasc Res, 2002. 54: p. 549-58.

63. Lee, R.T., Matrix metalloproteinase inhibition and the prevention of heart failure. Trends Cardiovasc Med, 2001. 11: p. 202-5.

64. Kocher, A.A., et al., Neovascularization of ischemic myocardium by human bone-marrow-derived angioblasts prevents cardiomyocyte apoptosis, reduces remodeling and improves cardiac function. Nat Med, 2001. 7: p. 430-6.

65. Orlic, D., et al., Bone marrow cells regenerate infarcted myocardium. Nature, 2001. 410: p. 701-5.

66. Jackson, K.A., et al., Regeneration of ischemic cardiac muscle and vascular endothelium by adult stem cells. J Clin Invest, 2001. 107: p. 1395-402.

Chapter 13

Experimental Models of MMP Activation: Ventricular Volume Overload

Baljit S. Walia, Stephen C. Jones, Jianming Hao, Ian M.C. Dixon
University of Manitoba, Winnipeg, Canada

1. Introduction

Ventricular dilatation is a hallmark of chronic volume overload manifest as eccentric hypertrophic enlargement of the affected ventricular chamber [1]. In these hearts, the ratio of myocyte length to width increases markedly with time resulting in a markedly dilated, relatively thin-walled chamber [1, 2]. Myocyte function is markedly depressed relative to that from animals with normal ventricular geometry [3].

The participation of non-myocytes is becoming clear as a key factor in the pathogenesis of many types of heart failure [4-6]. The extracellular collagen matrix connects muscle cells, bears mechanical load during systole and limits myocyte movement during diastole [4, 7]. Collagen degradation and the net loss of collagen *per se* may contribute to systolic contractile dysfunction in cardiovascular abnormalities such as ischemia [8, 9], stunned myocardium [10, 11], and volume overload hypertrophy [12]. However, cardiac collagen matrix metabolism in arterial-venous (AV) shunt-induced cardiac volume overload remains controversial. In a study of canine heart, cardiac volume overload by aortocaval (AC) shunting for 2 months resulted in no significant changes in the left ventricular (LV) collagen fraction [13]. Similarly, Michel et al. [14] found no change in LV collagen density after 1 or 3 months of cardiac volume overload by AC shunt in rats. In the porcine model of AV shunt, Harper et al. [15] noted a slight increase in collagen concentration at 4-6 weeks after induction of shunt. However, Ruzicka et al. [16] reported decreases in rat LV collagen after 4-10 weeks of volume overload by AC shunting. Variability in regarding the effects of volume overload on collagen turnover may due to species, severity and/or stage of the pathology.

Angiotensin II is implicated in the pathogenesis of cardiac hypertrophy and failure, and is known to increase synthetic activity of cardiac fibroblasts and myofibroblasts [17, 18]. These cells are the main cardiac source for the synthesis and secretion fibrillar collagens [19]. Increased cardiac renin activity as well as plasma and LV cardiac angiotensin I and angiotensin II levels occur in volume overload due to AV shunt [20] and in rats with AV shunt atrial hypertrophy is associated with increased angiotensin AT_1 receptor mRNA abundance in this tissue [21].

Cardiac fibrillar collagens are resistant to degradation by most proteinases with the exception of specific collagenases. Matrix metalloproteinases (MMPs) are specific for matrix collagens and other protein components and function in turnover of these target molecules [22, 23]. Altered expression and activity of MMPs have been identified in several pathological processes, including the development of severe heart failure, such as in experimental cardiomyopathy [24], myocardial infarction (MI) [25], pacing-induced heart failure [26], and clinical cardiomyopathy [27, 28]. MMP activation has been established as a mechanism of cardiac dilatation. Nevertheless, the role of MMPs in the pathogenesis of heart failure in chronic AV shunt (chronic volume overload) is not well-defined. A requisite for ventricular remodeling may be the disruption and degradation of collagen fibers [17]. We hypothesize that ventricular dilation in experimental volume overload is due to inappropriate collagen removal and alteration of the cardiac collagen weave via the activation of ventricular MMPs. As an increase or decrease in MMP activity may tip the balance in favor of net matrix accumulation or loss, respectively, this issue is relevant to the progression of ventricular dilatation and heart failure. The goal of this study is to determine whether hypertrophied and volume overloaded hearts are characterized by abnormal myocardial MMP activity and/or expression. Furthermore, the relationship among altered collagen concentration in volume overloaded hearts, collagen removal by MMPs, and cardiac AT_1 receptor activation is addressed.

2. Materials and Methods

2.1 Experimental animals

The experimental protocols for animal studies were approved by the Protocol Management and Review Committee of the University of Manitoba, following guidelines established by the Canadian Institutes of Health Research and the Canadian Council on Animal Care. Male Sprague-Dawley rats weighing 200-250 g were used for the study. Volume overload was produced by surgical creation of AC shunt as described previously by Garcia and Diebold [29]. The skin was sterilized with pivodine-iodine solution before a mid-line laparotomy was performed. After isofluorane anesthesia, the vena cava and abdominal aorta were exposed by opening the abdominal cavity via a mid-line incision. The intestines were displaced laterally using sterile guaze. Blunt dissection was used to remove the overlying adventitia and expose the vessels, taking care not to disrupt the tissue connecting the vessels, and both vessels were occluded proximal and distal to the intended puncture site. An 18 gauge needle was inserted into the exposed abdominal aorta and advanced through the medial wall into the vena-cava to create the fistula. The needle was then removed and the puncture point sealed by a drop of cyanoacrylate glue which conferred puncture patency. The abdominal musculature and skin incisions were closed by standard technique with absorbable suture and autoclips. Throughout the surgery, ventilation of the lung was maintained by positive pressure inhalation of 95% oxygen and 5% carbon dioxide mixed with isofluorane. Sham-operated rats were treated similarly except that no puncture was made. The animals were divided into three groups: sham operated animals, shunt animals and shunt animals treated with losartan (40 mg/kg/day, gastric gavage). Drug treatment was initiated 1 day after shunt surgery and animals were sacrificed by decapitation at either 1 week or 8 weeks subsequent to the surgery. Upon sacrifice, hearts were dissected *ex vivo* and great veins, extracardiac fat, and atria were trimmed and discarded. Right and left ventricular tissues were separated, weighed, snap-frozen in liquid nitrogen and stored at -70°C for further analysis.

2.2. Hemodynamic measurements

Mean arterial pressure (MAP) and left ventricular function of sham-operated control, shunt and losartan-treated shunt groups were measured following induction of AV shunt, as described previously [30]. Briefly, rats were anesthetized by intra-peritoneal injection of a ketamine:xylazine mixture (100mg/kg:10mg/kg). A micromanometer tipped catheter (2-0) (Millar SPR - 249) was advanced into the aorta to determine mean arterial pressure (MAP), and then further advanced to the left ventricular chamber to record left ventricular systolic pressure (LVSP), left ventricular end diastolic pressure (LVEDP), the maximum rate of isovolumic pressure development ($+dp/dt_{max}$) and the maximum rate of isovolumic pressure decay ($-dp/dt_{max}$). Hemodynamic data was computed instantaneously and displayed using a computer data acquisition workstation (Biopac, Harvard Apparatus Canada). In another series of experiments, left ventricular function and blood pressure of sham, shunt, and losartan treated rats were measured 8 weeks following induction of volume overload.

2.3. Protein extraction and assay

Ventricular samples were homogenized in cold 100 mM Tris pH 7.4

containing 1mM PMSF, 4 μM leupeptin, 1 μM pepstatin A, and 0.3 μM aprotinin and were then sonicated (Diamed Pro200, Mississauga) for 3 X 5 seconds. Crude membrane and cytosolic fractions were isolated according to the methods of Gettys et al. [31]. Briefly, samples were centrifuged for 3000 x g at 4°C for 10 min. to remove broken cells and nuclei. The supernatant was further subjected to centrifugation for 48,000 x g for 20 min. at 4°C. The cytosolic fraction (supernatant) was separated from the crude membrane pellet. Total protein concentration in cardiac samples was determined using the Bicinchoninic acid solution (BCA) kit (Sigma, St. Louis USA). These extracts were aliquoted and flash-frozen by immersion in liquid nitrogen, and stored at -80°C until the time of protein expression or enzyme activity assay.

2.4. Zymography: detection of cardiac matrix metalloproteinase activity

To identify changes in ventricular myocardial collagen removal in the progression of AV shunt induced cardiac hypertrophy and failure, MMP activity was determined from LV and RV myocardial extracts at 1 and 8 weeks of AV shunt [32]. Zymographic activities could be identified on the basis of calibrated molecular weight markers [Bio-Rad prestained broad range (catalogue # 161-0372), data not shown]. The zymograms were subjected to densitometric analysis in order to determine total proteolytic activity. Gelatin (300 bloom, final concentration 1 mg/ml) was added to a standard 7.5% SDS polyacrylamide and 30μg of protein was loaded per lane without reduction or boiling (to preserve MMP activity). Samples were run at 200 volts, maintaining a running buffer temperature of 4°C. After electrophoresis, gels were washed twice for 30 min each in 25 mM glycine (pH 8.3) containing 2.5% Triton X-100 with gentle shaking at 4°C in order to eliminate SDS. After incubating at 37 °C in incubation solution (50 mM Tris pH 8.0, 10 mM $CaCl_2$) for 18 h, gels were stained in 0.1% Coomassie Brilliant blue R-250 for 30 minutes and then destained in acetic acid/methanol. Gels were dried and scanned using a CCD camera densitometer (Bio-Rad imaging densitometer GS 670 Bio-Rad, CA, USA) to determine relative lytic activity.

2.5. Determination of total cardiac collagen

Right and left ventricular samples from different groups were ground into powder in liquid nitrogen and 100 mg (wet weight) cardiac tissue was then dried to constant weight. Tissue samples were digested in 6N HCl (6ml/100mg dry wt.) for 16 h at 105°C. Hydroxyproline was measured according to the method described by Chiariello et al. [33]. A stock solution containing 40mM of 4-hydroxyproline in 1mM HCl was used as a standard. Collagen concentration was calculated by multiplying hydroxyproline levels by a factor of 7.46, assuming that interstitial collagen contains an average of 13.4% hydroxyproline [33]. The data was expressed as μg collagen per mg dry tissue.

2.6. Masson's trichrome staining for total cardiac matrix

Serial cryostat cardiac sections were mounted on paraffin-fixed and coated slides. Slides were deparaffinized and hydrated with distilled water. Fixing was performed by immersing slides in Bouin's solution (saturated picric acid, 37-40% formaldehyde, glacial acetic acid) for 1 h at 60°C. After fixing, slides were cooled and washed with tap water for 5 min and then rinsed with distilled water. The slides were then placed in working hematoxylin solution (Hematoxylin, 95% ethanol, 29% aqueous ferric chloride, HCl) for 10 min and then rinsed as above. Slides were immersed in Biebrich-acid fuchsin solution (1% biebrich scarlet, 1% acid fuchsin, glacial acetic acid) for 15 minutes. Slides were rinsed and placed in phosphomolybdic-phosphotungstic acid solution (phosphomolybdic acid, phosphotungstic acid) for 10 min, transferred to aniline blue solution (aniline blue, glacial acetic acid) for 5 min and rinsed. Slides were placed in aqueous acetic acid solution for 3 min followed by dehydration and mounting in permount. As a result, nuclei were stained black; cytoplasm, keratin, muscle and intracellular fibers stained red; collagen and lipid stained blue.

2.7. Cardiac MMP protein detection

Western blot analysis was performed on tissue samples. Crude membrane and cytosolic fractions were isolated according to the methods of Gettys *et al.* [31]. Briefly, samples were centrifuged for 3000 x g at 4°C for 10 min to remove broken cells and nuclei. The supernatant was further subjected to centrifugation for 48,000 x g for 20 min at 4°C. The cytosolic fraction (supernatant) was separated from the crude membrane pellet. Total protein concentration of cytosolic fractions was measured using the BCA assay. Prestained low molecular weight marker and 30 μg of protein from samples was separated on 10% SDS-PAGE, after reduction (100°C, 5 min). Separated proteins were transferred to 0.45 μm polyvinylidene difluoride (PVDF) membrane which was then blocked overnight at 4°C or at room temperature for 1 h in Tris-buffered saline with 0.1% Tween-20 (TBS-T) containing 6% skim milk. After washing with TBS-T solution, membranes were probed with primary antibodies for 1 h at room temperature. Primary antibodies used included the MMP-1 (1:1000, The Binding Site, Birmingham, UK) and MMP-2 antibody

(1:500, Biogenesis, Poole, England). After washing, membranes were incubated with horseradish peroxidase (HRP) labeled secondary antibodies (horseradish peroxidase labeled goat Anti-Rabbit IgG, 1:10,000) for 1 h at room temperature. Target proteins were detected and visualized using enhanced chemiluminescence (ECL) "plus" according to manufacturer's instructions (Amersham Life Science Inc. Canada). Autoradiographs from the Westerns were quantified using a CCD camera imaging densitometer (Bio-Rad imaging densitometer GS670).

2.8. Statistical analysis

Assays were conducted in a completely randomized fashion in accordance with methods described in parametric statistics. All values were expressed as mean $\pm$ SEM. Differences between groups were assessed by one way analysis of variance (ANOVA) followed by Student-Newman-Keuls test (SigmaStat, Jandel Scientific) for significance of differences between data sets. Significant differences among groups were defined by a probability of less then 0.05.

3. Results

3.1. Hemodynamic changes

Animals were assessed for left ventricular function at 8 weeks post AV shunt surgery. The data revealed an increase in left ventricular end diastolic pressure (LVEDP) and a decrease in $\pm dP/dt_{max}$ relative to their controls (Table 1). Rats with shunt had significantly lower left ventricular systolic (LVSP) pressure at 8 weeks after creation of shunt. Systolic and diastolic pressures corresponded with changes in LVSP and were significantly lower in rats with 8 week shunt compared with controls. Heart rate (HR) showed no significant changes in the two groups (Table 1). Lung congestion was noted by the increase in lung weight in shunt animals compared to controls.

Losartan treatment (initiated 1 day after shunt surgery and maintained for 8 weeks) was associated with a trend to normalized LVEDP in treated shunt rats *vs* untreated values. LVSP was significantly increased with chronic losartan treatment vs. untreated shunt values, but remained significantly lower than control values (Table 1). Losartan treatment was associated with modest normalization of systolic and diastolic pressures; nonetheless these values remained significantly lower than the control group values. Losartan treatment for 8 weeks was also associated with significantly increased $\pm dP/dt_{max}$ vs. values from the untreated AV shunt values, and these were not different from control values.

Table 1. General and Hemodynamic characteristics of volume-overloaded rats with or without losartan treatment at 8 weeks following AV shunt surgery.

	Sham	AV Shunt	AV Shunt + Losartan
BW (g)	501.1 ± 13.7	493 ± 12.6	483.9 ± 17.9
HW (mg)	1307 ± 89	1898 ± 145*	1493 ± 105*#
LVW (mg)	1053 ± 77	1490 ± 102*	1205 ± 104*#
RVW (mg)	248 ± 10	408 ± 388*	289 ± 21*#
HW/BW (mg/100g)	261 ± 15	385 ± 34*	309 ± 258*#
LVW/BW (mg/100g)	211 ± 10	302 ± 23*	250 ± 16*#
RVW/BW (mg/100g)	50 ± 3.5	84 ± 7.5*	61 ± 5.2*#
Lung Weight (g)	1.78 ± 0.08	2.04 ± 0.06*	1.82 ± 0.13#
LVEDP (mmHg)	6.1 ± 0.57	13.9 ± 1.08*	9.1 ± 0.58
LVSP (mgHg)	134.8 ± 4.3	118.1 ± 3.8*	124.9 ± 5.5*#
HEART RATE	297 ± 15	321 ± 20	308 ± 14
+dP/dt (mmHg/sec)	5746 ± 329	4711 ± 130*	5360 ± 317#
-dP/dt (mmHg/sec)	5862 ± 356	4664 ± 134*	5793 ± 358#
ASP	125 ± 5.9	101 ± 2.6*	110 ± 6.5*#
ADP	97.8 ± 3.8	65.2 ± 3.1*	78.3 ± 5.2*#
APP	29.3 ± 2.1	48.6 ± 4.2*	33.0 ± 2.7#

Values are mean ± SEM (n=6-8 in each group). BW, body weight; HW, heart weight; LVW, left ventricular weight; RVW, right ventricular weight; LVEDP, left ventricular end diastolic pressure; LVSP, left ventricular systolic pressure; +dP/dt, maximum rate of isovolumic pressure development; -dP/dt, maximum rate of relaxation; ASP, arterial systolic pressure; ADP, arterial diastolic pressure; APP, arterial pulse pressure. $p < 0.05$ is expressed by * vs. sham and # vs. AV shunt values.

3.2. General characteristics and morphometry

Hearts of experimental animals were significantly hypertrophied, as reflected by increased LV mass and also by the increased LVW/BW and RVW/BW ratios in 8 week experimental animals compared to controls (Table 1). Liver and kidney weights remained unchanged in experimental hearts vs. control values (data not shown). Losartan treatment for 8 weeks was characterized by attenuation of the HW, LVW and RVW. However, these values remained significantly higher vs. control values. The ratios of heart weight/body weight (HW/BW), left ventricular weight/body weight (LVW/BW) and right ventricular weight/body weight (RVW/BW) were also significantly reduced in the treatment group but were still significantly higher vs. control animals. Lung weight was normalized relative to the 8 week losartan treatment group while losartan treatment was observed to have no effect on heart rate (Table 1).

3.3. Myocardial MMP activity and MMP-1 protein expression

A representative zymogram for 1 and 8 week shunt groups using gelatin as the proteolytic substrate is shown in Figure 1A and 1B. LV MMP-1 gelatinolytic activity was significantly increased in all AV shunt groups compared to sham operated control group values. Losartan treatment for 1 week was associated with significant improvement in MMP-1 activity. In right ventricular samples from AV shunt hearts, MMP-1 activity was significantly increased compared to controls. Treatment of experimental animals with losartan for 1 week had no effect on MMP-1 gelatinolytic activities in right ventricle, whereas losartan treatment for 8 weeks was associated with significant attenuation in both left and right ventricular MMP-1 activity compared to the untreated AV shunt group, although they remained significantly greater than controls (Figure 1A and 1B). We noted that MT-MMP (seen in the 8 week RV blot at 120 kDa; Figure 1A) appears in all of the AV shunt samples whereas it was rarely apparent in the control samples.

Western blot analysis of MMP-1 was performed using specific primary antibodies for interstitial collagenase (MMP-1; 54 kDa). The mean value of immunoreactive band density for left ventricular MMP-1 in volume overloaded heart samples was unchanged vs controls in either 1 or 8 week groups (Figure 2). Furthermore, densitometric scan results revealed no significant differences between untreated left ventricular AV shunt and losartan treated groups (Figure 2). Densitometric scanning revealed that control values for RV myocardial MMP-1 band intensity was similar to AV shunt values. Finally, no significant changes in specific MMP-2 or MMP-9 band density was observed between control, shunt, or losartan-treated shunt groups in either LV or RV at any point in the current study (data not shown).

3.4. Trichrome staining for total cardiac matrix

To determine the gross distribution of secreted fibrillar collagens in various groups, heart sections were stained with Masson's trichrome [34] in 8 wk AV shunt rats, losartan treated 8 wk AV shunt group treated with losartan and age-matched sham-operated controls. Representative cardiac sections from each group are shown in Figure 3. Collagen was reduced in AV shunt operated group compared to sham operated controls. Losartan treatment was associated with a modest increase in cardiac collagen. Furthermore, hypertrophy and dilation of the volume overloaded heart was clearly evident in cardiac sections as compared to those from sham operated controls. There was a marked decrease in hypertrophy and dilatation (Figure 3) compared to untreated volume overloaded hearts. This data closely parallels the cardiac mass data presented in Table 1.

A

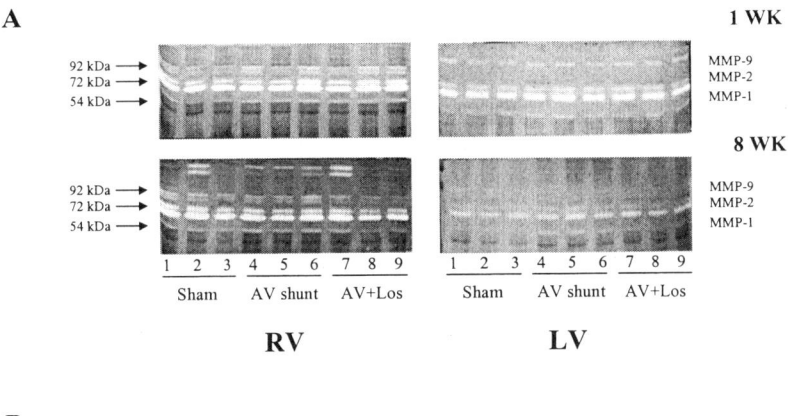

RV   LV

B

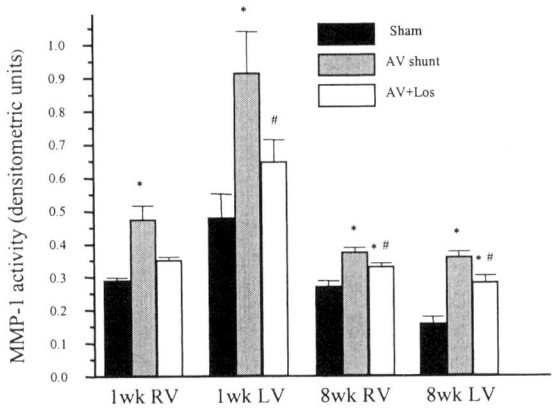

Figure 1. Zymographic analysis of matrix metalloproteinase (MMP) activity in 1 and 8 week AV shunt animals. Panel A. Representative zymograms from left and right ventricular (LV and RV) tissue samples from AV shunt animals, AV shunt animals treated with losartan (AV+Los) and age-matched sham-operated control animals (sham). MMP-1 (54kDa), MMP-2(72kDa) and MMP-9 (92kDa) bands are indicated on the basis of calibrated molecular weight markers included in each zymogram. Panel B. Histographic representations of MMP-1 data from multiple samples from groups in panel A, quantified by densitometric scanning. Data depicted is the mean $\pm$ SEM of 4-7 experiments. P<0.05 is expressed by * vs sham and # vs. AV shunt values.

A

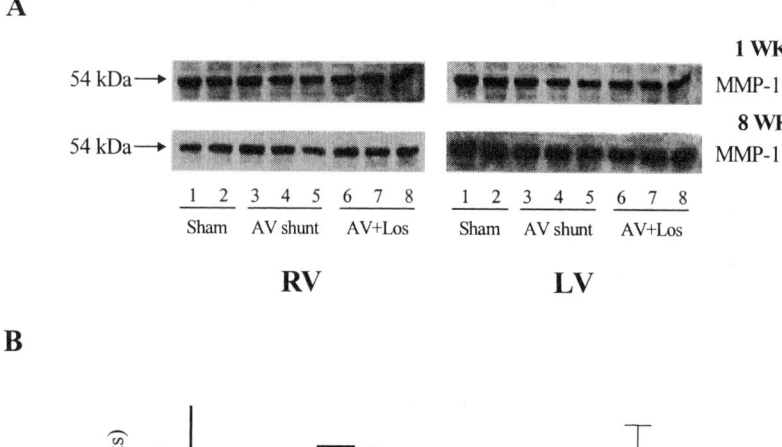

B

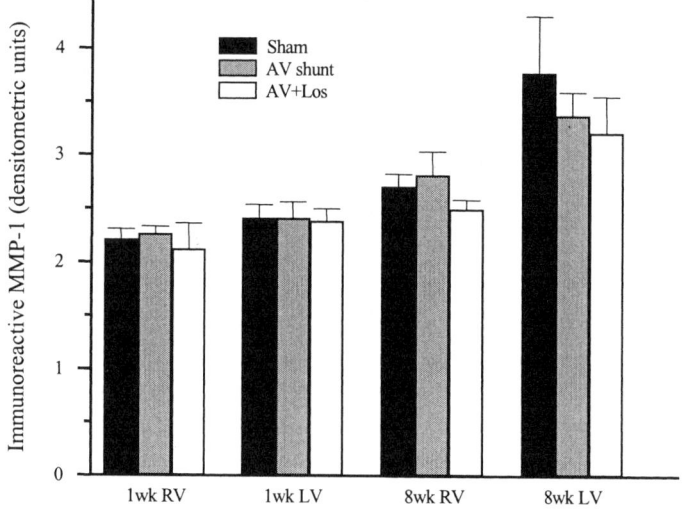

Figure 2. Western analysis of matrix metalloproteinase-1 in left and right ventricular (LV and RV) tissue from 1 and 8 week AV shunt animals. Panel A. Representative Western blots indicating the 54 kDa band specific for MMP-1 in 1 and 8 week AV shunt animals, AV shunt animals treated with losartan (AV+Los) and age-matched sham-operated control animals (Sham). Panel B. Histographic representation of data from multiple samples from groups in A, quantified by densitometric scanning. Data depicted is the mean ± SEM of 3-4 experiments. p<0.05 is expressed by * vs. sham and # vs. AV shunt values.

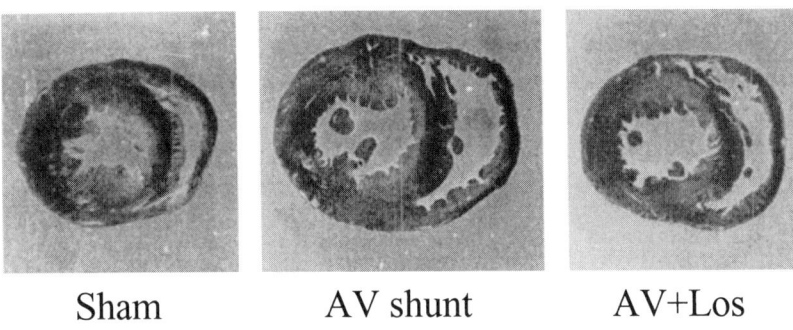

Sham   AV shunt   AV+Los

Figure 3. Cross-sections of trichrome stained hearts from 8 week sham-operated control animals (Sham), AV shunt animals and AV shunt animals treated for 8 weeks with losartan (AV+Los).

3.5. Quantitative assessment of cardiac collagen concentration

A unique feature of fibrillar collagen (as a protein) is the inclusion of a relative abundance of 4-hydroxyproline in the fibrillar α-chains. Fibrillar collagens types I and III share this trait and together these proteins constitute ~ 90% of the cardiac matrix protein content. Left and right ventricular samples were assessed for collagen concentration via 4-hydroxyproline detection and the results are shown as representative histograms in Figure 4. In the 1 week shunt group, a significant decrease in collagen concentration in the AV shunt group was observed vs control values in RV and LV samples. Losartan treatment for 1 week was associated with a significant elevation of collagen concentration in left ventricular samples while no difference of mean values was observed when compared to untreated right ventricular values. At 8 weeks, both LV and RV collagen concentration was significantly reduced vs control values, and losartan treatment for 8 weeks was associated with a partial reversal of these trends.

3.6. Myocardial TIMP activity

Tissue inhibitor of metalloproteinases (TIMP) were identified using reverse zymography (by their ability to inhibit gelatinase activity) and TIMP activity was reflected by relatively intense staining of bands at 28 kDa. Specific bands and were subjected to densitometric analysis in order to compare their relative inhibitory activities, and the results are seen in Figure 5. In this analysis, both RV and LV TIMP activity was unchanged in volume overloaded hearts at 1 and 8 weeks after AV shunt induction compared with values from control groups. Losartan treatment was not associated with any effect on myocardial TIMP activity early or late after overload induction (Figure 5).

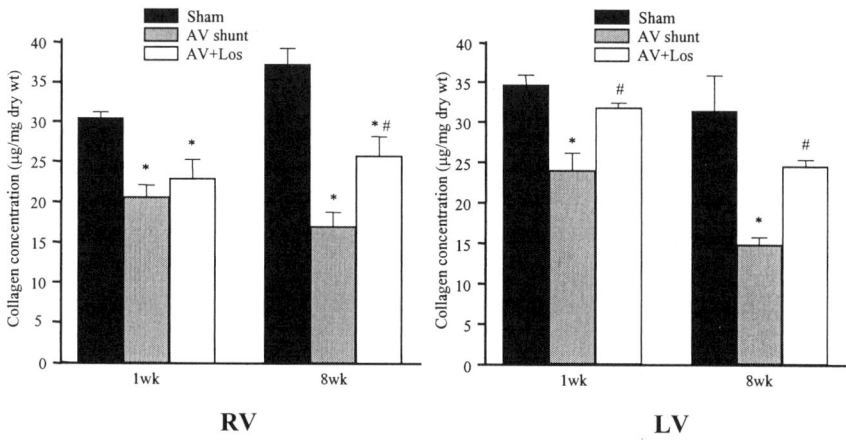

Figure 4. Cardiac collagen concentration (μg/mg dry weight) in left and right ventricular samples (LV and RV) from 1 and 8 week experimental animals (AV shunt), AV shunt animals treated with losartan (AV+Los) and age-matched sham-operated control animals (Sham). Data expressed as mean ± SEM from a total sample size of 4-6. $p < 0.05$ is expressed by * vs. sham and # vs. AV shunt values.

4. Discussion

The results of this study demonstrate positive correlations between increased end-diastolic loading of ventricular chambers (i.e., volume overload) and the ensuing myocardial remodeling process, resulting in marked ventricular dilatation and hypertrophy. Highlights of the current study include i) volume overload was associated with decreased myocardial collagen content in both the RV and LV chambers at 1 and 8 weeks subsequent to shunt induction, ii) myocardial LV and RV MMP-1 activity was significantly increased at 1 week after induction of volume overload and remained elevated at 8 weeks which implies enhanced removal of fibrillar collagens during that period, and that iii) losartan administration to experimental animals was associated with normalization of collagen accumulation and myocardial MMP-1 activity in both ventricles.

A

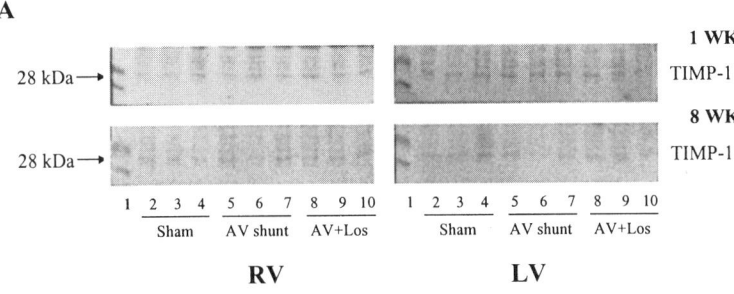

B

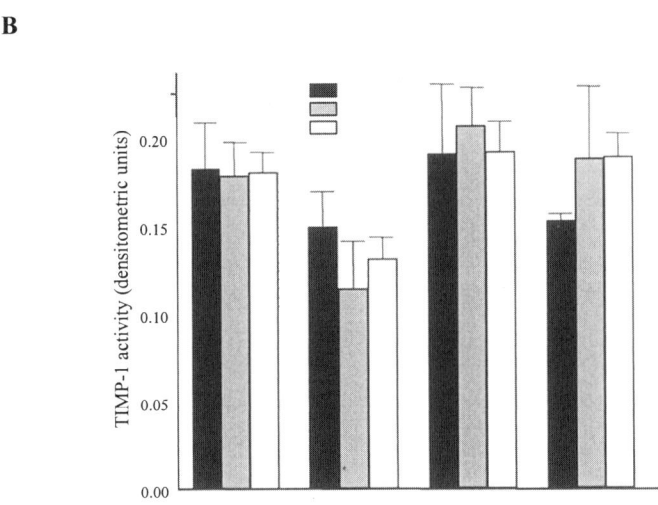

Figure 5. Reverse zymography for cardiac tissue inhibitor of matrix metalloproteinase-1 (TIMP-1) in 1 and 8 week AV shunt animals. Panel A. Representative reverse zymography gels showing TIMP-1 activity in 1 and 8 week left and right ventricular (LV and RV) tissue samples from AV shunt animals, AV shunt animals treated with losartan (AV+Los) and age-matched sham-operated controls (Sham). TIMP-1 (28 kDa) bands are indicated on the basis of calibrated molecular weight markers included in each zymogram (lane 1). Panel B. Histographic representation of data from multiple samples from groups in panel A, quantified by densitometric scanning. Data depicted is the mean $\pm$ SEM of 4-6 experiments. $p<0.05$ is expressed by * vs. sham and # vs AV shunt values.

Volume overload is associated with substantial morbidity and mortality in afflicted patients. Experimental models commonly employed to study volume overload have been mitral valve insufficiency, aortic valve insufficiency, and AV shunt [15, 16]. Each of these models is characterized by the production of volume overload and marked eccentric hypertrophy, but with notable differences in the hemodynamic variables [15, 16]. Mitral valve insufficiency results in a volume overload condition with relatively diminished afterload secondary to retrograde flow of blood into the compliant left atrium [3]. Aortic valve insufficiency is marked by diastolic retrograde flow into the left ventricle resulting in a volume overloaded ventricle with transiently elevated afterload on a beat-to-beat basis. However in both of these valvular conditions, net forward output is low and is limited by regurgitation. The AV shunt model produces a high output, bi-ventricular volume overload with normal systolic conditions, no retrograde flow, and a normal or slightly reduced afterload [16]. Thus despite numerous studies that have been carried out to understand cardiac pathologenesis subsequent to volume overload, data interpretation from various models requires a modicum of caution. We evaluated the role of MMPs in ventricular remodeling secondary to a sustained experimental volume overload using the rat model of AV shunt. The current model has the advantages of high reproducibility, low mortality, and the possibility to select the desired degree of volume overload.

4.1. LVEDP, cardiac hypertrophy, and failure in experimental AV shunt

Experimental AV shunt provides a reproducible model of volume overload with a consistent degree of cardiac hypertrophy. Eccentric LV hypertrophy has been shown to develop in response to an increase in LVEDP and wall stress [35]. In our hands, AV shunting was associated with significantly elevated LVEDP 8 weeks after surgery. Grossman et al. [1] found that end-diastolic wall stress was consistently increased in volume-overloaded ventricles independent of the degree of compensation or stage of ventricular remodeling. They proposed that increased end-diastolic wall stress is the stimulus for the eccentric hypertrophy and ventricular dilatation in this setting. It is possible that increased stretching of myocytes associated with elevated wall stress leads to angiotensin release [36], and our results indicate that AT_1 activation may contribute either directly or indirectly to hypertrophy in shunt animals. Furthermore, our results confirm that ventricular remodeling secondary to chronic volume overload in the AV fistula model is associated with gradual progression to heart failure.

4.2. Collagen protein concentration in volume overloaded hypertrophy

The present study demonstrated that interstitial collagen deposition was decreased in response to volume overload induced by AV shunt. This is consistent with an earlier report using a similar model in which Ruzika et al. [16] reported a decrease in left ventricular collagen concentration in response to volume overload. Similar results have been demonstrated in thyroxine induced cardiac hypertrophy where 8 weeks of daily thyroxine administration to juvenile rats resulted in a decrease in cardiac tissue collagen concentration [37]. Weber et al. [13] reported a modest decrease in the collagen volume fraction in a canine model of AV shunt. Similarly, Michel et al. [14] found a small decrease in LV collagen density in rat AV shunt. In a study of rat aortocaval shunt, upregulation of steady state mRNA levels of collagens type I and III was at times early after induction of volume overload although collagen protein concentration remained unchanged [38]. We suggest that the inconsistencies among these results could be due to differences in the severity of AV shunt/volume overload, as size of the shunt may be related to the severity of heart failure [20]. Nevertheless, the majority of data support the hypothesis that decreased collagen concentration is a component of the pathology of ventricular dilatation. As fibrillar collagens are essential for the maintenance of myocardial structural integrity, decreased collagen accumulation may lead to dysfunction of the collagen weave with attendant loss of cardiac geometry and function. These changes may contribute, at least in part, to progressive left ventricular dilatation observed during volume overload.

4.3. Myocardial MMP and TIMP activity

MMP-1 appears to be the rate-limiting metalloproteinase for the metabolism of collagen [38]. Elevated MMP-1 activity in volume overloaded hearts was sustained through 8 weeks (the end-point chosen for this study). Previous evidence supports the concept that increased MMP activity may contribute to the development of ventricular remodeling [24, 26]. Using a porcine model of pacing induced heart failure associated with ventricular dilatation, Spinale et al. [26] reported ~25% decrease in cardiac collagen content and an 80% increase in myocardial MMP activity 7 days after induction of supraventricular tachycardia. In these hearts, MMP activity increased to >100% over control baseline after 14 days. Data indicating activation of metalloproteinases in failing hearts is not limited to animal models; Tyagi et al. [39] reported increased collagenolytic activity in human dilated cardiomyopathy and in infarcted ventricular tissues. We suspect that a sustained increase in collagenase activity may contribute to degradation of collagen. Ventricular remodeling preceded by increased collagenase activity occurs in early and chronic stages of experimental myocardial infarction (MI) [34, 40]. In knockout mice lacking MMP-9 expression, induction of myocardial infarction was associated with significantly smaller increases in end-diastolic and end-systolic ventricular dimensions at both midpapillary and apical levels, compared with infarcted WT mice; these differences persisted at 15 days after MI [41]. In the present study, we found persistent upregulation in MMP-1 activity in both LV and RV, while MMP-2 activity was significantly increased only in 1 week post-AV shunt LV. This result is in agreement with the results of Armstrong et al. [42] in which MMP-2 did not change significantly during experimental pacing-induced ventricular dilatation and heart failure. Furthermore, we observed a significant increase in MMP-9 (vs. control activity) in 1 week tissue (both LV and RV) samples. This result is consistent with a previous report [42] of significantly increased MMP-9 activity after 1 wk of pacing induced CHF. Our data also indicate that increased zymogen activation rather than MMP protein expression is associated with altered cardiac collagen in volume overloaded hearts. We suggest that TIMP does not play a major role in modulation of MMP activation in these experimental hearts. Although altered net collagen synthesis and extracellular degradation by other proteinases cannot be ruled out as modulators in volume overloaded hearts, the results from this study would favor the hypothesis that increased net MMP activity is responsible for enhanced collagen removal and net matrix degradation.

4.4. Effect of AT_1 blockade on cardiac collagen concentration and MMP activity in volume overloaded hearts

Previous studies have suggested that the renin-angiotensin system (RAS) is involved in the remodeling of the myocardial collagen. Indeed, many cardiac RAS components including renin, angiotensin I, and angiotensin II expression have been shown to be elevated in volume overload due to AV shunt [20]. Furthermore, losartan treatment has been shown to be effective in the normalization of ventricular hypertropy associated with volume overload [16, 20]. In the present study, decreased cardiac left ventricular collagen concentration in the untreated animals at 1 and 8 weeks after shunt was partially normalized with losartan treatment. Ruzicka et al. [43] also found that 4-10 week losartan treatment was effective for attenuating decreased concentration of cardiac ventricular collagen in the rat model of volume overload and our results confirm this. We suggest that losartan treatment is effective in normalization of reduced collagen protein deposition in volume-overloaded hearts. The precise mechanism for the normalized MMP-1 activity subsequent to AT_1 blockade in the experimental model of volume overload is unclear. The lack of uniform response of different MMPs to losartan treatment suggests that a direct regulatory role by AT_1 activation may not be responsible for these changes. In this regard, a number of cytokines and growth factors have been shown to induce or stimulate the synthesis of MMPs including interleukin-1, platelet derived growth factor and tumor necrosis factor-α. Thus, it is possible that the effect of AT_1 blockade on MMP activity is mediated via the secondary activation of one or more of these cytokines and/or growth factors after activation of the AT_1 receptor.

5. Conclusions

In conclusion, these data indicate that early and sustained activation of cardiac MMP-1 in experimental volume overload may be responsible for loss of cardiac matrix collagen content during ventricular remodeling associated with volume overload. Losartan treatment was associated with normalization of MMP-1 activity and collagen content in volume overloaded hearts.

Acknowledgements

This work was supported via a grant-in-aid from the Canadian Institutes for Health Research (CIHR - IMCD). IMCD is a CIHR group scientist. We also thank Drs. Bin Ren, Xi Wang and Stephen C. Jones for their support, their assistance in creation of the experimental animal model, and for their kind assistance in the production of the manuscript.

References

1. Grossman, W., D. Jones, and L.P. McLaurin, Wall stress and patterns of hypertrophy in the human left ventricle. J Clin Invest, 1975. 56: p. 56-64.
2. Ross, J., Jr., et al., Diastolic geometry and sarcomere lengths in the chronically dilated canine left ventricle. Circ Res, 1971. 28: p. 49-61.
3. Urabe, Y., et al., Cellular and ventricular contractile dysfunction in experimental canine mitral regurgitation. Circ Res, 1992. 70: p. 131-47.
4. Borg, T. and M. Burgess, Holding it all together: Organization and functions(s) of the extracellular matrix of the heart. Heart Failure, 1993. 8: p. 230-238.
5. Weber, K.T., Cardiac interstitium in health and disease: the fibrillar collagen network. J Am Coll Cardiol, 1989. 13: p. 1637-52.
6. Factor, S.M. and T.F. Robinson, Comparative connective tissue structure-function relationships in biologic pumps. Lab Invest, 1988. 58: p. 150-6.
7. Factor, S., Role of the extracellular matrix in dilated cardiomyopathy. Heart Failure, 1994. 9: p. 260-268.
8. Takahashi, S., A.C. Barry, and S.M. Factor, Collagen degradation in ischaemic rat hearts. Biochem J, 1990. 265: p. 233-41.
9. Gerdes, A.M., et al., Structural remodeling of cardiac myocytes in patients with ischemic cardiomyopathy. Circulation, 1992. 86: p. 426-30.
10. Charney, R.H., et al., Collagen loss in the stunned myocardium. Circulation, 1992. 85: p. 1483-90.
11. Whittaker, P., et al., Stunned myocardium and myocardial collagen damage: differential effects of single and repeated occlusions. Am Heart J, 1991. 121: p. 434-41.
12. Iimoto, D.S., J.W. Covell, and E. Harper, Increase in cross-linking of type I and type III collagens associated with volume-overload hypertrophy. Circ Res, 1988. 63: p. 399-408.
13. Weber, K.T., et al., Fibrillar collagen and remodeling of dilated canine left ventricle. Circulation, 1990. 82: p. 1387-401.
14. Michel, J.B., et al., Morphometric analysis of collagen network and plasma perfused capillary bed in the myocardium of rats during evolution of cardiac hypertrophy. Basic Res Cardiol, 1986. 81: p. 142-54.
15. Harper, J., E. Harper, and J.W. Covell, Collagen characterization in volume-overload-

and pressure-overload-induced cardiac hypertrophy in minipigs. Am J Physiol, 1993. 265: p. H434-8.

16.   Ruzicka, M., B. Yuan, and F.H. Leenen, Effects of enalapril versus losartan on regression of volume overload-induced cardiac hypertrophy in rats. Circulation, 1994. 90: p. 484-91.

17.   Schelling, P., H. Fischer, and D. Ganten, Angiotensin and cell growth: a link to cardiovascular hypertrophy? J Hypertens, 1991. 9: p. 3-15.

18.   Aceto, J.F. and K.M. Baker, [Sar1]angiotensin II receptor-mediated stimulation of protein synthesis in chick heart cells. Am J Physiol, 1990. 258: p. H806-13.

19.   Eghbali, M., et al., Localization of types I, III and IV collagen mRNAs in rat heart cells by in situ hybridization. J Mol Cell Cardiol, 1989. 21: p. 103-13.

20.   Ruzicka, M., et al., The renin-angiotensin system and volume overload-induced cardiac hypertrophy in rats. Effects of angiotensin converting enzyme inhibitor versus angiotensin II receptor blocker. Circulation, 1993. 87: p. 921-30.

21.   Dostal, D. and Baker, K., Biochemistry, molecular biology and potential roles of the cardiac renin-angiotensin system. In: The Failing Heart, Dhalla, N.S., et al., Philadelphia, PA: Lippincott-Raven Publishers, 1995: p. 275-294.

22.   Woessner, J.F., Jr., Matrix metalloproteinases and their inhibitors in connective tissue remodeling. FASEB J, 1991. 5: p. 2145-54.

23.   Matrisian, L.M., Metalloproteinases and their inhibitors in matrix remodeling. Trends Genet, 1990. 6: p. 121-5.

24.   Gunja-Smith, Z., et al., Remodeling of human myocardial collagen in idiopathic dilated cardiomyopathy. Role of metalloproteinases and pyridinoline cross-links. Am J Pathol, 1996. 148: p. 1639-48.

25.   Dixon, I.M., et al., Effect of ramipril and losartan on collagen expression in right and left heart after myocardial infarction. Mol Cell Biochem, 1996. 165: p. 31-45.

26.   Spinale, F.G., et al., Time-dependent changes in matrix metalloproteinase activity and expression during the progression of congestive heart failure: relation to ventricular and myocyte function. Circ Res, 1998. 82: p. 482-95.

27.   Tyagi, S.C., et al., Differential gene expression of extracellular matrix components in dilated cardiomyopathy. J Cell Biochem, 1996. 63: p. 185-98.

28.   Tyagi, S.C., et al., Post-transcriptional regulation of extracellular matrix metalloproteinase in human heart end-stage failure secondary to ischemic cardiomyopathy. J Mol Cell Cardiol, 1996. 28: p. 1415-28.

29.   Garcia, R. and S. Diebold, Simple, rapid, and effective method of producing aortocaval shunts in the rat. Cardiovasc Res, 1990. 24: p. 430-2.

30.   Dixon, I.M., S.L. Lee, and N.S. Dhalla, Nitrendipine binding in congestive heart failure due to myocardial infarction. Circ Res, 1990. 66: p. 782-8.

31.   Gettys, T.W., et al., Characterization and use of crude alpha-subunit preparations for quantitative immunoblotting of G proteins. Anal Biochem, 1994. 220: p. 82-91.

32.   Tyagi, S.C., L. Matsubara, and K.T. Weber, Direct extraction and estimation of collagenase(s) activity by zymography in microquantities of rat myocardium and uterus. Clin Biochem, 1993. 26: p. 191-8.

33.   Chiariello, M., et al., A biochemical method for the quantitation of myocardial scarring after experimental coronary artery occlusion. J Mol Cell Cardiol, 1986. 18: p. 283-90.

34.   Cleutjens, J.P., et al., Regulation of collagen degradation in the rat myocardium after infarction. J Mol Cell Cardiol, 1995. 27: p. 1281-92.

35.   Carabello, B.A., et al., Left ventricular hypertrophy due to volume overload versus pressure overload. Am J Physiol, 1992. 263: p. H1137-44.

36.   Sadoshima, J., et al., Autocrine release of angiotensin II mediates stretch-induced hypertrophy of cardiac myocytes in vitro. Cell, 1993. 75: p. 977-84.

37.   Karim, M.A., et al., In vivo collagen turnover during development of thyroxine-induced left ventricular hypertrophy. Am J Physiol, 1991. 260: p. C316-26.

38.   Namba, T., et al., Regulation of fibrillar collagen gene expression and protein accumulation in volume-overloaded cardiac hypertrophy. Circulation, 1997. 95: p. 2448-54.

39.   Tyagi, S.C., et al., Matrix metalloproteinase activity expression in infarcted, noninfarcted and dilated cardiomyopathic human hearts. Mol Cell Biochem, 1996. 155: p. 13-21.

40.   Peterson, J.T., et al., Evolution of matrix metalloprotease and tissue inhibitor expression during heart failure progression in the infarcted rat. Cardiovasc Res, 2000. 46: p. 307-15.

41.   Ducharme, A., et al., Targeted deletion of matrix metalloproteinase-9 attenuates left ventricular enlargement and collagen accumulation after experimental myocardial infarction. J Clin Invest, 2000. 106: p. 55-62.

42.   Armstrong, P.W., et al., Structural remodelling in heart failure: gelatinase induction. Can J Cardiol, 1994. 10: p. 214-20.

43.   Ruzicka, M., F.W. Keeley, and F.H. Leenen, The renin-angiotensin system and volume overload-induced changes in cardiac collagen and elastin. Circulation, 1994. 90: p. 1989-96.

V. *PROSPECT FOR THERAPY OF THE FIBROSED HEART*

Chapter 14

Targets for Pharmacological Modulation of Cardiac Fibrosis

Lindsay Brown, Vincent Chan and Andrew Fenning
The University of Queensland, Australia

1. Introduction

The heart consists of the myocytes to produce force, the extracellular matrix to provide structural support and the blood vessels to supply nutrients and remove waste products. Studies on the mechanisms of cardiovascular disease, especially heart failure, have traditionally emphasised either deficiencies in myocyte function as the major cause of heart failure or inadequate perfusion as the major cause of ischaemic disease, for example atherosclerosis leading to coronary heart disease and myocardial infarction. The interest in the role of the third component, the extracellular matrix, in cardiovascular disease is much more recent and has emphasised the role of the collagens. The studies of Karl Weber and his colleagues [1-4] have played a key role in the realisation that excessive interstitial and perivascular collagen deposition, termed reactive fibrosis in contrast to reparative fibrosis (scar formation), is a critical component of cardiac remodelling in cardiovascular disease. Thus, Weber [4] argues that it is not the quantity but rather the quality of the myocardium that accounts for ventricular dysfunction in hypertension, the major risk factor for heart failure. This argues strongly that the extracellular matrix is a dynamic, rather than a static, component of the heart. This review will mention the biochemical processes leading to the synthesis and removal of collagens in the heart as the basis for understanding the targets for pharmacological intervention. The major emphasis of this review will be the drugs that may alter these biochemical processes involved in fibrosis. The ultimate aim of therapy with these drugs is to prevent or reverse deficits in cardiac function by controlling or reversing fibrosis to improve the quality of the myocardium.

The extracellular matrix is critically important in the normal function of the heart. This structural and protective framework of the heart connects myocytes, aligns contractile elements, prevents overextending and disruption of myocytes, transmits force and provides tensile strength to prevent rupture [1, 2]. It contains the fibrillar collagens type I and III, which constitute about 80% and 12 % respectively of the total cardiac collagen content [2], and fibronectin [5]. Collagen I is a heterotrimer that provides tensile strength from parallel rod-like fibres 50-150 nm in diameter. Collagen III is a homotrimer forming a fine network of fibrils. Fibronectin serves as a bridge between cells and the interstitial collagen network to influence cell growth, adhesion, migration and wound repair [5]. Excessive collagen deposition increases cardiac stiffness (or impairs cardiac compliance) and enhances the risk of adverse cardiovascular events such as diastolic and systolic ventricular dysfunction, myocardial infarction, heart failure, and arrhythmias [1, 4, 6].

Since fibrosis is an excessive accumulation of collagen, decreasing collagen synthesis or increasing collagen breakdown are the only ways to prevent or reverse this process. Understanding the complex process of collagen synthesis and degradation is therefore essential to utilise the possible points of attack of pharmacological therapy. Most collagen synthesis in the heart occurs in the fibroblasts. Using collagen I as an example, the mRNAs for the pro-α1 and pro-α2 chains are translated into prepro-α polypeptide chains in the nucleus that are extruded into the endoplasmic reticulum where a signal sequence is removed to give the pro-α chain. Selected proline and lysine residues are then hydroxylated in the presence of molecular oxygen and a reducing agent such as ascorbate. Some lysine residues are glycosylated with glucose or galactose. Three α-chains assemble and disulphide bonds form with the triple helix forming by zipper-like folding. This procollagen molecule is secreted from the Golgi vacuole into the extracellular space where N- and C-terminal propeptides are cleaved by procollagen peptidases. The resulting triple helix collagen molecule then undergoes self-assembly into the fibrils. In the long-term, collagens form glucose-dependent cross-links, the advanced glycation end-products (AGEs) [7].

Mature collagens are cleaved into two unequal fragments by matrix metalloproteinases (MMPs), especially MMPs-1, -2 and -8 (collagenases), produced as proMMPs by fibroblasts before activation by proteases including other MMPs; these fragments are then susceptible to cleavage by other proteases. There are at least 20 different MMPs which includes the collagenases, gelatinases, stromelysins and membrane-bound enzymes. MMP activity is regulated at several levels: by altering synthesis with growth factors, cytokines and corticosteroids, by regulating the activation of the inactive precursors such as the procollagenases and by blocking the enzyme by tissue inhibitors of metalloproteinases (TIMPs) [8]. These processes are altered by common cardiovascular diseases such as heart failure [9]. Fibronectins, like collagens, are degraded by MMPs, especially MMP-3 and -9 which are also inhibited by TIMPs [8].

This summary of collagen deposition and removal indicates that there are many possible targets for pharmacological intervention and some of these interventions are summarised in Figure 1.

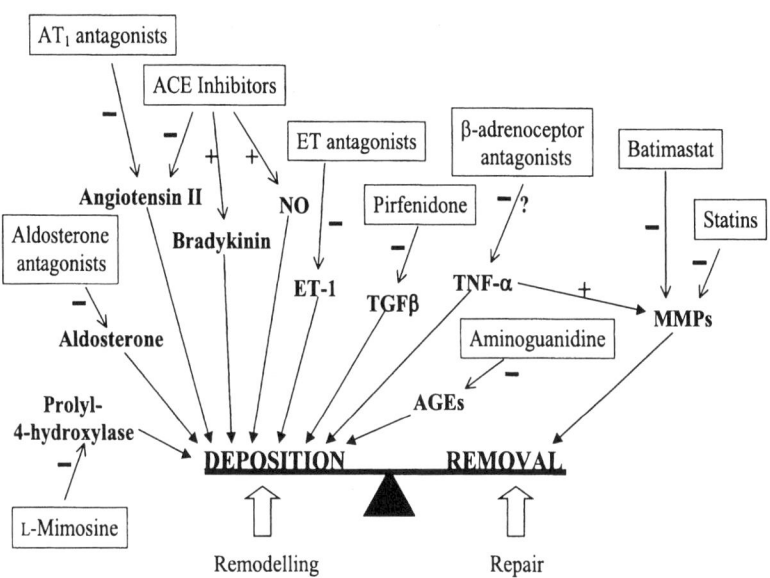

Figure 1. Some of the possible mechanisms to alter collagen deposition and removal in the heart, together with compounds shown to reduce fibrosis probably through these mechanisms.

Most antifibrotic compounds act primarily to decrease collagen expression by fibroblasts (inhibitors of the renin-angiotensin-aldosterone system, endothelin antagonists, calcium entry blockers, NO) or to modify cytokines such as transforming growth factor-β1 (TGF–β1) or tissue necrosis factor-α (TNF-α). Less researched targets include inhibitors of prolyl hydroxylase and lysyl oxidase, activators or inhibitors of MMPs, and modulators of AGE formation. A wide range of chemically unrelated structures has been shown to possess antifibrotic actions; some of the compounds are listed in Table 1.

Since these compounds have many physiological actions through receptors, ion channels, and cytokines distributed throughout the body, it is no surprise that they often have multiple actions on the cardiovascular system and possibly multiple points of attack on collagen synthesis and degradation.

278

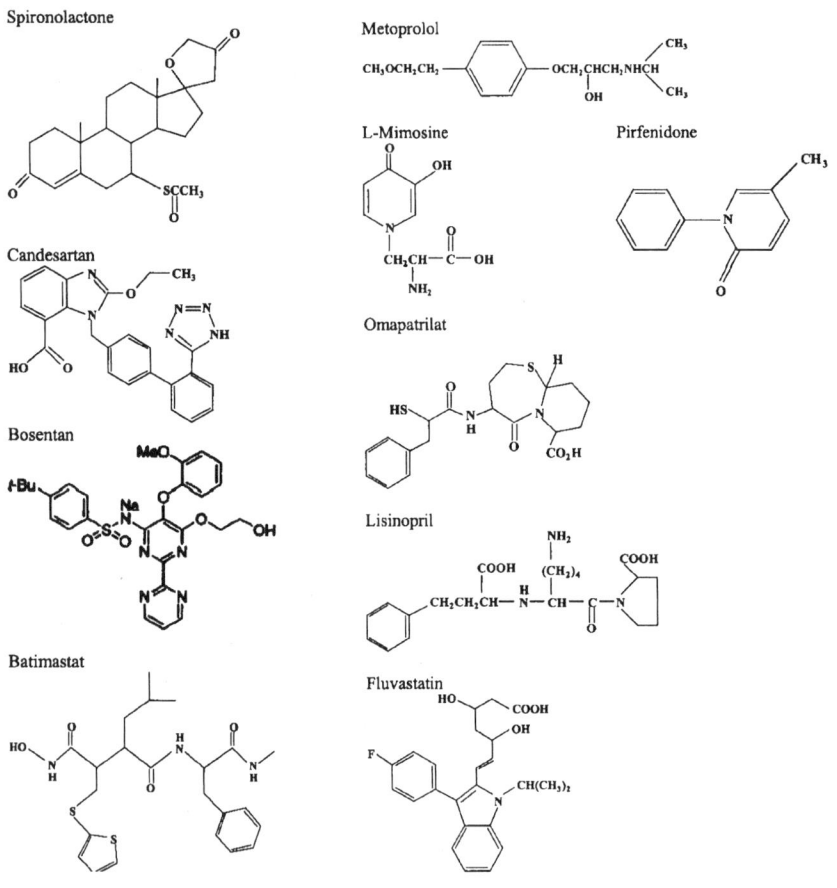

Figure 2. Compounds known to possess antifibrotic actions.

Experimental studies have induced cardiac fibrosis, usually in rats and mice, as a consequence of chronic hypertension, myocardial infarction, or gene deletions or additions. Most commonly used are the rat models of hypertension, in particular the ageing Spontaneous Hypertensive Rat (SHR) [10] or models with an altered renin-angiotensin-aldosterone system (for example, activation with renal hypertension or chronic angiotensin II infusions [11] or suppression in the DOCA-salt hypertensive rat [12]). Uncontrolled chronic hypertension, in both humans and SHR, leads to advanced hypertensive heart disease resulting in heart failure in the final quartile of life [10, 12, 13]. The SHR is widely used as an experimental model to mimic the progression of human essential hypertension although these rats are hypertensive in early adulthood, unlike most human hypertensive patients [12]. With this early development of hypertension, SHR hearts develop left ventricular hypertrophy accompanied by pathological cardiac remodelling observed as perivascular and interstitial fibrosis [10, 12, 14]. Since these models are also hypertensive and develop cardiac hypertrophy [10, 12, 13], antifibrotic actions may occur together with decreases in blood pressure and cardiac mass, making it harder to define selective antifibrotic effects. Early cardiac remodelling following myocardial infarction includes an increased collagen synthesis and deposition. Myocardial infarction is commonly modelled in rodents by acute ligation of the left coronary artery. This procedure will induce scar formation and heart failure in the relatively short period of 4-8 weeks, unlike the ageing SHR.

In addition, an increasing number of studies are using the techniques of molecular biology, especially cardiac-restricted gene expression, to define the actions of particular genes or products. One limitation of this approach is that cardiac fibrosis developing in humans as a chronic response to cardiovascular damage is unlikely to be the result of a defect in a single gene. Chronic cardiovascular disease is probably multigenic in cause, producing changes in the systems that control the cardiovascular system, especially the renin-angiotensin-aldosterone system, and in endothelial function, especially in the role of producing modulators such as endothelin and NO. Thus, many antifibrotic compounds target the mediators of these systems, such as angiotensin II, aldosterone, endothelin, and NO, rather than a single gene.

2. Inhibition of the Renin-Angiotensin-Aldosterone System

2.1 Inhibiting responses to angiotensin II

The chronic therapeutic management of cardiovascular symptoms in patients with hypertension, heart failure, myocardial infarction and diabetes relies on inhibition of the renin-angiotensin-aldosterone system [15, 16]. Classically, this system consists of liver-derived angiotensinogen being cleaved firstly by kidney-derived renin to inactive angiotensin I and then by angiotensin converting enzyme (ACE) during passage through the lungs to the active octapeptide, angiotensin II, which causes release of aldosterone from the adrenal glands. Organs such as the heart and brain may contain an intrinsic renin-angiotensin system independent of the circulating system. Angiotensin II acts on selective receptors, mostly of the angiotensin receptor Type 1 (AT1) subtype, to effect potent vasoconstriction, growth mediation and stimulation of collagen production; aldosterone is important in electrolyte control by the kidneys and also contributes to fibrosis by actions on non-epithelial cells. The actions of angiotensin II can be reduced by inhibition of its formation by ACE using compounds such as perindopril and enalapril, or by selective AT1 receptor antagonists such as irbesartan and candesartan. Aldosterone responses can be selectively antagonised by spironolactone.

Cardiac fibrosis characterises chronic cardiovascular diseases and the role of the renin-angiotensin-aldosterone system in the cellular process of fibrosis has been well established in both animal models of cardiovascular disease and in humans [11, 13, 17]. Thus, compounds which inhibit the production or actions of angiotensin II and aldosterone should be effective in controlling an increased blood pressure and both hypertrophy and fibrosis (cardiac remodelling). There is much evidence to support the effectiveness of ACE inhibitors, AT1 antagonists, and spironolactone in the control of cardiac remodelling so that angiotensin II could be described as a gatekeeper of the cascade of fibrosis by modifying the effects of many other possible mediators including NO, TGF-β1, eicosanoids, bradykinin, osteopontin, calcineurin, aldosterone and endothelin.

Angiotensin II rather than mechanical load has been shown to be the major stimulus for increased collagen deposition by cardiac fibroblasts leading to fibrosis [18]. In cultured adult cardiac fibroblasts, angiotensin II induced a dose-dependent synthesis of collagen which was selectively inhibited by the AT1 receptor antagonist, losartan, and unaffected by an angiotensin receptor type 2 antagonist [18, 19]. Local generation of angiotensin II and the novel idea of recruitable ACE have been proposed as mechanisms of local tissue repair. ACE binding density is low in the normal myocardium but high density binding is found in valve leaflets co-localized with angiotensin II and TGF-β1 receptors [20]. Valve leaflets are highly active in the remodelling of their structure particularly in the regulation of type I collagen. In an experimental model of myocardial infarction, high-density ACE binding was evident by day seven at the site of injury [20, 21]. Over the next eight weeks, ACE binding density continued to increase at the site of infarct as well as in other areas of the left ventricle remote to the injury [20, 21]. The recruitable ACE responsible for cardiac fibrosis following myocardial infarction was bound to macrophage and myofibroblast cell membranes and regulated local concentrations of angiotensin II involved in tissue repair [20]. ACE inhibition and AT1 receptor antagonism modulate fibrosis following myocardial infarction. Captopril, enalapril and losartan attenuated infarct size and expansion and inhibited the increase in hydroxyproline levels or collagen found at the infarct site [20, 22, 23].

The DOCA-salt rat is an established model of experimental hypertension characterised by extensive perivascular and interstitial cardiac fibrosis despite a suppressed renin-angiotensin system [24]. However, suppression of the renin-angiotensin system with captopril (100 mg/kg/day, ACE inhibitor), candesartan (2 mg/kg/day, AT1 receptor antagonist), or spironolactone (50 mg/kg/day, aldosterone antagonist) reversed and prevented further remodelling by reducing collagen expression and deposition [25]. Both captopril and spironolactone also decreased myocardial stiffness without decreasing systolic blood pressure or the degree of cardiac hypertrophy [25].

Most studies with SHR have investigated changes in young adult male rats where there is minimal cardiac fibrosis. As an example, 14 week old male SHR treated with the ACE inhibitor lisinopril (15 mg/kg/day for 12 weeks) showed reversal of the hypertrophy, fibrosis and hypertension with improved functional indices [26]. However, the reversal of established fibrosis in aged experimental models with mild to moderate cardiac failure is more relevant to the treatment of the human condition. Several studies have now examined reversal of existing fibrosis or the prevention of additional fibrosis in ageing SHR. Treatment of male 78 week old SHR with chronic hypertension, advanced ventricular hypertrophy and severe fibrosis with lisinopril (20 mg/kg/day) for 8 months normalised systolic blood pressure, reversed ventricular hypertrophy, attenuated fibrosis possibly by activating MMP-1 and improved diastolic stiffness [14]. Male 65 week old SHR treated with enalapril (30 mg/kg/day) for 12 weeks also showed decreased systolic blood pressure, left ventricular mass, and collagen content as well as improved coronary haemodynamics [27]. Treatment of male 12, 18 and 21 month old SHR with captopril (2g/L in the drinking water) until the age of 24 months prevented the characteristic decrease in the expression of alpha-myosin heavy chain and increase in the expression of pro-αI collagen and TGF–β1 [28] and restored inotropic responsiveness to β-adrenoceptor agonists [29]. Hypertension, hypertrophy, necrosis and fibrosis were prevented in male stroke-prone SHR treated from 1 month of age with ramipril (1 mg/kg/day) up to 15 months of age [30]. Treatment with the same dose of ramipril starting at 15 months of age significantly extended lifespan from 21 to 30 months, attenuated ventricular hypertrophy and endothelial dysfunction and enhanced NO release although collagen deposition was not measured [31].

Many clinical studies have consistently shown that ACE inhibitors prolong survival in human heart failure [32-34]. One reason could be a reversal of fibrosis leading to an improved systolic and diastolic function and fewer arrhythmias. Reversal of cardiac fibrosis has been shown in 18 patients with hypertension, left ventricular hypertrophy and diastolic dysfunction treated orally for 6 months with the ACE inhibitor, lisinopril (11.4±7.2 mg/day) [13]. This elegant study showed improved left ventricular dimensions and function by echocardiography as well as decreased collagen deposition in endomyocardial biopsies. Control patients received the diuretic, hydrochlorothiazide (45.6±9.8 mg/day), which controlled systolic blood pressure to the same extent as lisinopril but failed to improve collagen content or diastolic function [13]. This study highlights local effects of angiotensin II on cardiac myocytes and fibroblasts in humans that are independent of the blood pressure.

2.2. Inhibiting responses to other mediators: bradykinin, osteopontin, AcSDKP, calcineurin and aldosterone

Since bradykinin is a substrate for ACE, the increase in its concentrations following ACE inhibition could explain some of the positive responses to ACE inhibitors. Myocardial bradykinin concentrations in pigs were lower during pacing-induced heart failure while chronic treatment with the ACE inhibitor, benezaprilat (3.75 mg/day), normalised bradykinin concentrations and cardiac output [35]. Selective activation of bradykinin (B2) receptors produced vasodilatation, inhibition of cell growth, stimulation of NO synthase activity causing enhanced NO production [36, 37], and reductions in collagen I and III gene expression [38]. These decreases in collagen expression by fibroblasts were reversed by pre-treatment with indomethacin, a cyclo-oxygenase inhibitor, and mimicked by administration with beraprost, a stable prostacyclin analogue, indicating that increased bradykinin concentrations enhanced prostacyclin production which resulted in attenuation of collagen gene expression [38]. The role of bradykinin has been investigated using a selective B2 receptor antagonist, FR173657 (0.3mg/kg/day orally), in dogs with tachycardia-induced heart failure [39]. The B2 receptor antagonist worsened diastolic function, suppressed NO synthase and sarcoplasmic reticulum Ca^{2+}-ATPase expression and increased left ventricular collagen expression and deposition indicating that endogenous bradykinin participates in the cardioprotective effects of ACE inhibitors [39]. In contrast, co-administration of angiotensin II (150 ng/kg/min sc) and the B2 receptor antagonist, Hoe 140 (115 ng/kg/min sc), for 14 days completely prevented the reactive fibrosis which is characteristic of increased angiotensin II concentrations [40]. Further, in this study, oral indomethacin (2 mg/kg/day) attenuated perivascular collagen deposition suggesting that inhibition of activated myofibroblasts decreased collagen production [40].

The importance of bradykinin B2 receptors has been further defined using receptor gene knockout mice [41, 42]. These mice developed 5-fold higher myocardial fibrosis than control mice at 180 days of age, both as interstitial and perivascular fibrosis [41]. This collagen deposition was completely absent in knockout mice treated with an AT1 antagonist from conception [41]. However, collagen deposition was not different from control mice in another study of B2 receptor knockout mice of similar age either untreated or 12 weeks after coronary artery ligation [42]. The strains differed in that treatment with an ACE inhibitor or AT1 antagonist reduced the increased collagen deposition following infarction only in the control mice [42].

Another proposed mediator of the actions of angiotensin II is the adhesive glycophosphoprotein, osteopontin, an arginine-glycine-aspartic acid (RGD) containing protein which acts like a cytokine mediating cell adhesion, chemotaxis, and cell signalling. The expression of osteopontin was increased in the heart of SHR coincident with the development of heart failure [43]. There was an increased early expression of osteopontin in the left ventricle of rats transgenic for human renin, a model of angiotensin II-dependent left ventricular hypertrophy and failure [44]. The lack of osteopontin in a knockout mouse model was associated with an absence of an increased collagen following myocardial infarction and an approximately doubled post-infarction left ventricular chamber dilatation [45]. Angiotensin II stimulated osteopontin and TGF-β1 expression but not collagen I expression in fresh samples of human myocardium, suggesting both osteopontin and TGF-β1 as necessary mediators in the human heart [46]. However, there are as yet no studies demonstrating drug-induced inhibition of osteopontin expression leading to a prevention or reversal of cardiac fibrosis.

Plasma concentrations of the naturally occurring inhibitor of pluripotent haemopoietic stem cell proliferation, N-acetyl-seryl-aspartyl-lysyl-proline (AcSDKP) are increased during ACE inhibitor therapy since ACE cleaves AcSDKP to an inactive form [47]. Subcutaneous infusion of AcSDKP (800 μg/kg/day for 6 weeks) in aldosterone-salt hypertensive rats markedly prevented the development of both cardiac and renal fibrosis without affecting blood pressure or organ hypertrophy [47]. Rats treated with AcSDKP also showed fewer proliferating cells, probably fibroblasts, in both the heart and kidney [47]. In cultured rat cardiac fibroblasts, AcSDKP inhibited fibroblast proliferation, blocked endothelin-stimulated collagen synthesis and blunted the activation of p44/p42 MAP kinases [48]. Thus, AcSDKP may participate in the antifibrotic effects of ACE inhibitors by suppressing fibroblast proliferation and inhibiting collagen synthesis. AcSDKP inhibited the proliferation of neonatal rat ventricular fibroblasts and decreased phosphorylation and nuclear translocation of Smad2, a key step in the TGF-β1 pathway [49].

The Ca^{2+}-dependent protein phosphatase, calcineurin, an important signalling pathway component leading to cardiac hypertrophy, was activated following treatment of cardiac myocytes with angiotensin II [50]. Its role in cardiac fibrosis has been tested in Dahl salt-sensitive rats [51, 52], rats with abdominal aortic constriction [53] and aldosterone-salt-induced hypertensive rats [54] using the calcineurin inhibitor, FK506 (0.1 – 1 mg/kg/day). Fibrosis and hypertrophy are induced in the Dahl salt-sensitive rat by a high salt diet. Dosage with FK506 from 6 or 12 weeks of age attenuated the development of fibrosis without changing haemodynamic parameters [51]. Similar effects were measured with the AT1-receptor selective antagonist, candesartan (1 mg/kg/day) [52]. Ventricular calcineurin activity is also increased in the pressure-overloaded heart; treatment with FK506 prevented the increased wall thickening and perivascular fibrosis [53]. Mineralocorticoid excess increased expression of calcineurin and collagen and this was reduced either by AT1 receptor antagonism with losartan (10 mg/kg/day), or calcineurin inhibition with FK506 (0.5 mg/kg/day), or cyclosporine A (10 mg/kg/day) [54].

Activation of the renin-angiotensin system will also increase circulating aldosterone concentrations. Aldosterone acts through mineralocorticoid receptors on cardiac myocytes and endothelial cells and possibly on fibroblasts to induce both perivascular and interstitial fibrosis [55]. Aldosterone promoted fibrosis independent of blood pressure by activation of the transcription factors, AP-1 and NF-κB, and basic fibroblast growth factor in rats doubly transgenic for human renin and angiotensinogen genes [56]. Treatment with valsartan (10 mg/kg/day), an AT1 receptor antagonist, or spironolactone (20 mg/kg/day), an aldosterone antagonist, reduced both transcription factors and collagen [56]. The RALES (Randomized Aldactone Evaluation Study) clinical trial results have led to a re-evaluation of the role of spironolactone in human heart failure [57]. Low doses of spironolactone reduced the risk of death by 30% and improved the symptoms of heart failure. In a sub-group of RALES patients, high concentrations of serum markers for cardiac fibrosis were associated with poor outcome and these markers were decreased during spironolactone therapy [58]. In 46 patients with transmural infarction, the orally active aldosterone inhibitor, potassium canrenoate (50 mg/day), decreased postinfarction collagen synthesis defined by the serum concentration of the aminoterminal polypeptide of type III procollagen and also attenuated progressive left ventricular dilatation [59]. Increased myocardial expression of aldosterone synthase (CYP11B2) has been shown in the failing human heart and this expression and cardiac fibrosis were decreased in patients on spironolactone and ACE inhibitors [60]. In rats, chronic aldosterone-salt treatment increased blood pressure, ventricular hypertrophy and cardiac fibrosis; spironolactone (10 mg/kg/day) prevented collagen expression and deposition without affecting blood pressure or heart weight [61]. In streptozotocin-diabetic rats, spironolactone treatment (50 mg/kg/day for four weeks starting four weeks after streptozotocin) reversed the increased collagen deposition and also attenuated the increased ventricular stiffness of these rats [62].

Since ACE inhibitors may produce some of their beneficial effects by preventing the breakdown of the endogenous vasodilator, bradykinin, further benefit may be obtained by enhancing other endogenous vasodilators. The vasopeptidase inhibitors such as omapatrilat inhibit both ACE and neutral endopeptidase which further enhances NO and vasodilator prostaglandins and increases natriuretic peptides and adrenomedullin by blocking their metabolism. Omapatrilat (40 mg/kg/day for 10 weeks) significantly reduced both interstitial and perivascular collagen deposition as well as systolic blood pressure in 20 week old stroke-prone SHR [63]. However, a comparison of captopril (160 mg/kg/day) and omapatrilat (40 or 80 mg/kg/day) for 8 weeks starting immediately after myocardial infarction in rats showed that omapatrilat increased circulating atrial natriuretic peptide concentrations but did not result in further structural or functional improvement compared with captopril [64].

Plasma concentrations of circulating cell adhesion molecules such as vascular adhesion molecule-1 (VCAM-1), intercellular adhesion molecule-1 (ICAM-1) and E-selectin may predict adverse outcomes in patients at cardiovascular risk [65] while the chemokine monocyte chemotactic protein-1 (MCP-1) is expressed in atherosclerotic plaques [66]. Treatment with enalapril (10–20mg daily) but not losartan (50-100mg daily) lowered concentrations of cell adhesion molecules and MCP-1 in hypertensive patients [67]. ACE inhibitors reduced VCAM-1 concentrations following myocardial infarction [68], in heart failure [69] and in diabetics with borderline hypertension [70]. Since these responses appear selective to ACE inhibitors, the mechanism may involve an increase in NO leading to a decreased inflammatory cell infiltration or oxidative stress. Thus, blockade of angiotensin II formation or responses may lead to many other changes in signalling pathways leading to changes in collagen deposition. Defining the importance of each pathway, especially in different diseases, may lead to disease- or organ-specific control of fibrosis.

3. Modulation of the Endothelial Products Endothelin and NO

Vascular endothelial and smooth muscle cells control vascular tone and cellular proliferation by the synthesis of vasoconstrictors such as endothelin-1 and vasodilators such as NO. Endothelin-1 activates specific ET_A and ET_B receptors and both ET_A-receptor selective and non-selective antagonists have been shown to improve haemodynamics and symptoms in patients with congestive heart failure [71]. ET_A receptors mediate vasoconstriction; ET_B receptors may produce vasodilatation through release of NO. Non-selective endothelin receptor blockade with bosentan (100 mg/kg/day for 9 months) improved survival and decreased cardiac fibrosis, hypertrophy and dilatation in rats with heart failure following coronary artery ligation-induced myocardial infarction [72]. In the DOCA-salt hypertensive rat, bosentan (100 mg/kg/day for 6 weeks) decreased perivascular and subendocardial fibrosis with minimal effects on blood pressure showing that the different components of remodelling are controlled independently [73]. The ET_A-receptor selective antagonist, A-127722 (30 mg/kg/day), prevented the TGF-β1-dependent increase in cardiac collagen deposition induced by endothelin-1 over-expression in the heart also in the DOCA-salt hypertensive rat [74]. This study showed a separation of antifibrotic effects from effects on blood pressure or hypertrophy since systolic blood pressure was decreased to a small extent while hypertrophy was unchanged [74], a result also shown with renin-angiotensin-aldosterone system blockade in this model [25]. ET_A receptor stimulation stimulated collagen accumulation in infarct tissue in the rat since treatment with the selective antagonist, LU 135252 (30 mg/kg/day), decreased collagen and TGF-β1 gene expression and collagen deposition [75]. However, infarct expansion was increased and systolic function decreased when treatment was started 3 hours after coronary ligation [75]. Treatment with the non-selective antagonist, SB 209670 (6.25 mg/kg twice daily for 26 days starting 48 hours after coronary artery ligation), caused further dilatation of the left ventricle without changing collagen deposition or cross-linking, indicating that early intervention with endothelin antagonists may be harmful [76].

Nitric oxide (NO) is produced by endothelial cells from L-arginine by the action of NO synthase and regulates vascular tone, cardiac contractility, myocardial relaxation, diastolic function, and platelet aggregation [77, 78]. Compounds which release NO such as bradykinin negatively regulated cardiac fibroblast function to decrease collagen I and III expression, probably by increasing intracellular cGMP concentrations [79]. In addition, the NO donor DETA NONOate (100 μM) but not bradykinin decreased proliferation of fibroblasts [79]. In cultured rabbit vascular smooth muscle cells, NO-generating compounds such as S-nitroso-N-acetylpenicillamine and sodium nitroprusside showed reversible, haemoglobin-sensitive inhibition of collagen synthesis, implicating NO release, without damage to the cells [80]. Thus, NO from the endothelium may inhibit local collagen production in the heart and blood vessels. Further, NO suppresses the formation of plasminogen activator inhibitor-1 (PAI-1), which is critical in controlling endogenous fibrinolytic activity and also impairs matrix degradation [81, 82]. Genetically PAI-1-deficient mice treated with the NO synthase inhibitor, L-nitroarginine methyl ester (L-NAME), were protected against the development of coronary perivascular fibrosis indicating that inhibition of vascular PAI-1 activity may prevent fibrosis [83].

Both NO and the renin-angiotensin system are key regulators of vascular tone and there is significant cross-talk between these modulators. NO inhibited angiotensin converting enzyme activity and down-regulated AT1 receptors while angiotensin II stimulated NO synthesis and release [84]. In hypertensive humans, the ACE inhibitor enalapril dose-dependently increased serum nitrate/nitrite concentrations, indicating an increased NO production [85]. This mechanism could provide a mechanism whereby ACE inhibitors selectively suppress fibrosis by increasing NO in contrast to AT1 receptor antagonists which increase angiotensin II concentrations. On the other hand, AT1 antagonists could be more effective than ACE inhibitors in disease states with endothelial dysfunction, such as diabetes and hypertension.

Since NO has a very short half-life and its precursor, L-arginine, is orally active, the importance of NO in fibrosis may be tested by determining changes in cardiac structure and function following chronic L-arginine treatment. In male 12 month old SHR, 6 month treatment with L-arginine (1.2g/l in drinking water) reduced arterial pressure, peripheral resistance, left ventricular mass and collagen content and improved coronary haemodynamics [86]. Oral L-arginine treatment (2.25% in drinking water) reduced the cardiac pathology of myocarditis and improved survival in a mouse model of encephalomyocarditis virus-infected mice probably by reducing cellular infiltration and myocardial necrosis [87].

The relevance of NO has been further shown by studies in which NO production is markedly decreased using NO synthase inhibitors such as L-NAME. Deficiency of NO following chronic oral administration of this L-arginine derivative to rats induced hypertension, cardiac hypertrophy and fibrosis [88-91]. Inhibition of the synthesis of NO induces many changes that may be therapeutic targets to prevent the cardiovascular remodelling, specifically to inhibit fibrosis. In vivo, chronic inhibition of NO synthesis led to an upregulation of cardiac angiotensin II receptors [92]. Administration of candesartan, an orally active selective AT1 receptor antagonist, reversed the increased blood pressure, left ventricular wall thickness and collagen deposition of L-NAME-treated rats and normalised diastolic stiffness and cardiac function [89]. In L-NAME-treated rats, marked infiltration of leukocytes and fibroblast-like cells into the coronary vessels and myocardial interstitial areas occurred during the first week associated with expression of monocyte chemoattractant protein-1 [93]. The affected areas were replaced after 28 day treatment with vascular and myocardial remodelling. This suggests that early inhibition of inflammation, for example with corticosteroids or non-steroidal compounds, could prevent the subsequent development of fibrosis in this model. Cardiac PAI-1 expression was increased after 7 day L-NAME treatment; this increase was significantly prevented by the ACE inhibitor, imidapril, but not by candesartan, although both compounds inhibited collagen I expression [94]. While the products of the vascular endothelial cells have been primarily considered as modulators of vascular tone and proliferation, it is now clear that these products also control the synthesis and deposition of collagen in the heart.

4. Inhibition of Post-Translational Modifications

Enzymes which play key roles in the intracellular and extracellular maturation of collagen are obvious potential pharmacological targets in the control of cardiac fibrosis [95]. Since hydroxylation of the prolyl residues is a final common pathway in collagen synthesis, inhibition of this enzyme should be a major therapeutic target in reducing collagen production. Few studies have investigated prolyl 4-hydroxylase inhibitors but these have shown the potential of such compounds to prevent myocardial fibrosis and improve cardiac function. In neonatal rat cardiac fibroblasts, ascorbate deficiency led to decreased rates of prolyl hydroxylation without reducing procollagen mRNA levels [96].

Ascorbate-deficient fibroblasts showed increased intralysosomal degradation of newly synthesised procollagens, increased intracellular accumulation of Type I procollagen and decreased extracellular Type I collagen deposition [96]. The naturally occurring catechol analogue, L-mimosine, inhibited prolyl 4-hydroxylase in adult rat cardiac fibroblasts leading to increased intracellular accumulation of procollagens and diminished extracellular secretion with minimal cytotoxicity [97]. Treatment with L-mimosine also induced the activity of MMP-9 to increase the removal of fibrillar collagens [97]. L-Mimosine also reduced the secretion of hydroxyproline-containing proteins from smooth muscle cells obtained from human primary atherosclerotic and restenotic coronary arteries [98]. However, there are no reports investigating the effectiveness of L-mimosine *in vivo*. Treatment with the orally active prolyl 4-hydroxylase inhibitor, FG041 (100 mg/kg/day starting 48 hours after ligation), in female rats with myocardial infarction following coronary artery ligation prevented the substantial increase in the hydroxyproline/proline ratio in the infarcted hearts [97]. Further, there was partial recovery of left ventricular function as measured by echocardiography and haemodynamic measurements [99].

Procollagens are converted to fibrillar collagens by the removal of domains at the N-terminal by N-proteinases and at the C-terminal by C-proteinases allowing spontaneous self-assembly of the monomers. Thus, inhibition of these procollagen proteinases should block the deposition of collagen. Potent, non-peptide analogues of ornithine-derived sulfonamide hydroxamic acids have been shown to be inhibitors of the C-proteinase [100] although no results on cardiac fibrosis have been published. Peptide inhibitors of procollagen N-proteinase have been described [101] but no pharmacological results have been published. However, this could be an important mechanism to control cardiac fibrosis. Both the activity of procollagen C-proteinase (PCP) and its enhancer protein (PCPE) were stimulated by aldosterone in coordination with collagen production [102]. This recent study has shown that spironolactone prevented the upregulation of PCPE and collagen mRNAs following myocardial infarction in rats [102].

Hydroxylation of lysine is necessary for collagen cross-linking. Inhibition of lysyl oxidase would therefore be expected to alter collagen distribution and maturation. Treatment with β-aminopropionitrile (10 g/day orally), an active site irreversible inhibitor of lysyl oxidase, decreased left ventricular collagen deposition and collagen cross-linking and decreased myocardial stiffness when administered to normal adult pigs [103]. In rats treated chronically with 17α-methyltestosterone, β-aminopropionitrile prevented the decreased left ventricular compliance indicating that the increased stiffness with anabolic steroids may be due to increased cross-linking by lysyl oxidase rather than increased collagen formation [104]. Further studies are clearly needed to establish whether inhibition of lysyl oxidase prevents or reverses the chronic changes in cardiovascular disease. The control of post-translational modification of collagens is clearly an underused mechanism to control cardiac fibrosis.

5. β-Adrenoceptor Antagonists and Calcium Entry Blockers

Although hypertension and fibrosis are clearly independent variables as shown by antifibrotic actions without antihypertensive responses to ACE inhibitors, AT1 and endothelin receptor antagonists, it is feasible that other antihypertensive drugs act by mechanisms that additionally regulate collagen synthesis or degradation. Treatment with some β-adrenoceptor antagonists (carvedilol, metoprolol, and bisoprolol but not bucindolol or celiprolol) has been shown to improve survival in patients with heart failure [15, 16]. In patients treated with metoprolol, attenuated cardiac remodelling was shown as decreased left ventricular end-systolic and diastolic volume indices, decreased left ventricular mass index and an improved left ventricular ejection fraction [105]. However, studies on patients have not investigated whether this attenuated remodelling also involves a decreased deposition of collagen. Chronic β-adrenoceptor antagonism may indirectly decrease collagen deposition based on studies on cytokines in patients with dilated cardiomyopathy treated with metoprolol or bisoprolol [106] and on rat models of hypertension and heart failure treated with carvedilol or metoprolol [107-110]. The increased serum levels of interleukin-10, TNF-α and soluble TNF receptors in patients with dilated cardiomyopathy were significantly decreased during chronic treatment with β-adrenoceptor antagonists [106]. Since these cytokines have been implicated in inflammation and fibrosis, decreased serum levels may lead to a decreased collagen deposition in these patients. Similar effects have been reported in rats with increased TNF-α expression following large myocardial infarctions due to coronary artery ligation and attenuation with oral metoprolol administration (average dose 70.7 mg/kg/day) [108].

Animal studies have shown clear evidence of prevention of collagen deposition following administration of β-adrenoceptor antagonists although responses could be independent of β-adrenoceptor blockade. Administration of carvedilol to stroke-prone hypertensive rats on a high salt-fat diet decreased or prevented myocardial remodelling, in particular the increased hypertrophy, hyperplasia, inflammation, fibrosis and microinfarction without reducing blood pressure [107]. A comparison of equivalent β-adrenoceptor blocking doses of carvedilol (nonselective α- and β-adrenoceptor antagonist) and metoprolol (β_1-adrenoceptor selective antagonist) for 11 weeks showed that only carvedilol significantly reduced myocardial collagen in rats after coronary artery ligation-induced infarction [109]. This difference implies that β_1-adrenoceptors are not involved in a reduction of cardiac collagen. Consistent with these results, carvedilol but not metoprolol or prazosin reduced the increased collagen and fibronectin production in fibroblasts from rats with left ventricular hypertrophy following aortic banding [110]. The combined α- and β-adrenoceptor antagonist, labetalol, normalised blood pressure but did not regress myocardial fibrosis in rats with 8 week renovascular hypertension, in contrast to equieffective antihypertensive doses of the ACE inhibitor, zofenopril, or the calcium channel antagonist, nifedipine [111]. The relevance of antioxidant responses with carvedilol has not been satisfactorily resolved. Another mechanism to explain the limiting of remodelling and diastolic dysfunction with β-adrenoceptor antagonists may be modulation of MMP activity. Dogs infused with angiotensin II and given 48-hour tachycardia pacing showed increased MMP abundance and activity as well as increased chamber stiffness; these changes were prevented by almost complete β_1-adrenoceptor antagonism with atenolol treatment [112] .

Inhibition of voltage-dependent calcium entry into vascular smooth muscle cells is an accepted antihypertensive mechanism. Further, calcium is an important second messenger in myocytes and fibroblasts and is increased by hormones that cause fibrosis, such as angiotensin II and aldosterone. Thus, calcium channel blockade may have a role in preventing or attenuating cardiac fibrosis. Long-term verapamil for 45 weeks in SHR starting at 10 weeks of age decreased blood pressure and heart weight but did not change collagen concentration [113]. However, nifedipine treatment (30 mg/kg/day for 12 weeks) in renovascular hypertensive rats starting 8 weeks after induction significantly reduced left and right ventricular collagen deposition to a similar extent as the ACE inhibitor, zofenopril [111]. Blockade of T-type calcium channels with mibefradil (30 mg/kg/day for 2 weeks) significantly attenuated myocardial fibrosis in rats receiving either angiotensin II or aldosterone infusions [114]. Chronic treatment with mifebradil (10 mg/kg/day for 6 weeks) reduced interstitial and perivascular fibrosis and improved cardiac function following myocardial infarction in rats [115]. These results indicate that preventing calcium influx through calcium channels may be beneficial in reducing collagen deposition in chronic cardiovascular disease in addition to decreasing blood pressure and reducing anginal attacks.

6. Suppression of Autocrine and Paracrine Systems

The failing heart is characterised by the activation of humoral, autocrine and paracrine systems such as the renin-angiotensin-aldosterone, endothelin, and NO systems discussed above. The activity of other autocrine and paracrine factors in regulating the extracellular matrix provides further possible therapeutic targets to control or reverse collagen deposition. The most important mediators appear to be several pro-inflammatory cytokines such as the interleukins-1β, -6 and -8 and TNF-α as well as growth factors such as TGF-β, especially in cardiac fibroblasts, which are increased during the remodelling that follows myocardial infarction.

Expression of the interleukins-1β and -6 as well as TNF-α and TGF-β increased following coronary artery ligation in the rat heart [116, 117] and the expression of interleukin-1β correlated well with collagen deposition in the non-infarcted myocardium [116]. Treatment with anti-interleukin-1β antibody suppressed cardiac collagen expression and accumulation following myocardial infarction in mice [117]. However, wound healing mechanisms were delayed which led to left ventricular dilatation and an increased risk of ventricular rupture in these antibody-treated mice suggesting that interleukin-1β plays a protective role in the acute phase after myocardial infarction [118]. The role of an increased interleukin expression in the late phase has not been determined nor has the importance of an increased interleukin expression in other models of cardiac fibrosis such as the ageing SHR. One member of the interleukin-6 family of cytokines, leukaemia inhibitory factor (LIF), has multiple effects on collagen synthesis and degradation [119]. In mice cardiac fibroblasts, LIF inhibited differentiation into myofibroblasts, reduced collagen content and also reduced MMP activity [119].

TGF-β is a potent stimulus for matrix deposition by increasing the expression of collagen, decreasing the expression and activity of collagen degrading proteolytic enzymes such as MMP-2 and -9 and enhancing the expression of MMP inhibitors, the tissue inhibitors of MMPs (TIMP-1, -2 and -4) [117, 120-122]. TGF-β1 signalling occurs via ligand-induced heteromeric complex formation of type I and type II serine/threonine kinase receptors and downstream through the Smad protein family [123, 124]. In the mammalian heart, the Smad proteins are divided into the receptor-regulated Smads 2 and 3, the common mediator Smad 4 and the inhibitory Smads 6 and 7. After TGF-β1 receptor activation, the regulatory Smads are phosphorylated and form a dimer with Smad 4 which translocates to the nucleus to regulate gene transcription [123, 124]. Myocardial infarction leads to complex changes in this signal transduction pathway. Expression of TGF-β1 in isolated cardiac nonmyocytes was increased after 1 week before returning to baseline at 6 weeks [117]. In the whole rat heart, TGF-β1 expression was enhanced on day 2 after infarction and remained elevated for 28 days [125]. Smad 2, 3 and 4 proteins were significantly enhanced in border and scar tissues [126] while Smad 7 expression was decreased [127]. However, selective expression of TGF-β1 resulted in atrial but not ventricular fibrosis in transgenic mice, indicating that increased receptors or activating proteins are also necessary [128]. TGF-β1 may also increase the expression of other growth factors such as connective tissue growth factor

(CTGF) that can trigger cell proliferation, adhesion, migration and the synthesis of extracellular matrix [129]. In addition, TGF-β1 suppressed the activity or expression of NO synthase, especially the NOS2 isozyme expressed following stimulation with inflammatory compounds [130], which could also participate in increasing collagen synthesis and deposition. The increased collagen production in rat ventricular fibroblasts in culture may be due to an increased differentiation of fibroblasts to myofibroblasts that have a higher collagen production [131]. Thus, attenuation of these changes should lead to decreased collagen in the heart.

There are many studies showing that suppression or attenuation of the TGF-β1 pathway improves cardiac structure and function. Heterozygous TGF-β1 deficient mice showed decreased age-associated myocardial fibrosis and improved compliance which may have contributed to the improved survival [132]. Blockade of the actions of angiotensin II with the selective receptor antagonist, losartan, normalised Smad 2 and 4 over-expression and these changes were paralleled by modulation of the fibroproliferative events both in post-myocardial infarction rat hearts [133] and in Syrian hamsters at early and late stages of cardiomyopathy [134]. Combined blockade of angiotensin and endothelin receptors reduced TGF-β1 and collagen expression and improved ventricular function [135]. Suppression of an increased TGF-β1 expression by tranilast in hypertensive transgenic rats over-expressing human renin attenuated left ventricular hypertrophy and fibrosis without lowering blood pressure [136]. Daily administration of anti-TGF-β neutralising antibody in rats with pressure overload following aortic constriction inhibited fibroblast activation and subsequently collagen expression and myocardial fibrosis; diastolic dysfunction was reversed without affecting blood pressure, myocyte hypertrophy or systolic function [137].

Suppression of an increased TGF-β1 expression may be the mechanism for the antifibrotic actions of pirfenidone [138, 139]. Pirfenidone may also increase collagen breakdown by reducing the TGF-β1-induced inhibition of the degrading enzymes, the MMPs [140]. During chronic administration, pirfenidone consistently prevented collagen accumulation, for example in bleomycin-induced pulmonary fibrosis in hamsters [141] with attenuation of pulmonary functional deficits [142]. Pirfenidone reversed collagen deposition and reduced cardiac stiffness in streptozotocin-diabetic rats [62] and in DOCA-salt hypertensive rats [143].

In rat cardiac fibroblasts, TGF-β expression and production can be inhibited by hepatocyte growth factor (HGF) [144]. This study tested the role of HGF in cardiac fibrosis in cardiomyopathic hamsters treated with an ACE inhibitor or a selective angiotensin receptor antagonist. Angiotensin II blockade prevented myocardial fibrosis, accompanied by a significant increase in HGF, implying that local HGF expression may prevent myocardial injury [144]. In a further study, the human HGF gene was transfected into the heart of cardiomyopathic hamsters [145]. After 8 weeks, collagen density was decreased through activation of MMP-1 and inhibition of TGF-β expression; in addition, therapeutic angiogenesis was shown as an increased cardiac capillary density [145].

Cytokine activation is important in cardiovascular disease progression, in particular in heart failure [146, 147]. One of these cytokines, TNF-α, activates specific TNF-α receptors on all nucleated cells in the heart to change myocyte size and viability and up-regulate the different MMPs to induce variable proteolysis of extracellular matrix components. Since an increased TNF-α produces many responses and TNF-α concentrations are only one of many changes in heart failure, it is difficult to determine specific effects on cardiac collagen metabolism following TNF-α suppression in heart failure patients. One promising technique is the use of transgenic mice with over-expression of TNF-α that is restricted to the heart [148]. These mice showed an increase in MMP activity and a decrease in cardiac fibrillar collagen in the early stages followed by a significant decrease in MMP activity and increased collagen content as the mice aged. These changes in the ageing mice were associated with increased levels of both TIMP-1 and TGF-β. The relationship between TNF-α and MMPs has been investigated in dogs with evolving heart failure with 28-day chronic pacing given etanercept [149]. TNF-α block reduced or prevented pacing-induced changes in end-diastolic volume and MMP levels indicating that TNF-α acts by inducing specific MMPs [149]. Recent studies have shown that angiotensin II increased TNF-α expression, probably by a protein kinase C pathway, in the feline heart and in cultured cardiac myocytes [150]. This up-regulation was mediated by the AT1 receptor subtype which may explain why chronic blockade of these receptors reduced circulating TNF-α concentrations.

Despite these studies implicating TNF-α in cardiac fibrosis, there is little experimental or clinical evidence as yet that decreased TNF-α concentrations will decrease cardiac fibrosis and improve cardiac function. Etanercept, a recombinant TNF-α receptor antagonist that functionally inactivates TNF-α, improved ventricular function and remodelling in preliminary trials in heart failure patients [151]. However, reports state that the large RENEWAL clinical trial with etanercept has been prematurely stopped since interim analysis showed no likelihood of a difference between placebo and etanercept [152]. There are no reports measuring possible changes in cardiac collagen metabolism. Several drugs shown to block TNF-α expression such as prednisone or enhance mRNA degradation such as thalidomide or decrease TNF-α concentrations such as pentoxifylline have been shown to be beneficial in heart failure but their effects on cardiac collagen are unclear or unproven [147].

Brain natriuretic peptide (BNP) may be another locally-produced growth factor that acts as a myocyte-derived counter-regulatory mechanism to cardiac fibroblasts [153]. In this study, mice with targeted disruption of BNP showed marked fibrotic lesions without cardiac hypertrophy or systemic hypertension but with increased expression of ACE, TGF-β1 and pro-α1-collagen [148]. These BNP -/- mice also showed an increased fibrosis with an acute pressure overload, indicating that BNP moderates overload-induced progression of fibrosis [153].

The role of growth hormone and its major mediator, insulin-like growth factor-1 (IGF-1), have received little attention yet chronic excess of growth hormone in humans causes increased interstitial fibrosis as the major histological abnormality of the heart [154]. The major mediator may be an activated renin-angiotensin-aldosterone system. IGF-1 stimulated growth of neonatal rat cardiac fibroblasts which could be inhibited by either ACE inhibition or AT1 receptor antagonism [155]. The renin-angiotensin system is involved in the growth hormone-mediated modification of electrolyte and fluid homeostasis increasing angiotensinogen concentrations and angiotensin receptor density in the liver, kidney, and adrenals of dwarf rats supplemented with growth hormone [156], but similar effects on cardiac receptors have not been reported. Growth hormone has been reported as a treatment for heart failure in small trials, although a larger trial showed no benefit on cardiac structure or function [157]. The potential of growth hormone to worsen cardiac fibrosis should be considered in future trials.

Since cytokines are clearly involved in the progression of cardiac fibrosis and heart failure, the understanding of their multiple roles in the cardiovascular system is essential in understanding the possibilities for altering the progression of cardiovascular disease.

7. Inhibition of Inflammation, Free Radicals and Oxidative Stress

Interstitial fibrosis is accepted as a final common response to chronic inflammation, although fibrosis will become independent of the inflammatory process at some stage. Fibroblast activity and collagen deposition were closely related to the presence of lymphocytes and macrophages in the myocardium of ageing SHR [158]. In renovascular hypertensive rats, the correlation between macrophage density and plasma renin activity indicated that angiotensin II may be the initial signal which mobilised inflammatory cells [159]. Intercellular communication and intracellular signalling which confer an inflammatory phenotype to arteries have been reviewed [160]. The role of NO in inflammation remains unclear and it has been proposed that the physiological chemistry of NO may account for the differing responses [161]. Prevention of inflammation should prevent reactive cardiac fibrosis if this process is the response to inflammation.

Corticosteroids inhibit the synthesis of pro-inflammatory cytokines such as TNF-α. Chronic methylprednisolone (5 mg/kg/day for 21 days) prevented both interstitial and perivascular collagen deposition in the spared myocardium following myocardial infarction in rats [162]. In addition, baseline left ventricular function was improved by methylprednisolone treatment. These authors also showed that low-dose aspirin (25 mg/kg/day) reduced perivascular collagen deposition but this was not reflected in an improved ventricular diastolic function, possibly because interstitial collagen was unchanged [162]. This dose of aspirin had been shown previously to selectively inhibit platelet thromboxane production and lower plasma thromboxane concentrations without affecting left ventricular dysfunction in rats with myocardial infarction [163]. However, the key question remains whether the fibrotic process can be reversed by inhibition of inflammation.

Inflammation is also an important component of the atherosclerotic process. The major drugs used to reduce endogenous cholesterol biosynthesis to decrease atherosclerosis are the statins, orally active inhibitors of 3-hydroxy-3-methylglutaryl coenzyme A reductase. While many large trials have shown their beneficial effects in the prevention of coronary artery disease, recent evidence suggests that these compounds have important cholesterol-independent effects to restore endothelial function, enhance the stability of atherosclerotic plaques and decrease oxidative stress and vascular inflammation [164]. Many of the putative mechanisms may also lead to a reduction of cardiac collagen deposition, for example decreases in endothelin synthesis, reactive oxygen species, proinflammatory cytokines and MMP expression and secretion [164]. In a rat model of vascular remodelling following aortic banding, fluvastatin prevented the increased formation of superoxide anions and ICAM-1 expression in the aorta that was associated with an enhanced expression of endothelial nitric oxide expression and decreased perivascular fibrosis [165]. These antifibrotic changes probably also occur in the heart. Treatment of mice with coronary artery ligation-induced heart failure with fluvastatin (10 mg/kg/day for 4 weeks) reduced interstitial fibrosis and myocyte hypertrophy while improving left ventricular performance and survival [166]. These benefits were associated with an attenuation of the infarct-induced increase in left ventricular MMPs. Cholesterol-independent protective effects on the heart have been demonstrated in a double transgenic rat model with both the human renin and angiotensinogen genes. These rats develop severe cardiac and renal inflammatory injury as angiotensin II-induced end-organ damage and die at about 7 weeks of age if untreated [167]. Treatment with oral cerivastatin (0.5 mg/kg/day for 3 weeks) reduced mortality, blood pressure, cardiac hypertrophy, macrophage infiltration, and extracellular matrix (collagen, laminin and fibronectin) deposition [167]. Interstitial fibrosis together with cardiac hypertrophy and left ventricular dysfunction are some of the key characteristics of hypertrophic cardiomyopathy. These characteristics were recapitulated in rabbits with cardiac-restricted expression of β-myosin heavy chain-glutamine 403 (Q^{403}) [168]. In this model, treatment of adult rabbits with simvastatin (5mg/kg/day) for 12 weeks reduced collagen volume fraction and left ventricular mass and also improved left ventricular filling pressures [168]. One possible mechanism was a reduction in the activation of the predominant stress-responsive intracellular signalling kinase, ERK 1/2. These studies clearly show that the cholesterol-independent effects of the statins could be remarkably useful in improving the structure and function of the human heart in chronic cardiovascular disease by decreasing both hypertrophy and fibrosis.

Free radical-mediated cellular damage may be one possible cause of haemodynamic abnormalities leading to cardiac remodelling and dysfunction. Treatment with antioxidants may alleviate this oxidative stress and reduce cardiac damage. Myocardial infarction in rats has been shown to decrease concentrations of vitamin E in the left ventricle and liver and of vitamin A in the liver and kidney; dietary vitamin E supplementation led to an improved haemodynamic function [169]. However, vitamin E supplementation did not prevent cardiovascular events in a large trial of patients at high risk [170]. In mice with myocardial infarction following coronary artery ligation, hydroxyl radical concentrations were increased in the non-infarcted myocardium and the mice showed the symptoms of heart failure [171]. Treatment with the hydroxyl radical scavenger, dimethylthiourea (50 mg/kg/day ip for 4 weeks), attenuated the increased collagen deposition and myocardial MMP-2 activity while left ventricular function was significantly improved [171].

Probucol is an effective cholesterol-lowering compound with potent antioxidant properties. Oral probucol (61 mg/kg/day for 4 weeks) started 24 hours after coronary artery ligation in rats increased scar thickness and decreased cardiac fibrosis without altering ventricular hypertrophy or dilatation [172]. The positive effects on cardiac fibrosis may be due to decreased cardiac oxidative stress and expression of the proinflammatory cytokines, interleukin-1β and -6 [172]. Probucol (61 mg/kg/day for 80 days starting 20 days after infarction) improved left ventricular function in mice with heart failure following coronary artery ligation [173]. Cardiac fibrosis was decreased while left ventricular dilatation and wall thinning were prevented [173]. Oxidative stress may also be one factor in the development of cardiac interstitial fibrosis in renal failure as treatment of rats with subtotal nephrectomy with tocopherol (vitamin E; 2x1500 IE/kg/week for 12 weeks) attenuated but did not prevent interstitial fibrosis [174].

ACE inhibition may also be an effective antioxidant strategy by decreasing angiotensin II concentrations and increasing bradykinin and NO bioactivity [175, 176]. Oxidative stress begins in the vascular wall (endothelium, smooth muscle and fibroblasts) by enzymes that use NADH and NADPH as substrates for superoxide anion formation [175]. These enzyme systems are activated by angiotensin II and by a decrease in bradykinin levels leading to fibrosis, cell death, and necrosis [175]. Additionally superoxide reacts with NO to inactivate it, decreasing its bioavailability [176]. Overactive superoxide formation is also involved in the vascular pathology of diabetes mellitus [177]. Enalapril prevented oxidative stress in cells from streptozotocin-diabetic rats, inhibiting fibrosis and end-organ damage in the left ventricle, kidney and liver [177]. There is still much more to discover on the interrelationships between inflammation, oxidative stress and cardiac fibrosis.

8. Activation of Matrix Breakdown

The role of the MMPs, a family of at least 20 zinc-dependent enzymes responsible for myocardial matrix degradation, in the progression of cardiovascular disease is now being elucidated. The MMPs are regulated by many growth factors, cytokines and matrix fragments such as the matrikines [178, 179]. In addition, the endogenous physiological inhibitors of the MMPs, the tissue inhibitors of metalloproteinases (TIMPs), can also be regulated. Thus, the progression of the fibrotic process is determined by the interplay of MMPs, their inhibitors, and regulators, all of which may be altered in cardiovascular disease [178]. As an example, myocardial MMP-2 remained inactive during compensated left ventricular hypertrophy in Dahl salt-sensitive rats but was activated during the transition to heart failure [180]. Although TIMPs were also activated, the greater activation of MMP-2 may result in matrix breakdown and the progression of left ventricular dilatation following myocyte slippage. Thus, inhibition of MMPs would be a therapeutic target in the failing heart undergoing ventricular enlargement [181]. The actions of the nonselective MMP inhibitor, batimastat, on cardiac collagen, heart function, and survival have been measured in transgenic mice with cardiac-restricted overexpression of TNF-α [182]. In young mice, batimastat reduced collagen expression but increased insoluble collagen while myocardial hypertrophy and diastolic dysfunction were prevented and survival improved. However, no improvements were measured in old mice with established heart failure. Thus, MMP inhibition may be important in the treatment of heart failure but only early in its development [182]. Treatment for 4 months with PD166793 (5 mg/kg/day) of obese male spontaneously hypertensive heart failure (SHHF) rats attenuated the ventricular enlargement characteristic of the development of failure although collagen content was unchanged [183]. ACE inhibitors such as ramipril may also alter myocardial remodelling by MMPs in heart failure. In 16 week old SHR with heart failure following occlusion of the left coronary artery, ramipril (1 mg/kg/day for 6 weeks) reduced MMP-2 and collagen type 1 expression and increased TIMP-4 levels. These changes were associated with prevention of left ventricular dilatation, reduction of fibrosis, decreases in left ventricular end-diastolic pressure and mortality and increases in left ventricular pressure [184]. Although MMP inhibitors may be useful in the failing heart, these compounds may increase collagen deposition and ventricular stiffness in the non-failing heart [181].

One difficulty is to decide whether the aim of pharmacological modulation should be MMP activation or inhibition. Activation could be beneficial by allowing the removal of excessive collagen deposits. This is clinically attractive since it is essentially the reversal of an existing disease process, rather than prevention, but collagen removal could also cause progressive ventricular dilatation in the failing heart [179, 182]. Activation of MMPs with the serine protease, plasmin, acutely degraded collagen and decreased the elastic stiffness constant and viscosity constant in papillary muscles from hypertrophied hearts [179]. This study clearly shows that acute removal of collagen improved the function of hypertrophied myocardium.

9. Reduction of Cross-Linking

Collagen cross-linking occurs initially by the Maillard reaction of glucose with the amino groups of proteins to form a chemically reversible Schiff base adduct which rearranges to the more stable but still chemically reversible Amadori product, a ketoamine (Figure 2) [7]. AGEs are formed by further reactions of these Amadori products with amino groups on other proteins to form stable intermolecular cross-links. Several receptors which may mediate the responses of AGE have been identified. The best characterised receptor for AGE, known as RAGE, is increased in diabetes possibly to act as a scavenger and mediate intracellular signalling [7]. One possible signalling pathway leads to oxidant stress and activation of NF-κB to increase the generation of pro-inflammatory cytokines [7]. AGEs have also been shown to increase the expression of MMP-2, MMP-9 and MMP-13 in isolated rat cardiac fibroblasts which may alter cardiac remodelling [185].

Cross-linking of collagen fibres in the heart may be an important mechanism for the increased cardiac stiffness and more relevant than changes in collagen content [186]. This is supported by studies showing that therapeutic modulation of AGEs either by inhibiting AGE formation or breaking AGE cross-linkages improves cardiac function. The nucleophilic hydrazine, aminoguanidine (pimagedine), probably reacts with the reactive ketoamine Amadori product producing an unreactive product which leads to decreased cross-linking of collagens (Figure 2) [187]. Aminoguanidine has many possible mechanisms of action in the cardiovascular system, including selective inhibition of iNOS, quenching of hydroxyl radicals as well as inhibition of free radical formation, lipid peroxidation and oxidant-induced apoptosis [188].

As an example, aminoguanidine treatment of 13 month old normotensive Sprague Dawley rats for 9 months reduced blood pressure, improved glomerular filtration rate and renal plasma flow while reducing glomerular sclerosis but did not alter oxidative stress, lipid peroxidation, or immunostaining for AGEs, indicating that inhibition of iNOS was the most likely mechanism of action [189]. However, treatment of 6 month old Sprague Dawley and Fisher 344 rats for 18 months prevented the significant increases in AGE accumulation in the heart, aorta and kidney and also prevented age-linked vasodilatory impairment, indicating that interference with AGE accumulation by aminoguanidine may protect against cardiovascular and renal decline in ageing [190]. In normotensive male WAG/Rij rats, treatment with aminoguanidine for 6 months from 24 months of age prevented cardiac hypertrophy and arterial stiffening without changing collagen and elastin content [191]. In streptozotocin-diabetic rats, treatment with aminoguanidine for 4 months prevented both the increase in collagen cross-linking and the increased myocardial stiffness without changing the elevated blood glucose concentrations [192]. More potent inhibitors of the formation and accumulation of AGEs than aminoguanidine have been reported [193, 194]. However, the cardiovascular responses to chronic treatment have not yet been reported.

The breakage of established AGE cross-links is an additional potential mechanism to reverse the chronic effects of an increased collagen deposition, rather than to prevent collagen accumulation. The thiazolium derivative, ALT-711 (phenyl-4,5-dimethylthiazolium chloride), when given to aged dogs at 1 mg/kg daily for one month, reduced left ventricular stiffness by approximately 40% and increased stroke volume index since end-diastolic volume increased [195]. A possible mechanism has been described together with a review of cardiovascular studies of these compounds indicating their potential usefulness in ageing and diabetes [196]. Collagen cross-linking is an integral part of the chronic changes in cardiovascular disease so that either prevention or reversal of these processes holds promise for the improvement of cardiovascular function in chronic diabetes and hypertension.

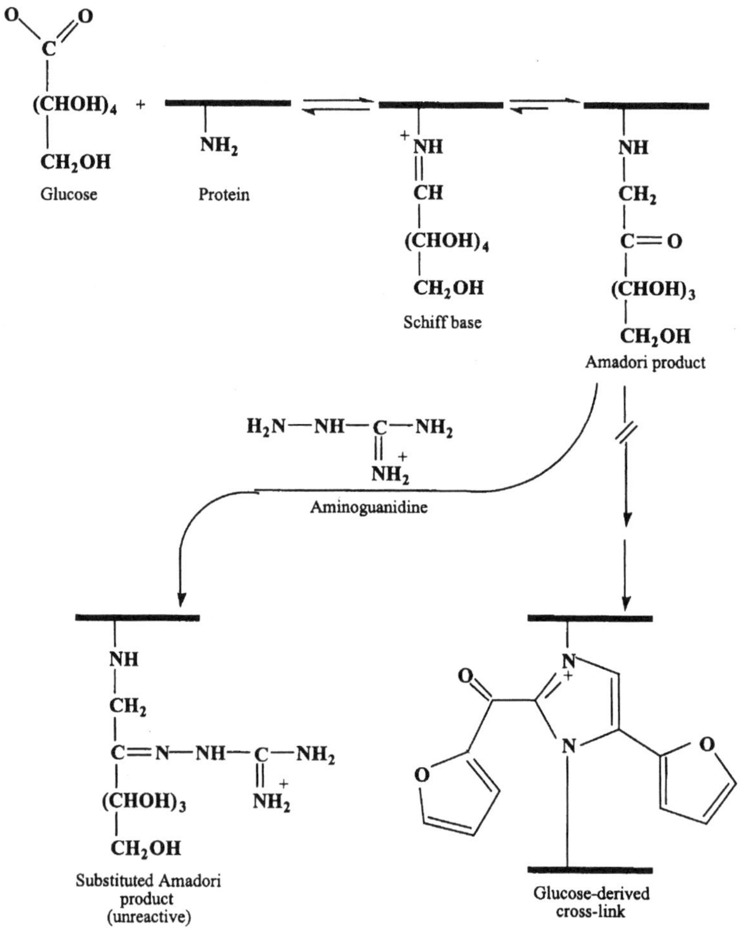

Figure 2. Mechanism of action of aminoguanidine.

10. Summary

These studies argue convincingly that the extracellular matrix, in particular collagen, is a dynamic component of the heart. Collagen deposition and removal are remarkably complex processes but this complexity provides many possible targets for pharmacological intervention. Many compounds, some in current therapeutic use especially in patients with hypertension, diabetes, and heart failure, have been shown to alter collagen content. While other compounds are unlikely to become therapeutic tools, they are allowing an investigation into possible mechanisms for the prevention or reversal of fibrosis. More importantly, these studies have shown us that controlling cardiac collagen is not simply a biochemical curiosity since many studies have now shown improvements in the functioning of the diseased heart. While most of these studies are in rodent models of human cardiovascular disease, those studies on humans are also positive. Research into cardiac collagen is still gaining momentum with almost all the studies cited in this review having been published in the last 10 years and many in the last 2-3 years. Thus, pharmacological control of collagen in the heart is likely to become a standard and successful component of the therapy of human cardiovascular disease.

References

1. Weber, K.T., et al., Collagen network of the myocardium: function, structural remodeling and regulatory mechanisms. J Mol Cell Cardiol, 1994. 26: p. 279-92.
2. Weber, K.T., Fibrosis and hypertensive heart disease. Curr Opin Cardiol, 2000. 15: p. 264-72.
3. Weber, K.T., C.G. Brilla, and J.S. Janicki, Myocardial fibrosis: functional significance and regulatory factors. Cardiovasc Res, 1993. 27: p. 341-8.
4. Weber, K.T., Cardioreparation in hypertensive heart disease. Hypertension, 2001. 38: p. 588-91.
5. Farhadian, F., et al., Fibronectin expression during physiological and pathological cardiac growth. J Mol Cell Cardiol, 1995. 27: p. 981-90.
6. Assayag, P., et al., Compensated cardiac hypertrophy: arrhythmogenicity and the new myocardial phenotype. I. Fibrosis. Cardiovasc Res, 1997. 34: p. 439-44.
7. Singh, R., et al., Advanced glycation end-products: a review. Diabetologia, 2001. 44: p. 129-46.
8. Dollery, C.M., J.R. McEwan, and A.M. Henney, Matrix metalloproteinases and cardiovascular disease. Circ Res, 1995. 77: p. 863-8.
9. Spinale, F.G., Matrix metalloproteinases: regulation and dysregulation in the failing heart. Circ Res, 2002. 90: p. 520-30.
10. Boluyt, M.O. and O.H. Bing, Matrix gene expression and decompensated heart failure: the aged SHR model. Cardiovasc Res, 2000. 46: p. 239-49.
11. Brilla, C.G., et al., Remodeling of the rat right and left ventricles in experimental hypertension. Circ Res, 1990. 67: p. 1355-64.
12. Doggrell, S.A. and L. Brown, Rat models of hypertension, cardiac hypertrophy and failure. Cardiovasc Res, 1998. 39: p. 89-105.
13. Brilla, C.G., R.C. Funck, and H. Rupp, Lisinopril-mediated regression of myocardial fibrosis in patients with hypertensive heart disease. Circulation, 2000. 102: p. 1388-93.
14. Brilla, C.G., L. Matsubara, and K.T. Weber, Advanced hypertensive heart disease in spontaneously hypertensive rats. Lisinopril-mediated regression of myocardial fibrosis. Hypertension, 1996. 28: p. 269-75.
15. Hunt, S.A., et al., ACC/AHA guidelines for the evaluation and management of chronic heart failure in the adult: executive summary. A report of the American College of

Cardiology/American Heart Association Task Force on Practice Guidelines (Committee to revise the 1995 Guidelines for the Evaluation and Management of Heart Failure). J Am Coll Cardiol, 2001. 38: p. 2101-13.

16. Remme, W.J. and K. Swedberg, Comprehensive guidelines for the diagnosis and treatment of chronic heart failure. Task force for the diagnosis and treatment of chronic heart failure of the European Society of Cardiology. Eur J Heart Fail, 2002. 4: p. 11-22.

17. Weber, K.T. and C.G. Brilla, Pathological hypertrophy and cardiac interstitium. Fibrosis and renin-angiotensin-aldosterone system. Circulation, 1991. 83: p. 1849-65.

18. Brilla, C.G., Renin-angiotensin system mediated mechanisms: cardioreparation and cardioprotection. Heart, 2000. 84: p. i18-9:discussion i50.

19. Brilla, C.G., et al., Collagen metabolism in cultured adult rat cardiac fibroblasts: response to angiotensin II and aldosterone. J Mol Cell Cardiol, 1994. 26: p. 809-20.

20. Weber, K.T. and Y. Sun, Recruitable ACE and tissue repair in the infarcted heart. J Renin Angiotensin Aldosterone Syst, 2000. 1: p. 295-303.

21. Sun, Y. and K.T. Weber, Infarct scar: a dynamic tissue. Cardiovasc Res, 2000. 46: p. 250-6.

22. Jugdutt, B.I., et al., Effect of enalapril on ventricular remodeling and function during healing after anterior myocardial infarction in the dog. Circulation, 1995. 91: p. 802-12.

23. De Carvalho Frimm, C., Y. Sun, and K.T. Weber, Angiotensin II receptor blockade and myocardial fibrosis of the infarcted rat heart. J Lab Clin Med, 1997. 129: p. 439-46.

24. Young, M.J. and J.W. Funder, The renin-angiotensin-aldosterone system in experimental mineralocorticoid-salt-induced cardiac fibrosis. Am J Physiol, 1996. 271: p. E883-8.

25. Brown, L., et al., Reversal of cardiac fibrosis in deoxycorticosterone acetate-salt hypertensive rats by inhibition of the renin-angiotensin system. J Am Soc Nephrol, 1999. 10: p. S143-8.

26. Brilla, C.G., J.S. Janicki, and K.T. Weber, Cardioreparative effects of lisinopril in rats with genetic hypertension and left ventricular hypertrophy. Circulation, 1991. 83: p. 1771-9.

27. Susic, D., J. Varagic, and E.D. Frohlich, Pharmacologic agents on cardiovascular mass, coronary dynamics and collagen in aged spontaneously hypertensive rats. J Hypertens, 1999. 17: p. 1209-15.

28. Brooks, W.W., et al., Captopril modifies gene expression in hypertrophied and failing hearts of aged spontaneously hypertensive rats. Hypertension, 1997. 30: p. 1362-8.

29. Brooks, W.W., et al., Altered inotropic responsiveness and gene expression of hypertrophied myocardium with captopril. Hypertension, 2000. 35: p. 1203-9.

30. Zimmermann, R., et al., Effect of long-term ACE inhibition on myocardial tissue in hypertensive stroke-prone rats. J Mol Cell Cardiol, 1999. 31: p. 1447-56.

31. Linz, W., et al., Late treatment with ramipril increases survival in old spontaneously hypertensive rats. Hypertension, 1999. 34: p. 291-5.

32. The CONSENSUS Trial Study Group, Effects of enalapril on mortality in severe heart failure. N Engl J Med, 1987. 316: p. 1429-1435.

33. Pfeffer, M.A., et al., Effect of captopril on mortality and morbidity in patients with left ventricular dysfunction after myocardial infarction. Results of the survival and ventricular enlargement trial. The SAVE Investigators. N Engl J Med, 1992. 327: p. 669-77.

34. Cleland, J., et al., Effect of ramipril on morbidity and mode of death among survivors of acute myocardial infarction with clinical evidence of heart failure. Eur Heart J, 1997. 18: p. 41-51.

35. Multani, M.M., et al., Long-term angiotensin-converting enzyme and angiotensin I-receptor inhibition in pacing-induced heart failure: effects on myocardial interstitial bradykinin levels. J Card Fail, 2001. 7: p. 348-54.

36.   Farhy, R.D., et al., Role of kinins and nitric oxide in the effects of angiotensin converting enzyme inhibitors on neointima formation. Circ Res, 1993. 72: p. 1202-10.

37.   Linz, W., G. Wiemer, and B.A. Scholkens, ACE-inhibition induces NO-formation in cultured bovine endothelial cells and protects isolated ischemic rat hearts. J Mol Cell Cardiol, 1992. 24: p. 909-19.

38.   Gallagher, A.M., H. Yu, and M.P. Printz, Bradykinin-induced reductions in collagen gene expression involve prostacyclin. Hypertension, 1998. 32: p. 84-8.

39.   Fujii, M., et al., Bradykinin improves left ventricular diastolic function under long-term angiotensin-converting enzyme inhibition in heart failure. Hypertension, 2002. 39: p. 952-7.

40.   Sigusch, H.H., S.E. Campbell, and K.T. Weber, Angiotensin II-induced myocardial fibrosis in rats: role of nitric oxide, prostaglandins and bradykinin. Cardiovasc Res, 1996. 31: p. 546-54.

41.   Madeddu, P., et al., Angiotensin II type 1 receptor blockade prevents cardiac remodeling in bradykinin B(2) receptor knockout mice. Hypertension, 2000. 35: p. 391-6.

42.   Yang, X.P., et al., Diminished cardioprotective response to inhibition of angiotensin-converting enzyme and angiotensin II type 1 receptor in B(2) kinin receptor gene knockout mice. Circ Res, 2001. 88: p. 1072-9.

43.   Singh, K., et al., Myocardial osteopontin expression coincides with the development of heart failure. Hypertension, 1999. 33: p. 663-70.

44.   Rothermund, L., et al., Early onset of chondroitin sulfate and osteopontin expression in angiotensin II-dependent left ventricular hypertrophy. Am J Hypertens, 2002. 15: p. 644-52.

45.   Trueblood, N.A., et al., Exaggerated left ventricular dilation and reduced collagen deposition after myocardial infarction in mice lacking osteopontin. Circ Res, 2001. 88: p. 1080-7.

46.   Kupfahl, C., et al., Angiotensin II directly increases transforming growth factor beta1 and osteopontin and indirectly affects collagen mRNA expression in the human heart. Cardiovasc Res, 2000. 46: p. 463-75.

47.   Peng, H., et al., Antifibrotic effects of N-acetyl-seryl-aspartyl-lysyl-proline on the heart and kidney in aldosterone-salt hypertensive rats. Hypertension, 2001. 37: p. 794-800.

48.   Rhaleb, N.E., et al., Effect of N-acetyl-seryl-aspartyl-lysyl-proline on DNA and collagen synthesis in rat cardiac fibroblasts. Hypertension, 2001. 37: p. 827-32.

49.   Pokharel, S., et al., N-acetyl-Ser-Asp-Lys-Pro inhibits phosphorylation of Smad2 in cardiac fibroblasts. Hypertension, 2002. 40: p. 155-61.

50.   Taigen, T., et al., Targeted inhibition of calcineurin prevents agonist-induced cardiomyocyte hypertrophy. Proc Natl Acad Sci U S A, 2000. 97: p. 1196-201.

51.   Shimoyama, M., et al., Calcineurin inhibitor attenuates the development and induces the regression of cardiac hypertrophy in rats with salt-sensitive hypertension. Circulation, 2000. 102: p. 1996-2004.

52.   Nagata, K., et al., AT1 receptor blockade reduces cardiac calcineurin activity in hypertensive rats. Hypertension, 2002. 40: p. 168-74.

53.   Shimoyama, M., et al., Calcineurin plays a critical role in pressure overload-induced cardiac hypertrophy. Circulation, 1999. 100: p. 2449-54.

54.   Takeda, Y., et al., Calcineurin inhibition attenuates mineralocorticoid-induced cardiac hypertrophy. Circulation, 2002. 105: p. 677-9.

55.   Lijnen, P. and V. Petrov, Induction of cardiac fibrosis by aldosterone. J Mol Cell Cardiol, 2000. 32: p. 865-79.

56.   Fiebeler, A., et al., Mineralocorticoid receptor affects AP-1 and nuclear factor-kappab activation in angiotensin II-induced cardiac injury. Hypertension, 2001. 37: p. 787-93.

57.   Pitt, B., et al., The effect of spironolactone on morbidity and mortality in patients with severe heart failure. Randomized Aldactone Evaluation Study Investigators. N Engl J Med, 1999. 341: p. 709-17.

58.   Zannad, F., et al., Limitation of excessive extracellular matrix turnover may contribute to survival benefit of spironolactone therapy in patients with congestive heart failure: insights from the randomized aldactone evaluation study (RALES). Rales Investigators. Circulation, 2000. 102: p. 2700-6.

59.   Modena, M.G., et al., Aldosterone inhibition limits collagen synthesis and progressive left ventricular enlargement after anterior myocardial infarction. Am Heart J, 2001. 141: p. 41-6.

60.   Satoh, M., et al., Aldosterone synthase (CYP11B2) expression and myocardial fibrosis in the failing human heart. Clin Sci, 2002. 102: p. 381-6.

61.   Robert, V., et al., Angiotensin AT1 receptor subtype as a cardiac target of aldosterone: role in aldosterone-salt-induced fibrosis. Hypertension, 1999. 33: p. 981-6.

62.   Miric, G., et al., Reversal of cardiac and renal fibrosis by pirfenidone and spironolactone in streptozotocin-diabetic rats. Br J Pharmacol, 2001. 133: p. 687-94.

63.   Pu, Q. and E.L. Schiffrin, Effect of ACE/NEP inhibition on cardiac and vascular collagen in stroke-prone spontaneously hypertensive rats. Am J Hypertens, 2001. 14: p. 1067-72.

64.   Lapointe, N., et al., Comparison of the effects of an angiotensin-converting enzyme inhibitor and a vasopeptidase inhibitor after myocardial infarction in the rat. J Am Coll Cardiol, 2002. 39: p. 1692-8.

65.   Ridker, P.M., et al., Plasma concentration of soluble intercellular adhesion molecule 1 and risks of future myocardial infarction in apparently healthy men. Lancet, 1998. 351: p. 88-92.

66.   Reape, T.J. and P.H. Groot, Chemokines and atherosclerosis. Atherosclerosis, 1999. 147: p. 213-25.

67.   Jilma, B., et al., Effects of enalapril and losartan on circulating adhesion molecules and monocyte chemotactic protein-1. Clin Sci, 2002. 103: p. 131-6.

68.   Soejima, H., et al., Angiotensin-converting enzyme inhibition reduces monocyte chemoattractant protein-1 and tissue factor levels in patients with myocardial infarction. J Am Coll Cardiol, 1999. 34: p. 983-8.

69.   Drexler, H., et al., Effect of chronic angiotensin-converting enzyme inhibition on endothelial function in patients with chronic heart failure. Am J Cardiol, 1995. 76: p. 13E-18E.

70.   Gasic, S., et al., Fosinopril decreases levels of soluble vascular cell adhesion molecule-1 in borderline hypertensive type II diabetic patients with microalbuminuria. Am J Hypertens, 1999. 12: p. 217-22.

71.   Spieker, L.E., et al., Endothelin receptor antagonists in congestive heart failure: a new therapeutic principle for the future? J Am Coll Cardiol, 2001. 37: p. 1493-505.

72.   Mulder, P., et al., Role of endogenous endothelin in chronic heart failure: effect of long-term treatment with an endothelin antagonist on survival, hemodynamics, and cardiac remodeling. Circulation, 1997. 96: p. 1976-82.

73.   Karam, H., et al., Respective role of humoral factors and blood pressure in cardiac remodeling of DOCA hypertensive rats. Cardiovasc Res, 1996. 31: p. 287-95.

74.   Ammarguellat, F., I.I. Larouche, and E.L. Schiffrin, Myocardial fibrosis in DOCA-salt hypertensive rats : Effect of Endothelin ET(A) receptor antagonism. Circulation, 2001. 103: p. 319-324.

75.   Fraccarollo, D., et al., Collagen accumulation after myocardial infarction: effects of ETA receptor blockade and implications for early remodeling. Cardiovasc Res, 2002. 54: p. 559-67.

76. Oie, E., et al., Early intervention with a potent endothelin-A/endothelin-B receptor antagonist aggravates left ventricular remodeling after myocardial infarction in rats. Basic Res Cardiol, 2002. 97: p. 239-47.

77. Vila-Petroff, M. and E. Lakatta, Nitric oxide: a multifaceted modulator of cardiac contractility. Asia Pacific Heart J, 1998. 7: p. 38-42.

78. MacCarthy, P. and A.M. Shah, The role of nitric oxide in the regulation of myocardial relaxation and diastolic function. Asia Pacific Heart J, 1998. 7: p. 29-37.

79. Kim, N.N., et al., Regulation of cardiac fibroblast extracellular matrix production by bradykinin and nitric oxide. J Mol Cell Cardiol, 1999. 31: p. 457-66.

80. Kolpakov, V., D. Gordon, and T.J. Kulik, Nitric oxide-generating compounds inhibit total protein and collagen synthesis in cultured vascular smooth muscle cells. Circ Res, 1995. 76: p. 305-9.

81. Bouchie, J.L., H. Hansen, and E.P. Feener, Natriuretic factors and nitric oxide suppress plasminogen activator inhibitor-1 expression in vascular smooth muscle cells. Role of cGMP in the regulation of the plasminogen system. Arterioscler Thromb Vasc Biol, 1998. 18: p. 1771-9.

82. Heymans, S., et al., Inhibition of plasminogen activators or matrix metalloproteinases prevents cardiac rupture but impairs therapeutic angiogenesis and causes cardiac failure. Nat Med, 1999. 5: p. 1135-42.

83. Kaikita, K., et al., Plasminogen activator inhibitor-1 deficiency prevents hypertension and vascular fibrosis in response to long-term nitric oxide synthase inhibition. Circulation, 2001. 104: p. 839-44.

84. Fernandez-Alfonso, M.S. and C. Gonzalez, Nitric oxide and the renin-angiotensin system. Is there a physiological interplay between the systems? J Hypertens, 1999. 17: p. 1355-61.

85. Di Girolamo, G., et al., The effect of enalapril on PGI(2) and NO levels in hypertensive patients. Prostaglandins Leukot Essent Fatty Acids, 2002. 66: p. 493-8.

86. Susic, D., A. Franischetti, and E.D. Frohlich, Prolonged L-arginine on cardiovascular mass and myocardial hemodynamics and collagen in aged spontaneously hypertensive rats and normal rats. Hypertension, 1999. 33: p. 451-5.

87. Hiraoka, Y., et al., Oral administration of L-arginine prevents congestive heart failure in murine viral myocarditis. J Cardiovasc Pharmacol, 2002. 40: p. 1-8.

88. Zatz, R. and C. Baylis, Chronic nitric oxide inhibition model six years on. Hypertension, 1998. 32: p. 958-64.

89. Brown, L., et al., Reversal of cardiovascular remodeling with candesartan. J Renin-angiotensin-aldosterone Sys, 2001: p. S141-S147.

90. Bernatova, I., et al., Regression of chronic L-NAME-treatment-induced left ventricular hypertrophy: effect of captopril. J Mol Cell Cardiol, 2000. 32: p. 177-85.

91. Takemoto, M., et al., Chronic angiotensin-converting enzyme inhibition and angiotensin II type 1 receptor blockade: effects on cardiovascular remodeling in rats induced by the long-term blockade of nitric oxide synthesis. Hypertension, 1997. 30: p. 1621-7.

92. Katoh, M., et al., Cardiac angiotensin II receptors are upregulated by long-term inhibition of nitric oxide synthesis in rats. Circ Res, 1998. 83: p. 743-51.

93. Tomita, H., et al., Inhibition of NO synthesis induces inflammatory changes and monocyte chemoattractant protein-1 expression in rat hearts and vessels. Arterioscler Thromb Vasc Biol, 1998. 18: p. 1456-64.

94. Katoh, M., et al., Differential effects of imidapril and candesartan cilexetil on plasminogen activator inhibitor-1 expression induced by prolonged inhibition of nitric oxide synthesis in rat hearts. J Cardiovasc Pharmacol, 2000. 35: p. 932-6.

95. Kagan, H.M., Intra- and extracellular enzymes of collagen biosynthesis as biological and chemical targets in the control of fibrosis. Acta Trop, 2000. 77: p. 147-52.

306

96.   Eleftheriades, E.G., et al., Prolyl hydroxylation regulates intracellular procollagen degradation in cultured rat cardiac fibroblasts. J Mol Cell Cardiol, 1995. 27: p. 1459-73.

97.   Ju, H., et al., Antiproliferative and antifibrotic effects of mimosine on adult cardiac fibroblasts. Biochim Biophys Acta, 1998. 1448: p. 51-60.

98.   McCaffrey, T.A., et al., Specific inhibition of eIF-5A and collagen hydroxylation by a single agent. Antiproliferative and fibrosuppressive effects on smooth muscle cells from human coronary arteries. J Clin Invest, 1995. 95: p. 446-55.

99.   Nwogu, J.I., et al., Inhibition of collagen synthesis with prolyl 4-hydroxylase inhibitor improves left ventricular function and alters the pattern of left ventricular dilatation after myocardial infarction. Circulation, 2001. 104: p. 2216-21.

100.   Dankwardt, S.M., et al., Amino acid derived sulfonamide hydroxamates as inhibitors of procollagen C-proteinase: solid-phase synthesis of ornithine analogues. Bioorg Med Chem Lett, 2001. 11: p. 2085-8.

101.   Morikawa, T., L. Tuderman, and D.J. Prockop, Inhibitors of procollagen N-protease. Synthetic peptides with sequences similar to the cleavage site in the pro alpha 1(I) chain. Biochemistry, 1980. 19: p. 2646-50.

102.   Kessler-Icekson, G., et al., Regulation of procollagenC-proteinase (PCP) and its enhancer protein (PCPE) in the remodeling myocardium. J Mol Cell Cardiol, 2002. 34: p. A33.

103.   Kato, S., et al., Inhibition of collagen cross-linking: effects on fibrillar collagen and ventricular diastolic function. Am J Physiol, 1995. 269: p. H863-8.

104.   LeGros, T., et al., The effects of 17 alpha-methyltestosterone on myocardial function in vitro. Med Sci Sports Exerc, 2000. 32: p. 897-903.

105.   Groenning, B.A., et al., Antiremodeling effects on the left ventricle during beta-blockade with metoprolol in the treatment of chronic heart failure. J Am Coll Cardiol, 2000. 36: p. 2072-80.

106.   Ohtsuka, T., et al., Effect of beta-blockers on circulating levels of inflammatory and anti-inflammatory cytokines in patients with dilated cardiomyopathy. J Am Coll Cardiol, 2001. 37: p. 412-7.

107.   Barone, F.C., et al., Carvedilol prevents severe hypertensive cardiomyopathy and remodeling. J Hypertens, 1998. 16: p. 871-84.

108.   Prabhu, S.D., et al., beta-adrenergic blockade in developing heart failure: effects on myocardial inflammatory cytokines, nitric oxide, and remodeling. Circulation, 2000. 101: p. 2103-9.

109.   Wei, S., L.T. Chow, and J.E. Sanderson, Effect of carvedilol in comparison with metoprolol on myocardial collagen postinfarction. J Am Coll Cardiol, 2000. 36: p. 276-81.

110.   Grimm, D., et al., Extracellular matrix proteins in cardiac fibroblasts derived from rat hearts with chronic pressure overload: effects of beta-receptor blockade. J Mol Cell Cardiol, 2001. 33: p. 487-501.

111.   Brilla, C.G., Regression of myocardial fibrosis in hypertensive heart disease: diverse effects of various antihypertensive drugs. Cardiovasc Res, 2000. 46: p. 324-31.

112.   Senzaki, H., et al., beta-blockade prevents sustained metalloproteinase activation and diastolic stiffening induced by angiotensin II combined with evolving cardiac dysfunction. Circ Res, 2000. 86: p. 807-15.

113.   Ruskoaho, H.J. and E.R. Savolainen, Effects of long-term verapamil treatment on blood pressure, cardiac hypertrophy and collagen metabolism in spontaneously hypertensive rats. Cardiovasc Res, 1985. 19: p. 355-62.

114.   Ramires, F.J., Y. Sun, and K.T. Weber, Myocardial fibrosis associated with aldosterone or angiotensin II administration: attenuation by calcium channel blockade. J Mol Cell Cardiol, 1998. 30: p. 475-83.

115. Sandmann, S., et al., The T-type calcium channel blocker mibefradil reduced interstitial and perivascular fibrosis and improved hemodynamic parameters in myocardial infarction-induced cardiac failure in rats. Virchows Arch, 2000. 436: p. 147-57.

116. Ono, K., et al., Cytokine gene expression after myocardial infarction in rat hearts: possible implication in left ventricular remodeling. Circulation, 1998. 98: p. 149-56.

117. Yue, P., et al., Cytokine expression increases in nonmyocytes from rats with postinfarction heart failure. Am J Physiol, 1998. 275: p. H250-8.

118. Hwang, M.W., et al., Neutralization of interleukin-1beta in the acute phase of myocardial infarction promotes the progression of left ventricular remodeling. J Am Coll Cardiol, 2001. 38: p. 1546-53.

119. Wang, F., et al., Regulation of cardiac fibroblast cellular function by leukemia inhibitory factor. J Mol Cell Cardiol, 2002. 34: p. 1309-16.

120. Brand, T. and M.D. Schneider, The TGF beta superfamily in myocardium: ligands, receptors, transduction, and function. J Mol Cell Cardiol, 1995. 27: p. 5-18.

121. Lijnen, P.J., V.V. Petrov, and R.H. Fagard, Induction of cardiac fibrosis by transforming growth factor-beta(1). Mol Genet Metab, 2000. 71: p. 418-35.

122. Seeland, U., et al., Myocardial fibrosis in transforming growth factor-beta(1) (TGF-beta(1)) transgenic mice is associated with inhibition of interstitial collagenase. Eur J Clin Invest, 2002. 32: p. 295-303.

123. Massague, J. and Y.G. Chen, Controlling TGF-beta signaling. Genes Dev, 2000. 14: p. 627-44.

124. Massague, J., How cells read TGF-beta signals. Nat Rev Mol Cell Biol, 2000. 1: p. 169-78.

125. Sun, Y. and K.T. Weber, Cardiac remodelling by fibrous tissue: role of local factors and circulating hormones. Ann Med, 1998. 30: p. 3-8.

126. Hao, J., et al., Elevation of expression of Smads 2, 3, and 4, decorin and TGF-beta in the chronic phase of myocardial infarct scar healing. J Mol Cell Cardiol, 1999. 31: p. 667-78.

127. Wang, B., et al., Decreased Smad 7 expression contributes to cardiac fibrosis in the infarcted rat heart. Am J Physiol, 2002. 282: p. H1685-96.

128. Nakajima, H., et al., Atrial but not ventricular fibrosis in mice expressing a mutant transforming growth factor-beta(1) transgene in the heart. Circ Res, 2000. 86: p. 571-9.

129. Chen, M.M., et al., CTGF expression is induced by TGF- beta in cardiac fibroblasts and cardiac myocytes: a potential role in heart fibrosis. J Mol Cell Cardiol, 2000. 32: p. 1805-19.

130. Vodovotz, Y., Control of nitric oxide production by transforming growth factor-beta1: mechanistic insights and potential relevance to human disease. Nitric Oxide, 1997. 1: p. 3-17.

131. Petrov, V.V., R.H. Fagard, and P.J. Lijnen, Stimulation of collagen production by transforming growth factor-beta1 during differentiation of cardiac fibroblasts to myofibroblasts. Hypertension, 2002. 39: p. 258-63.

132. Brooks, W.W. and C.H. Conrad, Myocardial fibrosis in transforming growth factor beta(1)heterozygous mice. J Mol Cell Cardiol, 2000. 32: p. 187-95.

133. Hao, J., et al., Interaction between angiotensin II and Smad proteins in fibroblasts in failing heart and in vitro. Am J Physiol, 2000. 279: p. H3020-30.

134. Dixon, I.M., et al., Effect of chronic AT(1) receptor blockade on cardiac Smad overexpression in hereditary cardiomyopathic hamsters. Cardiovasc Res, 2000. 46: p. 286-97.

135. Tzanidis, A., et al., Combined angiotensin and endothelin receptor blockade attenuates adverse cardiac remodeling post-myocardial infarction in the rat: possible role of transforming growth factor beta(1). J Mol Cell Cardiol, 2001. 33: p. 969-81.

136. Pinto, Y.M., et al., Reduction in left ventricular messenger RNA for transforming growth factor beta(1) attenuates left ventricular fibrosis and improves survival without lowering blood pressure in the hypertensive TGR(mRen2)27 Rat. Hypertension, 2000. 36: p. 747-54.

137. Kuwahara, F., et al., Transforming growth factor-beta function blocking prevents myocardial fibrosis and diastolic dysfunction in pressure-overloaded rats. Circulation, 2002. 106: p. 130-5.

138. Iyer, S.N., G. Gurujeyalakshmi, and S.N. Giri, Effects of pirfenidone on procollagen gene expression at the transcriptional level in bleomycin hamster model of lung fibrosis. J Pharmacol Exp Ther, 1999. 289: p. 211-8.

139. Iyer, S.N., G. Gurujeyalakshmi, and S.N. Giri, Effects of pirfenidone on transforming growth factor-beta gene expression at the transcriptional level in bleomycin hamster model of lung fibrosis. J Pharmacol Exp Ther, 1999. 291: p. 367-73.

140. Suga, H., et al., Preventive effect of pirfenidone against experimental sclerosing peritonitis in rats. Exp Toxicol Pathol, 1995. 47: p. 287-91.

141. Iyer, S.N., et al., Dietary intake of pirfenidone ameliorates bleomycin-induced lung fibrosis in hamsters. J Lab Clin Med, 1995. 125: p. 779-85.

142. Schelegle, E.S., J.K. Mansoor, and S. Giri, Pirfenidone attenuates bleomycin-induced changes in pulmonary functions in hamsters. Proc Soc Exp Biol Med, 1997. 216: p. 392-7.

143. Mirkovic, S., et al., Attenuation of cardiac fibrosis by pirfenidone and amiloride in DOCA-salt hypertensive rats. Br J Pharmacol, 2002. 135: p. 961-8.

144. Taniyama, Y., et al., Potential contribution of a novel antifibrotic factor, hepatocyte growth factor, to prevention of myocardial fibrosis by angiotensin II blockade in cardiomyopathic hamsters. Circulation, 2000. 102: p. 246-52.

145. Taniyama, Y., et al., Angiogenesis and antifibrotic action by hepatocyte growth factor in cardiomyopathy. Hypertension, 2002. 40: p. 47-53.

146. Bradham, W.S., et al., Tumor necrosis factor-alpha and myocardial remodeling in progression of heart failure: a current perspective. Cardiovasc Res, 2002. 53: p. 822-30.

147. Baumgarten, G., P. Knuefermann, and D.L. Mann, Cytokines as emerging targets in the treatment of heart failure. Trends Cardiovasc Med, 2000. 10: p. 216-23.

148. Sivasubramanian, N., et al., Left ventricular remodeling in transgenic mice with cardiac restricted overexpression of tumor necrosis factor. Circulation, 2001. 104: p. 826-31.

149. Bradham, W.S., et al., TNF-alpha and myocardial matrix metalloproteinases in heart failure: relationship to LV remodeling. Am J Physiol, 2002. 282: p. H1288-95.

150. Kalra, D., N. Sivasubramanian, and D.L. Mann, Angiotensin II induces tumor necrosis factor biosynthesis in the adult mammalian heart through a protein kinase C-dependent pathway. Circulation, 2002. 105: p. 2198-205.

151. Bozkurt, B., et al., Results of targeted anti-tumor necrosis factor therapy with etanercept (ENBREL) in patients with advanced heart failure. Circulation, 2001. 103: p. 1044-7.

152. Pugsley, M.K., Etanercept. Immunex. Curr Opin Investig Drugs, 2001. 2: p. 1725-31.

153. Ogawa, Y., et al., Brain natriuretic peptide appears to act locally as an antifibrotic factor in the heart. Can J Physiol Pharmacol, 2001. 79: p. 723-9.

154. Colao, A., et al., Growth hormone and the heart. Clin Endocrinol (Oxf), 2001. 54: p. 137-54.

155. van Eickels, M., H. Vetter, and C. Grohe, Angiotensin-converting enzyme (ACE) inhibition attenuates insulin-like growth factor-I (IGF-I) induced cardiac fibroblast proliferation. Br J Pharmacol, 2000. 131: p. 1592-6.

156. Wyse, B., M. Waters, and C. Sernia, Stimulation of the renin-angiotensin system by growth hormone in Lewis dwarf rats. Am J Physiol, 1993. 265: p. E332-9.

157. Smit, J.W., et al., Six months of recombinant human GH therapy in patients with ischemic cardiac failure does not influence left ventricular function and mass. J Clin Endocrinol Metab, 2001. 86: p. 4638-43.

158. Hinglais, N., et al., Colocalization of myocardial fibrosis and inflammatory cells in rats. Lab Invest, 1994. 70: p. 286-94.

159. Nicoletti, A., et al., Inflammatory cells and myocardial fibrosis: spatial and temporal distribution in renovascular hypertensive rats. Cardiovasc Res, 1996. 32: p. 1096-107.

160. Nicoletti, A. and J.B. Michel, Cardiac fibrosis and inflammation: interaction with hemodynamic and hormonal factors. Cardiovasc Res, 1999. 41: p. 532-43.

161. Grisham, M.B., D. Jourd'Heuil, and D.A. Wink, Nitric oxide. I. Physiological chemistry of nitric oxide and its metabolites:implications in inflammation. Am J Physiol, 1999. 276: p. G315-21.

162. Van Kerckhoven, R., et al., Altered cardiac collagen and associated changes in diastolic function of infarcted rat hearts. Cardiovasc Res, 2000. 46: p. 316-23.

163. Kalkman, E.A., et al., Chronic aspirin treatment affects collagen deposition in non-infarcted myocardium during remodeling after coronary artery ligation in the rat. J Mol Cell Cardiol, 1995. 27: p. 2483-94.

164. Takemoto, M. and J.K. Liao, Pleiotropic effects of 3-hydroxy-3-methylglutaryl coenzyme A reductase inhibitors. Arterioscler Thromb Vasc Biol, 2001. 21: p. 1712-9.

165. Katoh, M., et al., Fluvastatin inhibits O2- and ICAM-1 levels in a rat model with aortic remodeling induced by pressure overload. Am J Physiol, 2001. 281: p. H655-60.

166. Hayashidani, S., et al., Fluvastatin, a 3-hydroxy-3-methylglutaryl coenzyme a reductase inhibitor, attenuates left ventricular remodeling and failure after experimental myocardial infarction. Circulation, 2002. 105: p. 868-73.

167. Dechend, R., et al., Amelioration of angiotensin II-induced cardiac injury by a 3-hydroxy-3-methylglutaryl coenzyme A reductase inhibitor. Circulation, 2001. 104: p. 576-81.

168. Patel, R., et al., Simvastatin induces regression of cardiac hypertrophy and fibrosis and improves cardiac function in a transgenic rabbit model of human hypertrophic cardiomyopathy. Circulation, 2001. 104: p. 317-24.

169. Palace, V.P., et al., Mobilization of antioxidant vitamin pools and hemodynamic function after myocardial infarction. Circulation, 1999. 99: p. 121-6.

170. Yusuf, S., et al., Vitamin E supplementation and cardiovascular events in high-risk patients. The Heart Outcomes Prevention Evaluation Study Investigators. N Engl J Med, 2000. 342: p. 154-60.

171. Kinugawa, S., et al., Treatment with dimethylthiourea prevents left ventricular remodeling and failure after experimental myocardial infarction in mice: role of oxidative stress. Circ Res, 2000. 87: p. 392-8.

172. Sia, Y.T., et al., Improved post-myocardial infarction survival with probucol in rats: effects on left ventricular function, morphology, cardiac oxidative stress and cytokine expression. J Am Coll Cardiol, 2002. 39: p. 148-56.

173. Sia, Y.T., et al., Beneficial effects of long-term use of the antioxidant probucol in heart failure in the rat. Circulation, 2002. 105: p. 2549-55.

174. Amann, K., et al., Effect of antioxidant therapy with dl-alpha-tocopherol on cardiovascular structure in experimental renal failure. Kidney Int, 2002. 62: p. 877-84.

175. Griendling, K.K., D. Sorescu, and M. Ushio-Fukai, NAD(P)H oxidase: role in cardiovascular biology and disease. Circ Res, 2000. 86: p. 494-501.

176. Munzel, T. and J.F. Keaney, Jr., Are ACE inhibitors a "magic bullet" against oxidative stress? Circulation, 2001. 104: p. 1571-4.

177. de Cavanagh, E.M., et al., Enalapril attenuates oxidative stress in diabetic rats. Hypertension, 2001. 38: p. 1130-6.

178.   Li, Y.Y., C.F. McTiernan, and A.M. Feldman, Interplay of matrix metalloproteinases, tissue inhibitors of metalloproteinases and their regulators in cardiac matrix remodeling. Cardiovasc Res, 2000. 46: p. 214-24.

179.   Creemers, E.E., et al., Matrix metalloproteinase inhibition after myocardial infarction: a new approach to prevent heart failure? Circ Res, 2001. 89: p. 201-10.

180.   Iwanaga, Y., et al., Excessive activation of matrix metalloproteinases coincides with left ventricular remodeling during transition from hypertrophy to heart failure in hypertensive rats. J Am Coll Cardiol, 2002. 39: p. 1384-91.

181.   Spinale, F.G., et al., Myocardial matrix degradation and metalloproteinase activation in the failing heart: a potential therapeutic target. Cardiovasc Res, 2000. 46: p. 225-38.

182.   Li, Y.Y., et al., MMP inhibition modulates TNF-alpha transgenic mouse phenotype early in the development of heart failure. Am J Physiol, 2002. 282: p. H983-9.

183.   Peterson, J.T., et al., Matrix metalloproteinase inhibition attenuates left ventricular remodeling and dysfunction in a rat model of progressive heart failure. Circulation, 2001. 103: p. 2303-9.

184.   Seeland, U., et al., Effect of ramipril and furosemide treatment on interstitial remodeling in post-infarction heart failure rat hearts. J Mol Cell Cardiol, 2002. 34: p. 151-63.

185.   Daoud, S., et al., Advanced glycation endproducts: activators of cardiac remodeling in primary fibroblasts from adult rat hearts. Mol Med, 2001. 7: p. 543-51.

186.   Norton, G.R., et al., Myocardial stiffness is attributed to alterations in cross-linked collagen rather than total collagen or phenotypes in spontaneously hypertensive rats. Circulation, 1997. 96: p. 1991-8.

187.   Nilsson, B.O., Biological effects of aminoguanidine: an update. Inflamm Res, 1999. 48: p. 509-15.

188.   Giardino, I., et al., Aminoguanidine inhibits reactive oxygen species formation, lipid peroxidation, and oxidant-induced apoptosis. Diabetes, 1998. 47: p. 1114-20.

189.   Reckelhoff, J.F., et al., Chronic aminoguanidine attenuates renal dysfunction and injury in aging rats. Am J Hypertens, 1999. 12: p. 492-8.

190.   Li, Y.M., et al., Prevention of cardiovascular and renal pathology of aging by the advanced glycation inhibitor aminoguanidine. Proc Natl Acad Sci U S A, 1996. 93: p. 3902-7.

191.   Corman, B., et al., Aminoguanidine prevents age-related arterial stiffening and cardiac hypertrophy. Proc Natl Acad Sci U S A, 1998. 95: p. 1301-6.

192.   Norton, G.R., G. Candy, and A.J. Woodiwiss, Aminoguanidine prevents the decreased myocardial compliance produced by streptozotocin-induced diabetes mellitus in rats. Circulation, 1996. 93: p. 1905-12.

193.   Rahbar, S., et al., Novel inhibitors of advanced glycation endproducts. Biochem Biophys Res Commun, 1999. 262: p. 651-6.

194.   Rahbar, S., et al., Novel inhibitors of advanced glycation endproducts (part II). Mol Cell Biol Res Commun, 2000. 3: p. 360-6.

195.   Asif, M., et al., An advanced glycation endproduct cross-link breaker can reverse age-related increases in myocardial stiffness. Proc Natl Acad Sci U S A, 2000. 97: p. 2809-13.

196.   Vasan, S., P.G. Foiles, and H.W. Founds, Therapeutic potential of AGE inhibitors and breakers of AGE protein cross-links. Expert Opin Investig Drugs, 2001. 10: p. 1977-87.

Chapter 15

Biological and Functional Effects of Chronic Mechanical Support Induced by Left Ventricular Assist Devices on Failing Human Myocardium

Guillermo Torre-Amione, Cynthia K. Wallace, O.H. Frazier
Texas Heart Institute at St. Luke's Episcopal Hospital, Houston,Texas U.S.A.

1. Introduction

The use of ventricular assist devices (VADs) to provide mechanical circulatory support for end-stage heart patients has become a standard therapeutic option to bridge patients to cardiac transplantation [1-4]. Recently, a randomized clinical trial, REMATCH, demonstrated that mechanical support was superior to medical therapy in patients ineligible for transplant [5]. Partially based on the results of this trial, the Heartmate LVAS XVE system was approved in the United States as a destination therapy for end-stage heart failure patients. In addition to these indications, there is a growing interest in the use of mechanical support as a strategy to unload the heart with the hope of achieving enough myocardial recovery to allow removal of the device with at least partial resolution of the heart failure state. While VADs are primarily used to support left-sided circulation (LVADs), they can also be used to support the failing right ventricle (RVADs) or both ventricles (BIVADs).

Currently approved LVADs are pulsatile devices that receive blood from the apex of the left ventricle and mechanically pump it into the aorta, bypassing the aortic valve (Figure 1). Newer pumps under investigation employ axial rotational systems that provide a continuous flow resulting in a pulseless state when they are used to fully support the circulation. When used for partial support, axial flow pumps create a state of partial unloading leading to maintenance of left ventricular function, flow through the aortic valve, and, thus, pulsatility. For both systems, the pumps and cannulae are fully implantable, but they must be connected via a percutaneous cable to an external console for programming and power. Ongoing research into newer generation LVADs promise advanced features, such as size, portability, and improved power supply.

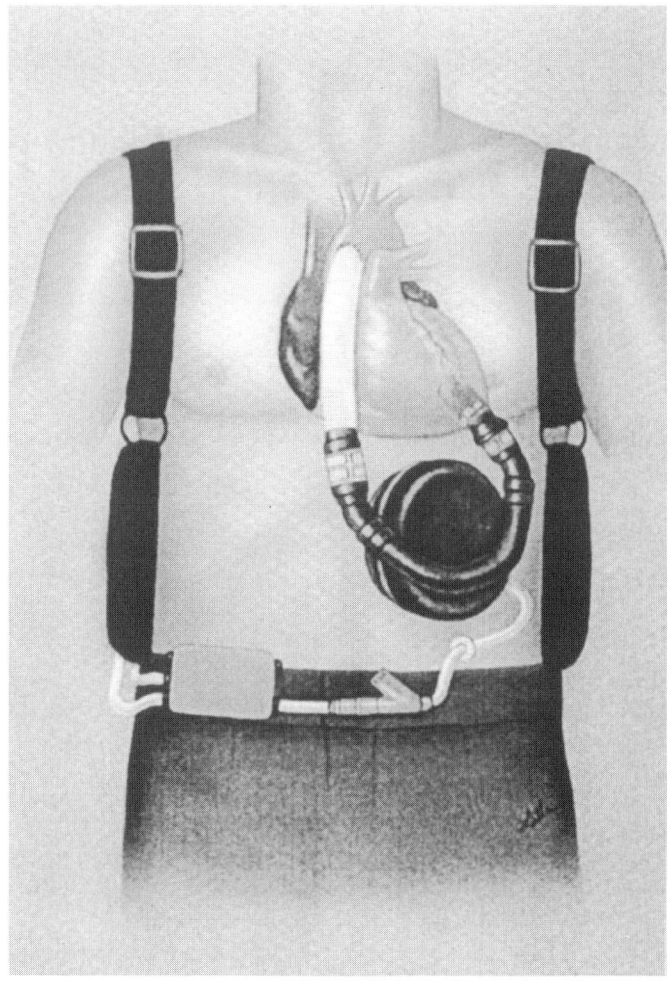

Figure 1. Illustration of a left ventricular assist device (LVAD).

The widely accepted indications for LVAD placement are based on the ability of these devices to maintain sufficient cardiac output in the face of fulminate heart failure. This results in increased survival to transplant for the most critically ill transplant candidates [3, 4] and improved outcomes in patients with end-stage heart failure who are not transplant candidates [5]. The use of LVADs to treat patients with refractory heart failure has permitted the study of human failing myocardium at various time points. Failing myocardial tissue can be obtained when the device is implanted and when the heart and the device are removed for cardiac transplantation. The ability to obtain paired human myocardium at these two times has facilitated the detailed analysis of the effect of chronic mechanical unloading on the expression of various genes and proteins that typify the failing phenotype. In this chapter, we will discuss the effects of LVAD support on failing myocardium at the cellular and functional level.

2. End-Stage Heart Failure, Hypertrophy, and Fibrosis

End-stage cardiac failure can be initiated by various insults. The most common insults are myocardial infarction and hypertension, but other causes include viral infection, exposure to toxins, chemotherapeutic agents, or muscle abnormalities triggered by genetic defects. Regardless of the etiology, however, end-stage myocardial dysfunction is characterized by cardiac hypertrophy, interstitial fibrosis, and decreased contractile function. This is secondary to a loss of myocytes and muscle function that requires the surviving muscle to compensate for the injured area. Local and systemic compensatory responses continue over years until the portion of the muscle not initially affected is eventually damaged from the persistent overactivity of the compensatory responses.

In response to the "heart failure state," there is activation of the inflammatory system that results in the production and release of proinflammatory cytokines, activation of the complement system, production of autoantibodies, and over expression of major histocompatibility complex (MHC II) molecules, as well as adhesion molecules that may perpetuate the inflammatory state. Thus, the inflammatory response to failing myocardium may be crucially important to the establishment of fibrosis and hypertrophy. Myocytes become enlarged and tissue damage ensues, thereby perpetuating the inflammatory state and resulting in scarring of the normal myocardium.

3. LVADs and Repair of the Failing Heart

In the setting of chronic end-stage heart failure, LVAD implantation was initially viewed as a vehicle by which to bridge severely decompensated heart failure patients to transplantation. However, experience with LVAD implantation has demonstrated that some end-stage heart failure patients supported by LVADs recover cardiac function sufficiently to be weaned from the devices [1, 6-17]. The mechanisms through which this recovery occurs is an area of extensive research, both with an eye toward identifying the factors that may predict which patients can be successfully weaned as well as to the potential of developing new treatment strategies for chronic heart failure.

The underlying hypothesis that unites much of the research into LVAD-mediated cardiac recovery is that hemodynamic unloading of the failing heart allows reversal of the compensatory and stress responses of the overloaded myocardium, resulting in structural and functional remodeling of the tissue. These issues have been studied at various levels in an attempt to parse out the critical determinants of both myocardial injury and recovery.

4. Mechanisms of Reverse Remodeling

4.1. Decreases in Myocyte Size and Myocardial Collagen Content

Hypertrophy and fibrosis are characteristic findings in the failing myocardium and result both in systolic and diastolic dysfunction. Several studies have evaluated the impact of mechanical unloading on myocyte size as well as well as on myocardial collagen content.

Increased myocyte size is associated with decreased cardiac performance [18, 19]. Myocyte size has been shown to be increased in the failing heart requiring LVAD support as compared both to normal controls [20, 21] and to heart failure patients treated medically [21]. Direct analysis of myocyte cellular diameters in paired pre- and post-LVAD myocardial samples indicates that myocyte size is significantly reduced in the unloaded ventricle after prolonged mechanical support [20-25] and becomes statistically indistinguishable from the myocyte size of normal controls (Figure 2, top panel) [21]. The magnitude of the reduction significantly correlated with length of LVAD support in one study population [20]. While cardiomyocyte hypertrophy occurs in the failing left and right ventricles, the reduction in myocyte size is more prominent in the left ventricle when an LVAD is used to support the circulation. These data demonstrate that the magnitude of unloading, greater on the left side, has a direct impact on the cellular response [21].

The impact of ventricular unloading on myocardial collagen content has also been examined. The importance of defining the effect of LVAD support on collagen content stems from the fact that increases in collagen content are associated with abnormal LV stiffness and that regression of collagen content is associated with improvements in myocardial function [26, 27]. Collagen I is the major collagen component that establishes the myocyte-collagen matrix relationship. Accordingly, changes in collagen I content may specifically influence LV size and stiffness [28]. Collagen III contributes to elasticity [29] but also plays a minor role in the myocyte-collagen interaction.

Immunohistochemistry techniques have been used to evaluate total collagen, collagen I, and collagen III levels in paired myocardial samples before and after LVAD support. Collagen content, as measured by total collagen, collagen I, or collagen III levels, was increased in the failing myocardium before

LVAD implantation as compared to normal controls. At the time of LVAD removal, all three measured collagen levels were statistically indistinguishable from the normal controls (Figure 2, middle panel). No association was found between length of support and magnitude of change in collagen levels [20]. Differential gene expression analysis using oligonucleotide microarrays has supported the finding that collagen production is down-regulated following LVAD support when compared to pre-implant gene expression (Table 1) [30].

Observations regarding changes in myocyte size and collagen content after ventricular unloading demonstrate that regression in hypertrophy with LVAD support is due to a decrease both in cellular size and interstitial fibrosis. Since mechanical stretch is one of the major signals for activation of genes responsible for the heart failure phenotype [31], it follows that reduction in mechanical stretch by unloading may lead to a deactivation of these genes.

4.2. Reversal of the Pro-inflammatory State

At the cellular level, the failing myocardium is known to exist in a pro-inflammatory state. Studies looking at differential gene expression pre- and post-LVAD support have found a reduction in the expression of important pro-inflammatory cytokines after periods of mechanical support. One such cytokine, tumor necrosis factor-α (TNF-α), is an inflammatory cytokine that is produced by the failing myocardium but that is not seen in the normal heart [32]. When present in the myocardium, TNF-α stimulates cardiac growth and produces cardiac enlargement, heart failure, and death in experimental animals [33].

Intracardiac TNF-α expression has been shown to be significantly increased in the failing myocardium of individuals requiring LVAD support as compared to normal myocardium [34-36] and may be higher than in heart failure patients not requiring LVAD support [30]. Importantly, serial comparisons of pre- and post-implant myocardial TNF-α levels have shown TNF-α expression significantly decreased with chronic ventricular unloading [30, 34, 35, 37] (Figure 2, bottom panel; Table 1). Interestingly, in a small study in which cardiac TNF-α concentrations were determined in a group of patients who had an LVAD removed due to functional improvement and in another group of patients who underwent cardiac transplantation, it was found that the reduction in intracardiac TNF-α was greater among those patients who had the device explanted because they showed signs of myocardial recovery [35]. Observations from that study were twofold: first, it demonstrated that the response of the failing myocardium to chronic mechanical unloading was diverse and second, although reductions in cardiac TNF-α may not be responsible for myocardial recovery, the findings are consistent with the hypothesis that approximation of a more "normal myocardial state" can lead to cardiac recovery.

316

Table 1. Gene Expression Before and After Mechanical Unloading

Gene Product	Function	Pre-LVAD levels[1]	Post-LVAD levels[2]	Technique	Patient Population	Ref
TNFα	Pro-inflammatory cytokine	Elevated	Normalized	Immunostaining	Non-IsCM	35
		Not compared	↓1.6-fold	Affymetrix Genechip	Both	30
		↑29.8-fold[3]	↓33.6%[3]	RT-PCR	Both	37
IL-8	Pro-inflammatory cytokine	Not compared	↓9.3-fold	Affymetrix Genechip	Both	38
N–terminal Dystrophin	Cell-ECM interactions	Decreased[4]	Increased[5]	Immunostaining/ Western blot	Both	43
gp130	RTK	↑3-fold[6]	Decreased, but not normalized	RT-PCR	Both	37
Her2/neu	RTK	No difference	Increased[6]	RT-PCR	Both	37
Her4	RTK	↑3-fold	↑2.5-fold[6]	RT-PCR	Both	37
Total Collagen	Structural/fibrosis	Elevated	Normalized	Immunostaining	Both	20
		Not compared	↓3-fold	Affymetrix genechip	Both	30
Collagen I	Structural/fibrosis	Elevated	Normalized	Immunostaining	Both	20
Collagen III	Structural/fibrosis	Elevated	Normalized	Immunostaining	Both	20

Non-IsCM=non-ischemic cardiomyopathy
Both=ischemic cardiomyopathy & non-ischemic cardiomyopathy
[1]Compared to normal control
[2]Compared to pre-LVAD sample
[3]Average reduction among samples with reduced expression (24/36 samples)
[4]In 5/6 patients
[5]In 4/5 patients with reduced expression pre-LVAD
[6]In TNF-α responders

Circulating levels of the pro-inflammatory cytokines interleukin-6 and interleukin-8 have been compared before and after LVAD implantation in patients with acute circulatory collapse; levels of these cytokines were significantly diminished after LVAD support [38]. Additionally, IL-8 levels in the left ventricular wall of end-stage cardiomyopathy patients were compared pre- and post- implant and found to be significantly reduced [30] (Table 1).

Reversal of the pro-inflammatory state in the myocardium may play a critical role in normalization of myocardial structure and recovery of cardiac function by interrupting the cycle of inflammation, necrosis, and fibrosis.

Pre LVAD **Post LVAD**

Size

TC

TNF

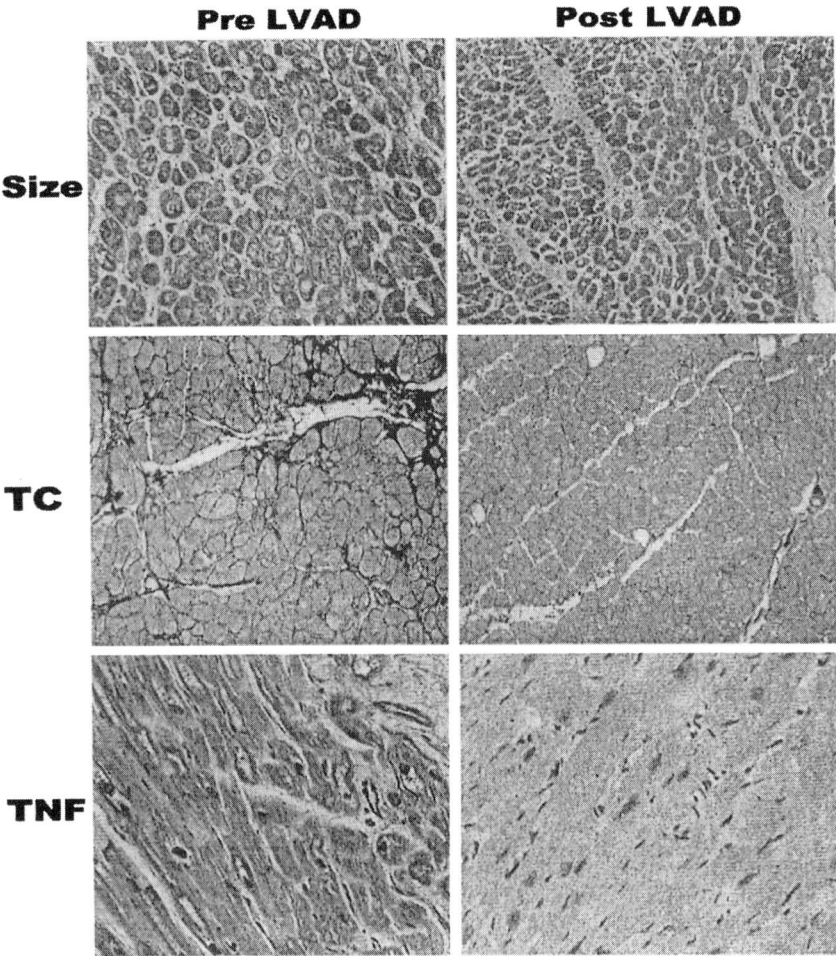

Figure 2. Top panel, Myocyte size pre (left) and post-LVAD (right). Middle panel, total collagen (TC) staining using Masson's trichrome. Lower panel, TNF-α staining pre (left) and post-LVAD implant (right).

4.3. Normalization of Structural Proteins: the Effect on Dystrophin

Dystrophin is a structural protein thought to provide support for the myocyte and the cardiomyocyte membrane by linking actin at the N-terminus with the dystrophin-associated protein complex and sarcolemma at the C-terminus. Defects in the gene for the structural protein dystrophin have been identified as responsible for X-linked dilated cardiomyopathy [39-42] and are generally manifested as mutations in the N-terminus of dystrophin. Because of the importance of dystrophin in the contractile apparatus of the myocardial cell and the functional consequences of its abnormal expression, the integrity of dystrophin expression in acquired forms of cardiomyopathies was investigated.

N-terminus dystrophin expression was evaluated in paired pre- and post-LVAD myocardial samples from a group of patients with non-X-linked dilated cardiomyopathy, using immunohistochemistry and Western blot [43]. Expression of the normal N-terminus dystrophin domain was reduced in most of the patients with dilated cardiomyopathy before LVAD implant as compared to a normal control, while immunostaining and Western blot analysis of the mid-rod and C-terminus domains revealed no discernible differences. After LVAD support, the majority of the patients with reduced N-terminus expression before implant exhibited an increase in expression (Table 1). These findings suggest that the mechanical stress of myocyte hypertrophy might disrupt normal N-terminus dystrophin activity and that this disruption is potentially reversible through unloading, when not a manifestation of an underlying genetic disorder.

4.4. Myocyte Survival, Growth, and Apoptotic Signaling

The transmission of hypertrophic and survival signals in the myocardial cell is mediated in part through receptor tyrosine kinases (RTK). In animal models, the cardiac survival factor cardiotrophin-1 was shown to increase neonatal cardiac myocyte survival and proliferation through interactions with the RTK, gp-130 [44]. The interactions of neuregulin growth factors with the RTKs, Her2/neu, and Her4 (ErbB2 and ErbB4) have been shown to promote survival and inhibit apoptosis in rat cardiac myocytes *in vitro* [45].

Expression of these RTKs has been retrospectively evaluated in paired pre- and post-LVAD myocardial samples through the quantification of mRNA transcripts in the tissue samples using reverse-transcriptase polymerase chain reaction (RT-PCR) techniques [37]. Subgroup analysis of those paired samples that exhibited suppression of TNF-α or atrial natriuretic factor (ANF) after unloading showed a 2.5 to 3-fold increase in gp130 expression before LVAD implant when compared to normal controls. After unloading, a significant decrease in gp130 expression occurred in the subgroup that also exhibited suppression of TNF-α (Table 1).

Her2/neu expression in LVAD recipients before implant was found to be indistinguishable from normal controls. The same subgroup analysis of TNF-α and ANF responders showed an upregulation of her2/neu expression after unloading (Table 1). Her4 expression was significantly higher in pre-LVAD myocardial samples as compared to normal controls and was up-regulated further after unloading in the TNF-α and ANF responders (Table 1).

Interference with the her2/neu signaling pathway has been hypothesized to contribute to heart failure in breast cancer patients treated with Herceptin, an inhibitory antibody that targets her2/neu [46]. It is therefore not surprising to find this receptor down-regulated in the failing heart before LVAD implantation. The discordant changes in the expression of gp130, and her2/neu and her4, may point to complex interactions between these RTKs in the coordination of myocyte survival and growth and in the facilitation of cardiac recovery after mechanical unloading.

Apoptosis is an inconsistent finding in heart failure [47]; however, this is thought to be due to interruption midway through the complex apoptotic pathway, with apoptosis itself prevented but functional damage occurring due to upstream steps [48]. There is growing evidence that unloading tends to normalize apoptotic pathways in the failing heart, perhaps contributing to reverse remodeling.

The effect of mechanical unloading of the failing heart on several regulators of apoptosis has been studied. Evaluation of transcript and protein expression of the anti-apoptotic proteins Bcl-x_L and FasExo6Del/Fas in pre- and post-LVAD myocardial samples from patients with dilated cardiomyopathy has been evaluated. Expression of both proteins was reduced before LVAD implant and exhibited a time-dependent increase during LVAD support [49]. Interestingly, a related anti-apoptotic protein, Bcl-2, was found to be overexpressed in the failing heart, with unloading resulting in decreased (normalized) expression [47]. Additionally, phosphorylation-mediated regulation of mitogen-activated protein kinases (MAPK), which affect both myocyte hypertrophy and apoptosis, was investigated. Unloading was shown to induce a differential regulation of MAPK, resulting in reduced myocyte apoptosis and decreased myocyte size [24].

4.5. Calcium Homeostasis

Myocardial contraction relies on precise regulation of calcium homeostasis. The sarcoplasmic reticular Ca^{2+} ATPase (SERCA2a) plays a vital role in this homeostasis by mediating the re-uptake of Ca^{2+} into the sarcoplasmic reticulum (SR) after contraction, thereby replenishing SR stores for the next contraction and enabling myocyte relaxation. Reduced SR Ca^{2+} release due to diminished SR Ca^{2+} stores is directly related to diminished SERCA2a activity in failing cardiac myocytes [50]. Decreased SERCA2a activity in human failing myocytes contributes to elevated diastolic calcium concentrations and decreased contractility at higher rates of stimulation [51]. This is coupled with an increased Na^+-Ca^{2+} exchanger expression in failing myocytes [49] that may serve to further increase cytoplasmic Ca^{2+} concentrations and worsen myocyte contractility. In the hypertrophied left ventricle, abnormal calcium handling translates into slow and dyssynchronous left ventricular isovolumic relaxation after left ventricular unloading and diastolic left ventricular aftercontractions [52].

Initial investigations of paired myocardial samples from 7 LVAD-supported failing hearts taken at time of LVAD implant and cardiac transplant indicated that SERCA2a protein levels were significantly increased after ventricular unloading [53]. The effect of mechanical unloading on abnormal calcium homeostasis was further evaluated in paired pre- and post-LVAD samples from 20 failing hearts from patients bridged to transplant. Northern blot analysis showed increased SERCA2a mRNA after unloading, with increased protein concentrations confirmed by Western blot. This increase in SERCA2a translated functionally into an increase in Ca^{2+} uptake by the SR of the LVAD-supported hearts [54].

Improvement in the impaired calcium handling of the failing heart is a necessary precondition for functional improvement of ventricular contractility. LVAD-mediated unloading appears to alter gene expression and protein concentrations of critical components of calcium homeostasis, thereby paving the way for myocardial recovery with regression of the failing phenotype.

5. Functional Improvements with Mechanical Unloading

Based on the above discussion, it is clear that chronic mechanical unloading improves the molecular and cellular alterations that characterize the failing myocardium. The impact of these changes on myocardial function has not been fully defined; however, significant improvements in anatomic, hemodynamic, and electrophysiological properties are known to occur.

5.1. Anatomic and Hemodynamic Improvements

The failing heart is characterized by increased mass and dilation, with resultant increases in both end diastolic volume (EDV) and end systolic volume (ESV) and a decrease in left ventricular ejection fraction (LVEF). Dilated cardiomyopathy specifically results in a rightward shift of the pressure volume curve due both to the increased end diastolic volume (EDV) and to dysfunctionally increased compliance of the myocardium [55]. Several studies have investigated the impact of mechanical unloading on the gross structural properties of the failing heart.

The physical characteristics of the trabeculae in LVAD supported hearts have been compared to those of failing hearts managed medically [21]. Although the sample sizes were small and the difference not significant, the LVAD-supported hearts on average exhibited reduced left ventricular trabeculae length, diameter, and mass as compared to the medically managed hearts. Despite not reaching the level of statistical significance, these observations are consistent with findings that show gross reductions in left ventricular mass, diameter, and volumes after periods of mechanical unloading.

In a study that compared LV size in post-LVAD, normal, and medically managed hearts by assessing ventricular volume necessary to reach 30 mm Hg (V_{30}), the LVAD supported hearts had decreased V_{30} when compared to the medically managed hearts. In addition, hearts supported for more than 30 days had a lower V_{30} than hearts supported for less than 30 days [21].

Retrospective assessment of echocardiographs from patients bridged to transplant with LVAD support has also provided early evidence of functional improvement in supported hearts [16]. After a period of LVAD support, echocardiographs showed significant decreases in LV end diastolic diameter (6.81 to 5.39) with corresponding increases in ejection fraction (11% to 22%).

More recently, echocardiographic studies of hearts supported by pulsatile-type LVADs were done pre-implant and at the time of explant to further investigate the effect of unloading on structural characteristics of the left ventricle. After maximal LVAD support, there was a significant decrease in LV mass, LV end-diastolic dimension, LV end-diastolic volume, LV end-systolic volume, left atrial volume, and LV ejection fraction by echocardiograph (Table 2) [56]. Gross improvement in these hearts is observable on the chest radiographs (Figure 3).

5.2. Electrophysiologic Improvements

Electrophysiologic disturbances are important manifestations of the heart failure state, resulting in part from abnormalities in myocyte action potential shape and duration [57]. The prolongation of the action potential seen in heart failure may be a compensatory response that initially results in increased Ca^{2+} flux with a positive inotropic result, but later becomes maladaptive due to a decreased Ca^{2+} response [58]. The question of whether the cellular and structural changes observed with LVAD-unloading impact the electrophysiologic responses of the heart has not been extensively investigated. Comparisons of the electrophysiologic characteristics of myocytes from the explanted hearts of transplant patients who had had prior LVAD support with myocytes from unsupported transplant patients indicated improved function in the LVAD-supported hearts [59]. The magnitude of myocyte contraction was significantly higher in the LVAD-supported hearts while the time to maximal contraction and the action potential duration at 50% repolarization (APD_{50}) were decreased, indicating superior electrophysiologic function

322

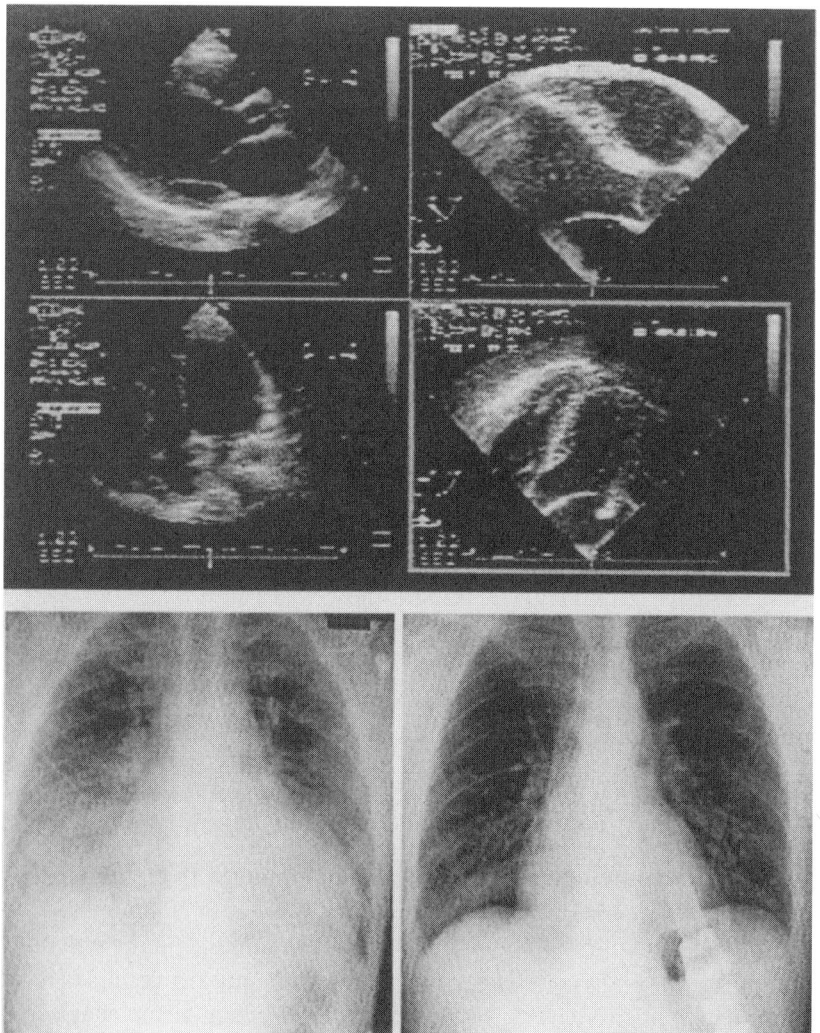

Figure 3. Cardiac images of pre- (left) and post- implantation (right) of left ventricular assist device. Top panel: echocardiograph; Bottom panel: X-ray.

More recently, a retrospective analysis of LVAD-supported patients was undertaken that compared paired ECG findings before (pre), immediately after (post), and at least one week after (delayed) LVAD implantation [60]. The delayed ECGs exhibited significantly slower heart rates (107 vs. 91 bpm) than the pre-implant tracings, consistent with the expected increase in cardiac output with LVAD support. There was also a decrease in QT intervals corrected for heart rate from the pre-implant ECG to the delayed ECG (479±10 vs. 445±9 ms).

Table 2. Echocardiographic structural changes

	Pre-LVAD	Post-LVAD	%Change	p-value
LV mass (gms)	231.3	183.5	-20.7	0.001
LVEDD (cm)	6.3	4.2	-33.3	0.000
LVEDV (ml)	223.2	131.2	-41.2	0.000
LVESV (ml)	175.1	74.3	-57.6	0.000
LAV (ml)	121.2	7.22	-40.4	0.000
LVEF (%)	20.9	42.5	+103.3	0.000
RVFAC (%)	22.5	30.0	+33.3	0.007

LV = left ventricle
LVEDD = left ventricular end-diastolic dimension
LVEDV = left ventricular end diastolic volume
LVESV = left ventricular end systolic volume
LAV = left atrial volume
LVEF = left ventricular ejection fraction
RVFAC = right ventricular fractional area of contraction

Table 3. LVAD Removal After Cardiac Recovery

No of Patients	Length of Follow-up	Outcomes			Patient Population	Ref
		Trp-free	Trp	Death		
28	1 mo - 5.5 yrs	16	8	4	Non-IsCM	61
22	1.2 - 10 yrs	17	3	2	Non-IsCM	6
5	2 - 35 mos	4	0	1	Both	62
4	Not reported	4	0	0	Not reported	13
4	Not reported	4	0	0	Not reported	63
3	Not reported	3	0	0	Not reported	64
2	1 yr, 3 yrs	2	0	0	Non-IsCM	65
2	Not reported	2	0	0	Not reported	66
2	Not reported	0	2	0	Not reported	67
1	1 yr	1	0	0	Non-IsCM	68
1	Not reported	1	0	0	Not reported	69
1	Not reported	1	0	0	Not reported	70

Non-IsCM=non-ischemic cardiomyopathy
Both=ischemic cardiomyopathy and non-ischemic cardiomyopathy
Trp=transplant

Interestingly, the immediate impact of unloading was to significantly increase the QT interval (479 ± 10 vs. 504 ± 11 ms). A comparison of myocytes from LVAD-supported hearts with myocytes from unsupported hearts showed that the APD_{50} was markedly lower in the previously supported cells (863 ± 37 vs. 529 ± 154 ms).

Importantly, the findings from this analysis indicated that the immediate and delayed effects of unloading were distinct in both magnitude and direction, lending credence to the idea that long-term ventricular unloading induces cellular and structural changes in the myocardium that have a direct and beneficial effect on cardiac function.

6. Device Removal for Myocardial Recovery/Improvement

Perhaps the best evidence for improvement of failing myocardium in patients with end-stage heart failure necessitating mechanical support to prevent death is the ability to remove the device and maintain circulation without the need of further mechanical or intravenous inotropic support. The largest series of patients confirming this phenomenon was reported by the German Heart Institute and included 95 patients with non-ischemic, idiopathic dilated cardiomyopathy who received an LVAD while awaiting heart transplantation. These patients were followed up and treated aggressively in an attempt to remove the device if recovery or myocardial improvement was observed. Of the original 95 patients, 28 recovered cardiac function sufficiently to be weaned from their devices, and 16 of these individuals maintained "normal" heart function for a significant period of time (follow-up, 1 month to 5.5 years) [61]. The total number of reported cases of patients with chronic refractory heart failure who improved function sufficiently to be weaned from an LVAD is small—approximately 75 (Table 3). However, these findings are important because they show that improvement in myocardial function is possible in some patients. Therefore, our future goals should, perhaps, be aimed at determining signals needed to enhance recovery or designing better devices to permit optimal physiological recovery.

7. Summary

The use of mechanical support for patients with refractory, end-stage heart failure will continue to increase. The indications may also expand as the technology improves and we are able to avoid long-term mechanical failures and reduce the complications associated with the implantation of these devices. However, what has become clear is the failing heart, which appears to be "end-stage," may have the remarkable ability to recover at least at the cellular and structural levels. The translation of these changes into functional myocardial improvement has been demonstrated by the experience of a small number of investigators who have proven that device explantation is possible in some patients with advanced refractory non-ischemic cardiomyopathy. Others and we are actively pursuing new strategies to modify and enhance the probability of myocardial recovery or improvement following LVAD support.

References

1. Loebe, M., et al., Long-term mechanical circulatory support as a bridge to transplantation, for recovery from cardiomyopathy, and for permanent replacement. Eur J Cardiothorac Surg, 1997. 11: p. S18-24.
2. Oz, M.C., et al., Bridge experience with long-term implantable left ventricular assist devices. Are they an alternative to transplantation? Circulation, 1997. 95: p. 1844-52.
3. Arabia, F.A., et al., Success rates of long-term circulatory assist devices used currently for bridge to heart transplantation. ASAIO J, 1996. 42: p. M542-6.
4. Frazier, O.H., et al., Improved survival after extended bridge to cardiac transplantation. Ann Thorac Surg, 1994. 57: p. 1416-22.
5. Rose, E.A., et al., The REMATCH trial: rationale, design, and end points. Randomized Evaluation of Mechanical Assistance for the Treatment of Congestive Heart Failure. Ann Thorac Surg, 1999. 67: p. 723-30.
6. Farrar, D.J., et al., Long-term follow-up of Thoratec ventricular assist device bridge-to-recovery patients successfully removed from support after recovery of ventricular function. J Heart Lung Transplant, 2002. 21: p. 516-21.
7. Kumpati, G.S., P.M. McCarthy, and K.J. Hoercher, Left ventricular assist device as a bridge to recovery: present status. J Card Surg, 2001. 16: p. 294-301.
8. Hetzer, R., et al., Midterm follow-up of patients who underwent removal of a left ventricular assist device after cardiac recovery from end-stage dilated cardiomyopathy. J Thorac Cardiovasc Surg, 2000. 120: p. 843-53.
9. DeRose, J.J., Jr., et al., Implantable left ventricular assist devices provide an excellent outpatient bridge to transplantation and recovery. J Am Coll Cardiol, 1997. 30: p. 1773-7.
10. Koul, B., et al., HeartMate left ventricular assist device as bridge to heart transplantation. Ann Thorac Surg, 1998. 65: p. 1625-30.
11. Pietsch, L., et al., Recovery from end-stage ischemic cardiomyopathy during long-term LVAD support. Ann Thorac Surg, 1998. 66: p. 555-7.
12. Mancini, D.M., et al., Low incidence of myocardial recovery after left ventricular assist device implantation in patients with chronic heart failure. Circulation, 1998. 98: p. 2383-9.
13. Sun, B.C., et al., 100 long-term implantable left ventricular assist devices: the Columbia Presbyterian interim experience. Ann Thorac Surg, 1999. 68: p. 688-94.
14. Muller, J., et al., Weaning from mechanical cardiac support in patients with idiopathic dilated cardiomyopathy. Circulation, 1997. 96: p. 542-9.
15. Levin, H.R., et al., Transient normalization of systolic and diastolic function after support with a left ventricular assist device in a patient with dilated cardiomyopathy. J Heart Lung Transplant, 1996. 15: p. 840-2.
16. Frazier, O.H., et al., Improved left ventricular function after chronic left ventricular unloading. Ann Thorac Surg, 1996. 62: p. 675-81.
17. Delgado, R., 3rd, et al., Neurohormonal changes after implantation of a left ventricular

assist system. ASAIO J, 1998. 44: p. 299-302.

18. Marian, A.J., Pathogenesis of diverse clinical and pathological phenotypes in hypertrophic cardiomyopathy. Lancet, 2000. 355: p. 58-60.

19. Zile, M.R., et al., Role of microtubules in the contractile dysfunction of hypertrophied myocardium. J Am Coll Cardiol, 1999. 33: p. 250-60.

20. Bruckner, B.A., et al., Regression of fibrosis and hypertrophy in failing myocardium following mechanical circulatory support. J Heart Lung Transplant, 2001. 20: p. 457-64.

21. Barbone, A., et al., Comparison of right and left ventricular responses to left ventricular assist device support in patients with severe heart failure: a primary role of mechanical unloading underlying reverse remodeling. Circulation, 2001. 104: p. 670-5.

22. Baba, H.A., et al., Reversal of metallothionein expression is different throughout the human myocardium after prolonged left-ventricular mechanical support. J Heart Lung Transplant, 2000. 19: p. 668-74.

23. Khan, T., et al., Dobutamine stress echocardiography predicts myocardial improvement in patients supported by left ventricular assist devices (LVADs): hemodynamic and histologic evidence of improvement before LVAD explantation. J Heart Lung Transplant, 2003. 22: p. 137-146.

24. Flesch, M., et al., Differential regulation of mitogen-activated protein kinases in the failing human heart in response to mechanical unloading. Circulation, 2001. 104: p. 2273-6.

25. Grabellus, F., et al., Reversible activation of nuclear factor-kappaB in human end-stage heart failure after left ventricular mechanical support. Cardiovasc Res, 2002. 53: p. 124-30.

26. Krayenbuehl, H.P., et al., Left ventricular myocardial structure in aortic valve disease before, intermediate, and late after aortic valve replacement. Circulation, 1989. 79: p. 744-55.

27. Brilla, C.G., B. Maisch, and K.T. Weber, Renin-angiotensin system and myocardial collagen matrix remodeling in hypertensive heart disease: in vivo and in vitro studies on collagen matrix regulation. Clin Investig, 1993. 71: p. S35-41.

28. Zafeiridis, A., et al., Regression of cellular hypertrophy after left ventricular assist device support. Circulation, 1998. 98: p. 656-62.

29. Bishop, J.E., et al., Enhanced deposition of predominantly type I collagen in myocardial disease. J Mol Cell Cardiol, 1990. 22: p. 1157-65.

30. Blaxall, B.C., et al., Differential gene expression and genomic patient stratification following left ventricular assist device support. J Am Coll Cardiol, 2003. 41: p. 1096-106.

31. Schneider, M.D., R. Roberts, and T.G. Parker, Modulation of cardiac genes by mechanical stress. The oncogene signalling hypothesis. Mol Biol Med, 1991. 8: p. 167-83.

32. Torre-Amione, G., et al., Tumor necrosis factor-alpha and tumor necrosis factor receptors in the failing human heart. Circulation, 1996. 93: p. 704-11.

33. Feldman, A.M., et al., Selective changes in cardiac gene expression during compensated hypertrophy and the transition to cardiac decompensation in rats with chronic aortic banding. Circ Res, 1993. 73: p. 184-92.

34. McCarthy, P.M., et al., Structural and left ventricular histologic changes after implantable LVAD insertion. Ann Thorac Surg, 1995. 59: p. 609-13.

35. Torre-Amione, G., et al., Decreased expression of tumor necrosis factor-alpha in failing human myocardium after mechanical circulatory support : A potential mechanism for cardiac recovery. Circulation, 1999. 100: p. 1189-93.

36. Nagueh, S.F., et al., Decreased expression of tumor necrosis factor-alpha and regression of hypertrophy after nonsurgical septal reduction therapy for patients with hypertrophic

obstructive cardiomyopathy. Circulation, 2001. 103: p. 1844-50.

37. Uray, I.P., et al., Left ventricular unloading alters receptor tyrosine kinase expression in the failing human heart. J Heart Lung Transplant, 2002. 21: p. 771-82.

38. Goldstein, D.J., et al., Circulatory resuscitation with left ventricular assist device support reduces interleukins 6 and 8 levels. Ann Thorac Surg, 1997. 63: p. 971-4.

39. Towbin, J.A., et al., X-linked dilated cardiomyopathy. Molecular genetic evidence of linkage to the Duchenne muscular dystrophy (dystrophin) gene at the Xp21 locus. Circulation, 1993. 87: p. 1854-65.

40. Muntoni, F., et al., Brief report: deletion of the dystrophin muscle-promoter region associated with X-linked dilated cardiomyopathy. N Engl J Med, 1993. 329: p. 921-5.

41. Milasin, J., et al., A point mutation in the 5' splice site of the dystrophin gene first intron responsible for X-linked dilated cardiomyopathy. Hum Mol Genet, 1996. 5: p. 73-9.

42. Ortiz-Lopez, R., et al., Evidence for a dystrophin missense mutation as a cause of X-linked dilated cardiomyopathy. Circulation, 1997. 95: p. 2434-40.

43. Vatta, M., et al., Molecular remodelling of dystrophin in patients with end-stage cardiomyopathies and reversal in patients on assistance-device therapy. Lancet, 2002. 359: p. 936-41.

44. Sheng, Z., et al., Cardiotrophin-1 displays early expression in the murine heart tube and promotes cardiac myocyte survival. Development, 1996. 122: p. 419-28.

45. Zhao, Y.Y., et al., Neuregulins promote survival and growth of cardiac myocytes. Persistence of ErbB2 and ErbB4 expression in neonatal and adult ventricular myocytes. J Biol Chem, 1998. 273: p. 10261-9.

46. Ewer, M.S., et al., Cardiotoxicity in patients receiving transtuzumab (Herceptin): primary toxicity, synergistic or sequential stress, or surveillance artifact? Semin Oncol, 1999. 26: p. S96-101.

47. Francis, G.S., et al., Apoptosis, Bcl-2, and proliferating cell nuclear antigen in the failing human heart: observations made after implantation of left ventricular assist device. J Card Fail, 1999. 5: p. 308-15.

48. Haider, N., N. Narula, and J. Narula, Apoptosis in heart failure represents programmed cell survival, not death, of cardiomyocytes and likelihood of reverse remodeling. J Card Fail, 2002. 8: p. S512-7.

49. Bartling, B., et al., Myocardial gene expression of regulators of myocyte apoptosis and myocyte calcium homeostasis during hemodynamic unloading by ventricular assist devices in patients with end-stage heart failure. Circulation, 1999. 100: p. SII216-23.

50. Pieske, B., L.S. Maier, and S. Schmidt-Schweda, Sarcoplasmic reticulum Ca2+ load in human heart failure. Basic Res Cardiol, 2002. 97: p. SI63-71.

51. Del Monte, F., et al., Defects in calcium control. J Card Fail, 2002. 8: p. S421-31.

52. Paulus, W.J., M.A. Goethals, and S.U. Sys, Failure of myocardial inactivation: a clinical assessment in the hypertrophied heart. Basic Res Cardiol, 1992. 87: p. 145-61.

53. Takeishi, Y., et al., Alterations in Ca2+ cycling proteins and G alpha q signaling after left ventricular assist device support in failing human hearts. Cardiovasc Res, 2000. 45: p. 883-8.

54. Heerdt, P.M., et al., Chronic unloading by left ventricular assist device reverses contractile dysfunction and alters gene expression in end-stage heart failure. Circulation, 2000. 102: p. 2713-9.

55. Barbone, A., et al., Normalized diastolic properties after left ventricular assist result from reverse remodeling of chamber geometry. Circulation, 2001. 104: p. SI229-32.

56. Thohan, V., et al., Cellular, structural and hemodynamic responses of failing myocardium to continuous flow mechanical circulatory support: comparative analysis with pulsatile-type devices. In Press.

57. Tomaselli, G.F. and E. Marban, Electrophysiological remodeling in hypertrophy and

heart failure. Cardiovasc Res, 1999. 42: p. 270-83.

58.   Wickenden, A.D., et al., The role of action potential prolongation and altered intracellular calcium handling in the pathogenesis of heart failure. Cardiovasc Res, 1998. 37: p. 312-23.

59.   Dipla, K., et al., Myocyte recovery after mechanical circulatory support in humans with end-stage heart failure. Circulation, 1998. 97: p. 2316-22.

60.   Harding, J.D., et al., Electrophysiological alterations after mechanical circulatory support in patients with advanced cardiac failure. Circulation, 2001. 104: p. 1241-7.

61.   Hetzer, R., et al., Bridging-to-recovery. Ann Thorac Surg, 2001. 71: p. S109-13.

Chapter 16

Molecular Strategies for the Prevention of Cardiac Fibrosis

Ramareddy V. Guntaka, Karl T. Weber
University of Tennessee Health Science Center, Memphis, Tennessee. U.S.A.

1. Introduction

Pathophysiologic mechanisms underlying chronic cardiac failure, a major health problem, have received considerable interest. An adverse accumulation of extracellular matrix proteins in the heart's interstitial space, fibrillar type I and III collagens in particular and defined as cardiac fibrosis, is now recognized as a major determinant of ventricular dysfunction [1]. Normally the turnover of these structural proteins is gradual and carried out by interstitial fibroblasts. Accordingly, the half-life of collagen is 80–120 days [2]. In the diseased heart, phenotypically transformed fibroblast-like cells, termed myofibroblasts, are responsible for synthesizing and maintaining collagen turnover (reviewed in 3). A variety of injuries and insults, including myocardial infarction and activation of the circulating renin-angiotensin-aldosterone system (RAAS), induce deleterious structural remodeling within the matrix (reviewed in 4). This includes not only tissue repair found at a site of injury, but within the interstitial space remote to it, a reactive fibrosis, where the progressive accumulation of heavily cross-linked type I fibrillar collagen with the tensile strength of steel serves to adversely raise the stiffness of myocardial tissue leading to ventricular dysfunction and ultimately symptomatic heart failure (reviewed in 5). The prevention of cardiac fibrosis and thereby ventricular dysfunction is an important management strategy and focus of ongoing research.

Given the role of circulating hormones in regulating myofibroblast collagen turnover (vis-à-vis hemodynamic factors), promising results with angiotensin converting enzyme (ACE) inhibitors, antagonists to angiotensin II, aldosterone and endothelin-1 receptors, and inhibitors of transforming growth factor-β_1 (TGF-β_1) induced signaling pathways have each been reported as effective in attenuating fibrous tissue formation in diverse forms of experimental injury [3-8]. These promising findings notwithstanding, there is the need for additional strategies (e.g., drugs or other antidotes) that can prevent this reactive cardiac fibrosis. In this brief report, we review the prevention of cardiac fibrosis associated with various pharmacologic and nonpharmacologic interventions that interfere with collagen turnover by targeting ligand-receptor interactions, signaling pathways and transcriptional activation of various genes.

2. ACE Inhibitors

Several structurally heterogenous inhibitors of ACE have been developed and are in clinical use for the management of heart failure and hypertension [6, 9]. These inhibitors, which may differ in their ability to enter and bind cells, act by binding to the active site of ACE [10, 11]. Various well-known clinical trials have established the efficacy and safety of ACE inhibitors in reducing morbidity and mortality in asymptomatic and symptomatic patients with left ventricular dysfunction [12-16]. Experimental studies in diseased organs have demonstrated a role for circulating angiotensin II and that formed *de novo* at sites of repair in regulating the expression of a fibrogenic cytokine, TGF-β_1, and its influence on fibrous tissue formation [4, 17, 18]. Accordingly, ACE inhibitor treatment of rodents with diverse forms of injury involving various tissues has proven effective in attenuating subsequent organ fibrosis (reviewed in 21).

3. Antagonism of Receptor-Ligand Binding

As shown in Figure 1, there are several potential targets in the pathway leading to myofibroblast production of fibrillar collagen that leads to cardiac fibrosis. Endocrine properties of circulating angiotensin II, aldosterone and endothelin-1 induce signaling pathways by binding to their cognate receptors located in either the cell membrane or cytosol and lead to an overexpression of matrix genes, especially type I and III collagens. In addition, auto/paracrine properties of these substances generated *de novo* at sites of injury contribute to myofibroblast collagen synthesis (reviewed in 3, 22). Whether derived from the circulation or locally within injured tissue, these stimulators of tissue repair regulate one another to create a redundant, failsafe system of repair (see Figure 2). They are normally opposed by inhibitors of repair that include bradykinin, nitric oxide and prostaglandins. It is this reciprocal regulation that maintains a steady state in collagen turnover. When there exists a predominance of stimulators, either by an absolute increase in their formation or in relative proportion to reduced inhibitors, collagen formation ensues. This topic has been reviewed elsewhere [19].

Drugs that antagonize ligand-receptor interaction are effective in preventing fibrosis in injured heart and kidneys [3-6, 20]. The relative importance of this mode of action in the overall clinical efficacy of these agents remains uncertain. However,

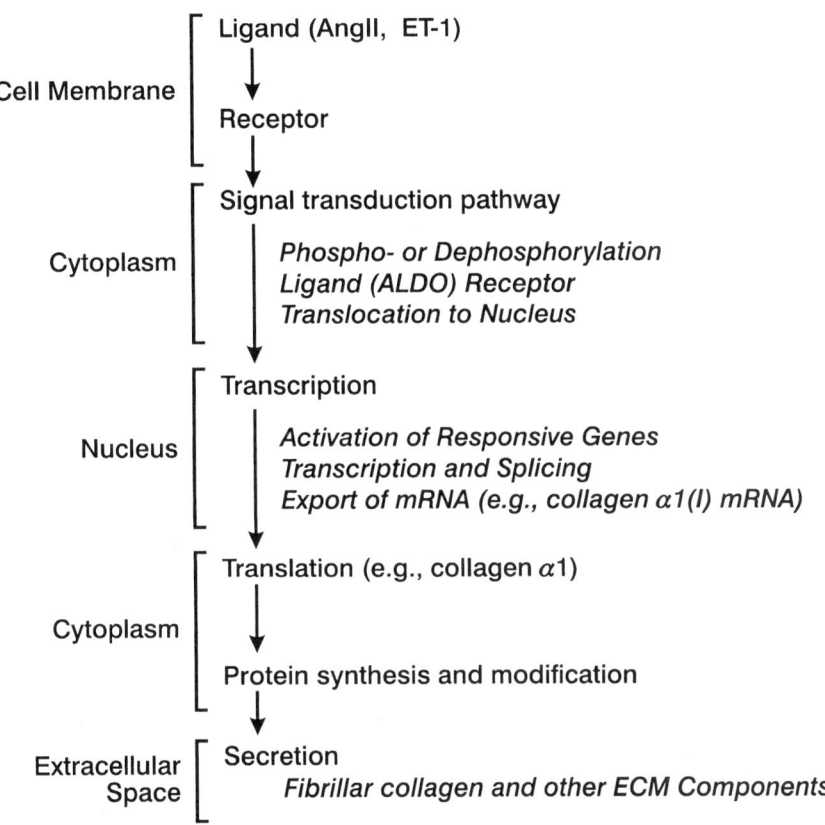

Figure 1. Diagram illustrating potential targets for therapeutic molecules in the signaling pathways present in different compartments of a cell.

clinical trials are useful in highlighting the potential for receptor antagonists to modify the course of developing tissue fibrosis. Results recently reported from a clinical trial (RALES), conducted in 19 countries on 5 continents in over 1660 patients, the use of an aldosterone receptor antagonist (spironolactone) in combination with angiotensin converting enzyme inhibitor and a loop diuretic was found to significantly reduce risk of all-cause and cardiac mortality and morbidity [21]. In a substudy to this trial, where serologic markers of collagen synthesis were monitored, a marked reduction in their plasma levels in the spironolactone treatment group was associated with survival benefit not evident in patients randomized to placebo [22].

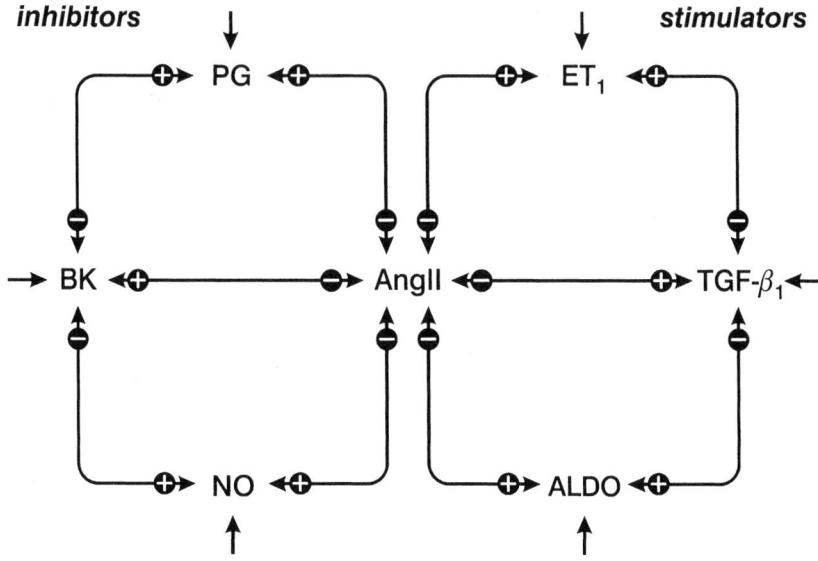

Figure 2. A theoretical paradigm representing several of the stimulators and inhibitors of fibrous tissue formation. Not shown are catecholamines, as stimulators, nor natriuretic peptides, as inhibitors. AngII, angiotensin II; ALDO, aldosterone; ET-1, endothelin; BK, bradykinin; NO, nitric oxide; PG, prostaglandins. Reproduced with permission from Weber KT.

4. Potential Inhibitors of Signaling Pathways

The binding of angiotensin II, aldosterone or endothelin-1 to their cognate receptors elicits a response based on a stimulation of intracellular and nuclear pathways that then eventuate in the transcription of fibrillar collagens (see Figure 1). There is, therefore, a vast potential to develop therapeutic molecules that target specific enzymatic reactions in signaling pathways. In this connection, aldosterone and angiotensin II each stimulate endothelin-1 gene transcription. In rats treated with aldosterone (by implanted mini-pump), the accompanying reactive fibrosis that appears throughout both atria and both ventricles can be prevented by either spironolactone or an ET-1 receptor antagonist [23]. The ET-1 gene promoter contains transcription factor AP-1 responsive elements. Activation of the nuclear hormone receptors, peroxisome-proliferator-activated receptor α, by fenofibrate suppresses ET-1 production to inhibit activation of the AP-1 transcription factor [24]. The ET-1-induced collagen synthesis could be inhibited by PD98059, a compound that specifically inhibits mitogen-activated protein kinase (MAP kinase), p44/p42 [25]. Another ET-1 receptor antagonist, LU135253, has been shown to inhibit ET-1 collagen accumulation and enhance expression of matrix metalloproteinases MMP-13 and MMP-2 at the site of myocardial infarction [26].

As mentioned above, TGF-β_1 is a potent profibrotic factor because it activates several genes involved in matrix buildup. It acts through two types of receptors (types I and II), which mediate their response directly through intracellular Smad proteins and indirectly through the p38 MAP kinase [27]. Smad proteins contribute to the progressive cardiac fibrosis seen in the infarcted rat heart [28]. Two inhibitors that effectively inhibit matrix components have been developed [29]. One of them, SB431542, specifically inhibits phosphorylation of Smad3 by ALK5 (activin receptor-like kinase). This phosphorylation inhibits TGF-β_1-induced nuclear localization of Smad3. The p38 mitogen-activated protein kinase inhibitor (SB-242235) selectively inhibits TGF-β_1-induced collagen $\alpha 1$(I) transcription [29], whereas both inhibitors blocked TGF-β_1-induced fibronectin mRNA synthesis. Further evidence indicates that biochemical blockade of p38, but not other MAP kinases, abolished TGF-β_1-induced $\alpha 2$(I) collagen mRNA expression [30]. These examples demonstrate the feasibility of developing antifibrotic agents by targeting several different steps involved in signaling pathways.

5. Type I Collagen Gene Promoter as Target for Controlling Fibrosis

Several different transcription factors (trans-acting factors) bind to distinct cis-acting motifs in the promoter of eukaryotic genes. At least 6 factors have been shown to interact with the promoter sequence (-1 to -300, relative to the transcription start site) of the $\alpha 1$(I) collagen gene. These include NF-1, SP-1, AP-1, c-Krox-1, YB-1, and CBF (reviewed by [18]). Each is a potential target by itself as these factors are required for efficient transcription of the $\alpha 1$(I) gene. For example, as described above, AP-1 is required for both ET-1 synthesis and type I collagen synthesis and AP-1 is activated by several external stimuli, such as thrombin and aldosterone. Therefore, induction of AP-1 likely results in fibrosis. An alkaloid drug, halofuginone (RU-19110), which was originally isolated from the plant *Dichroa febrifuga* and used as an antifungal agent, or a coccidiostat for chickens and turkeys, has been shown to inhibit fibrosis by blocking $\alpha 1$(I) gene transcription, in various tissues and in several different animal models of fibrosis. It has been reported to be effective against scleroderma, lung fibrosis and restenosis (reviewed in [31]). The mechanism of inhibition of collagen gene transcription is not known. Recently it has been shown that at least in fibroblasts, halofuginone appears to inhibit $\alpha 2$(I) collagen promoter probably by blocking TGF-β-induced phosphorylation and subsequent activation of Smad3 [31]. Although it has potential to become an antifibrotic agent, so far, to our knowledge, it has not been approved and released into clinical practice.

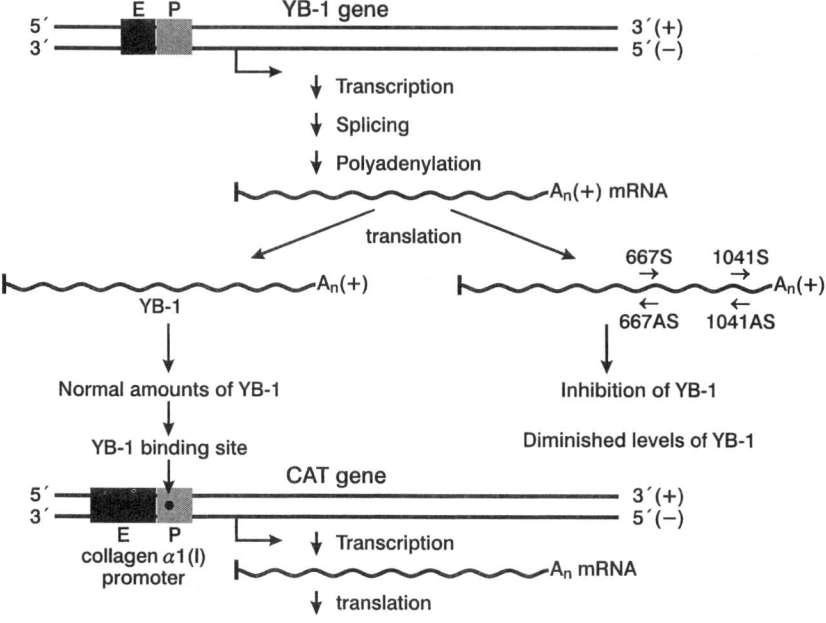

Figure 3. Inhibition of rat α1(I) collagen promoter-driven transcription of AS ODNs. In the upper portion, we have outlined the antisense strategy to specifically inhibit YB-1 protein synthesis. In the lower portion, a schematic illustration to monitor inhibition of collagen promoter, which drives expression of the reporter gene (CAT), is shown. Note that YB-1 binds to the promoter and activates transcription. When the AS ODNs inhibit YB-1 synthesis, levels of YB-1 are diminished, resulting in reduced promoter activity.

Our own work indicates that YB-1, a multifunctional protein, is a potential target for controlling fibrosis. It inhibits cell growth as well as collagen synthesis. Depending on its intracellular concentrations, YB-1 regulates type I collagen synthesis both positively [32] and negatively [33]. At normal levels YB-1 positively regulates α1(I) transcription by binding to a polypyrimidine sequence present at −140 to −170 of the promoter, as evidenced by using antisense oligonucleotides as well as by using retroviral vectors expressing antisense RNA specific for YB-1 [32]. However, at higher concentrations, it represses transcription by binding to the sequence from −83 to −59 [33].

Gene expression can be modulated at the level of translation (Figures 3 and 4) by using antisense oligodeoxyribonucleotides (AS ODNs) [34, 35] or at the level of transcription (Figure 5) by antigene triplex-forming oligonucleotides (TFOs) [36-38]. At least one AS ODN was approved by the FDA for the treatment of cytomegalovirus-induced retinitis. More than 70 to 80 AS ODNs, targeting various

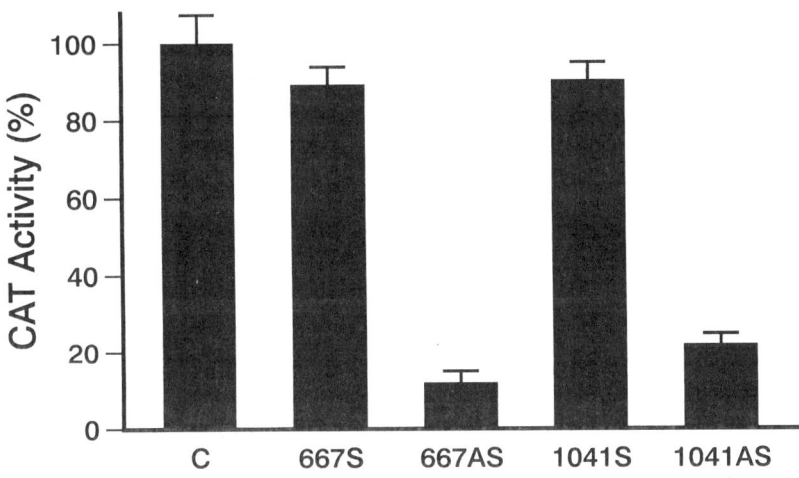

Figure 4. Inhibition of CAT activity by AS ODNs. Note that both AS ODNs (667AS and 1041AS, the location of these sites are shown in Figure 3) drastically inhibit CAT activity indicating that YB-1 positively activates the collagen promoter.

genes, are in different stages of clinical trials. In this approach, an AS ODN specific for a gene of interest is delivered into the cells where it hybridizes to mRNA corresponding to the gene and prevents translation of mRNA (Figure 3). As a result, protein levels are reduced because the hybridized AS ODN interferes with translation of the mRNA. In some instances it has been shown that cellular nucleases destroy the mRNA by specifically cleaving the RNA at the site of hybrid formation. Regardless of the mechanism the net result is inhibition of protein synthesis corresponding to the gene that has been targeted [35]. One example of this approach is the use of an AS ODN to specifically down-regulate the c-MYC oncogene, which has been shown to play a crucial role in the proliferation of vascular cells in atherosclerotic lesions and plaque fibrous tissue formation. Two clinical trials using a 15-mer AS ODN against c-MYC has been reported [39, 40]. Results indicate that although this AS ODN is well tolerated by patients it is not effective in blocking restenosis following vascular injury. In the future, AS ODNs specific for other regions of the c-MYC gene or a targeting of other genes, such as YB-1, need to be developed.

In our laboratory we have used a similar approach to inhibit the Y-box binding protein, which has been shown to play a major role in cell cycle and activation of several genes including type I collagen gene [32, 41]. Since fibrosis is due to abnormal accumulation of type I and type III collagens in damaged tissue, it should be possible to block its expression by targeting collagen genes. We have found that when YB-1 synthesis is inhibited by AS ODNs targeted to YB-1 (see

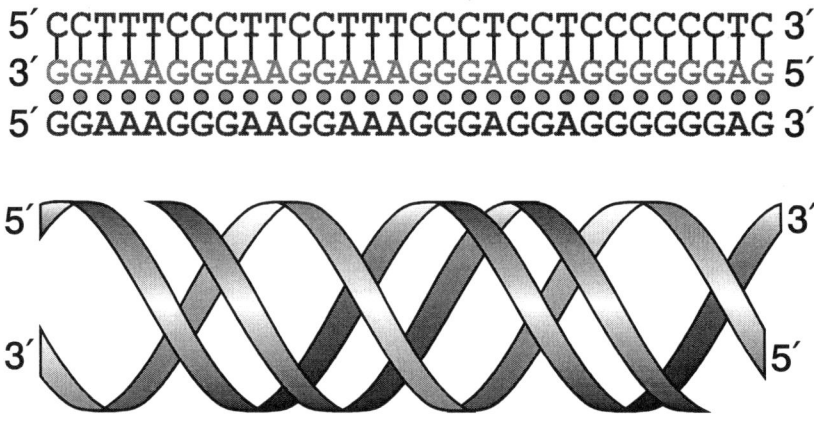

Figure 5. Triplex formation by collagen promoter-specific TFO. The target duplex DNA sequence present at −170 to −140 in the promoter of α1(I) gene is shown at the top. The TFO strand's polarity is not indicated in the helix. The polypurine TFO strand is shown red. Note that upon triplex formation, factors that normally bind to the duplex DNA can no longer bind to the triplex and hence inhibition of transcription.

Figure 3), the levels of α1(I) collagen promoter-mediated reporter gene product (chloramphenicol acetyl transferase, CAT) are greatly reduced (Figure 4). These results were further confirmed by cloning a fragment of the YB-1 cDNA in a retrovirus vector in such a way that the vector makes YB-1-specific antisense RNA. When we infected rat fibroblasts with this retrovirus expressing antisense YB-1 RNA, we have observed significant reduction in the levels of both α1(I) and α2(I) mRNAs suggesting that YB-1 exerts its effect on both promoters [32]. These results provide direct evidence that YB-1 activates collagen gene transcription.

Although fibrosis is largely due to abnormal deposition of type I collagen, synthesized by myofibroblasts, inhibition of cell growth or depletion of myofibroblasts by apoptosis should, in principle, control fibrosis. Recently we have demonstrated unequivocally that disruption of one allele of the YB-1 gene in a pre-B cell line, DT-40, caused multiple defects in cell cycle by blocking some steps in the G2/M phase of cell cycle [41]. We have also found that within 3 days after myocardial infarction, we could detect increased levels of YB-1 in infarcted compared to noninfarcted tissue. Collagen also begins to accumulate by day 3 following infarction [3]. Thus, antisense oligonucleotides to YB-1 offer a potential approach, as this gene function is required for both cardiac myofibroblast growth and collagen synthesis. A recent report provides evidence that an AS ODN, targeted to a cyclin kinase inhibitor gene, p21Waf1/Cip1, implicated in cell cycle progression, markedly reduced the production of matrix proteins in vascular smooth muscle cells suggesting that inhibition of growth of cells responsible for collagen synthesis could also block fibrosis [42].

Triplex-forming oligonucleotides (TFOs) have gained recent prominence because of their potential applications in antigene therapy (reviewed by 43). The $\alpha 1$(I) collagen gene promoter contains a unique polypyrimidine/polypurine tract of about 30 base pairs, located between -170 to -141 upstream of the transcription start site (Figure 5). We have used this strategy to down-regulate collagen gene-specific transcription by targeting $\alpha 1$(I) collagen promoter [43-45]. Although we have not yet used the TFO to prevent cardiac fibrosis, our results in preventing liver fibrosis in rats are encouraging. For example, in experiments of hepatic injury, induced by dimethylnitrosamine (DMN), the collagen gene-specific TFO, when administered simultaneously along with DMN, prevented hepatic fibrosis and preserved normal liver function. This effect of the TFO was observed even when administered 2 weeks after the initiation of DMN treatment. Since oligonucleotides are rapidly distributed into various tissues, including the heart, upon intravenous or intraperitoneal administration, we believe the collagen gene-specific TFO will be effective in controlling cardiac fibrosis.

6. Collagen-Modifying Enzymes as Targets for Controlling Fibrosis

Following transcription of collagen genes, corresponding mRNAs are translated into proteins. Type I collagen is a heterotrimer, made of two $\alpha 1$(I) and one $\alpha 2$(I) polypeptide chains that form a triple helix. Post-translational modification by hydroxylation of proline residues is absolutely necessary for the structural integrity and stability of collagen and is carried out by an enzyme, prolyl 4-hydroxylase. This enzyme can be a target to control fibrosis as the unmodified collagen is unstable and is likely to be rapidly degraded in the endoplasmic reticulum. Several inhibitors have been synthesized and tested for their activity on collagen synthesis [46]. Although several compounds progressed to clinical trials, they were subsequently withdrawn due to their toxicity and before their therapeutic potential could be assessed [46]. However, there are several potential molecules that are in various stages of development. FG-041, an inhibitor of prolyl hydroxylase, has been shown to block vascular restenosis and cardiac fibrosis in rat models. Recent studies demonstrated prevention of interstitial fibrosis and partial recovery of left ventricular function after myocardial infarction [47]. These and other recently developed phenanthrolinone derivatives, which effectively inhibit collagen synthesis by interfering with the prolyl 4-hydroxylase, offer promise for controlling fibrosis [46]. These are being evaluated as potential agents for the prevention of human dermal scarring.

7. Future Directions

In this chapter we have cited examples of specific steps in ligand-receptor interactions, in signaling pathways and transcription activation of various genes, which can be targeted in order to prevent cardiac fibrosis. Drugs that block receptors have been in use for some time and others are in various stages of development. However, we believe the future lies in targeting various genes that play a critical role in cardiac remodeling and fibrosis following injury. This is based on the proposition that by directly attacking gene function incriminated in fibrosis, we will have a better chance of controlling fibrosis without adversely affecting other functions in the cell. For instance, if we inhibit the function of an enzyme, such as p38 that is involved in TGF-β signaling, we may solve one problem but may likely create others, as this protein is required for many other events in cell metabolism. Similarly, since each transcription factor is probably involved in the transcriptional regulation of many genes, inhibition of these transcription factors may likely affect many other genes. On the other hand, if we selectively modulate expression of type I collagen and other matrix components (the final step in the pathways leading to fibrosis) by targeting the *cis*-acting elements in their promoters, we will be down-regulating only one gene. We have already established proof-of-concept by down-regulating the expression of the α1(I) collagen gene promoter by a promoter-specific TFO as described in Figure 5. Our future plans call for organ-specific delivery of this TFO to the injured heart, liver or kidney.

Another example of a *cis*-acting target is c-MYC, which plays a critical role in cell proliferation. Recently, a novel quadruplex structure formation in the promoter of the c-MYC gene has been described [48]. The sequence located at -142 to -115 upstream from the promoter of c-MYC is able to assume a non-B DNA structure called G-quadruplex structure (reviewed in 43). The cationic porphyrin TMPyP4 represses transcription of the c-MYC gene by stabilizing the G-quadruplex structure [48]. Therefore, this porphyrin can be selectively delivered into cells, such as myofibroblasts in injured myocardium or smooth muscle cells in atherosclerotic plaques, to block the production of matrix components. Similar approaches with other genes involved in proliferation of fibroblasts, trans-differentiation of precursor cells into myofibroblasts and synthesis of fibrillar collagens, make them ideal targets for controlling organ fibrosis.

Acknowledgements

This work was supported in part by National Institutes of Health (NHLBI) grant R01-HL62229.

References

1.   Weber, K.T., Cardiac interstititum. In Heart Failure: Scientific Principles and Clinical Practice, Poole-Wilson, P.A., Colucci, W.S., Massie, B.M., Chatterjee, K., Coats, A.J.S. eds. New York NY: Churchill Livingstone, 1997: p. 13-31.
2.   Laurent, G.J., Dynamic state of collagen: pathways of collagen degradation in vivo and their possible role in regulation of collagen mass. Am J Physiol, 1987. 252: p. C1-9.
3.   Sun, Y., et al., Fibrous tissue and angiotensin II. J Mol Cell Cardiol, 1997. 29: p. 2001-12.
4.   Border, W.A. and N.A. Noble, Transforming growth factor beta in tissue fibrosis. N Engl J Med, 1994. 331: p. 1286-92.
5.   Greenberg, B., Treatment of heart failure: state of the art and prospectives. J Cardiovasc Pharmacol, 2001. 38 Suppl 2: p. S59-63.
6.   Zaman, M.A., S. Oparil, and D.A. Calhoun, Drugs targeting the renin-angiotensin-aldosterone system. Nat Rev Drug Discov, 2002. 1: p. 621-36.
7.   Rocha, R., et al., Aldosterone: a mediator of myocardial necrosis and renal arteriopathy. Endocrinology, 2000. 141: p. 3871-8.
8.   Modena, M.G., et al., Aldosterone inhibition limits collagen synthesis and progressive left ventricular enlargement after anterior myocardial infarction. Am Heart J, 2001. 141: p. 41-6.
9.   Leonetti, G. and C. Cuspidi, Choosing the right ACE inhibitor. A guide to selection. Drugs, 1995. 49: p. 516-35.
10.   Fabris, B., et al., Increased cardiac angiotensin-converting enzyme in rats with chronic heart failure. Clin Exp Pharmacol Physiol, 1990. 17: p. 309-14.
11.   Fabris, B., et al., Characterization of cardiac angiotensin converting enzyme (ACE) and in vivo inhibition following oral quinapril to rats. Br J Pharmacol, 1990. 100: p. 651-5.
12.   Anon., Effect of enalapril on survival in patients with reduced left ventricular ejection fractions and congestive heart failure. N Engl J Med, 1991. 325: p. 293-302.
13.   Anon., Effect of enalapril on mortality and the development of heart failure in asymptomatic patients with reduced left ventricular ejection fractions. N Engl J Med, 1992. 327: p. 669-677.
14.   Pfeffer, M.A., et al., Effect of captopril on mortality and morbidity in patients with left ventricular dysfunction after myocardial infarction. Results of the survival and ventricular enlargement trial. The SAVE Investigators. N Engl J Med, 1992. 327: p. 669-77.
15.   St John Sutton, M., et al., Cardiovascular death and left ventricular remodeling two years after myocardial infarction: baseline predictors and impact of long-term use of captopril: information from the Survival and Ventricular Enlargement (SAVE) trial. Circulation, 1997. 96: p. 3294-9.
16.   Konstam, M.A., et al., Effects of the angiotensin converting enzyme inhibitor enalapril on the long-term progression of left ventricular dysfunction in patients with heart failure. SOLVD Investigators. Circulation, 1992. 86: p. 431-8.
17.   Weber, K.T., Extracellular matrix remodeling in heart failure: a role for de novo angiotensin II generation. Circulation, 1997. 96: p. 4065-82.
18.   Weber, K.T., et al., Angiotensin II and extracellular matrix homeostasis. Int J Biochem Cell Biol, 1999. 31: p. 395-403.
19.   Weber, K.T., Angiotensin II and connective tissue: homeostasis and reciprocal regulation. Regul Pept, 1999. 82: p. 1-17.
20.   Brilla, C.G., L.S. Matsubara, and K.T. Weber, Anti-aldosterone treatment and the prevention of myocardial fibrosis in primary and secondary hyperaldosteronism. J Mol Cell Cardiol, 1993. 25: p. 563-75.

21.   Pitt, B., et al., The effect of spironolactone on morbidity and mortality in patients with severe heart failure. Randomized Aldactone Evaluation Study Investigators. N Engl J Med, 1999. 341: p. 709-17.

22.   Zannad, F., et al., Limitation of excessive extracellular matrix turnover may contribute to survival benefit of spironolactone therapy in patients with congestive heart failure: insights from the randomized aldactone evaluation study (RALES). Rales Investigators. Circulation, 2000. 102: p. 2700-6.

23.   Park, J.B. and E.L. Schiffrin, Cardiac and vascular fibrosis and hypertrophy in aldosterone-infused rats: role of endothelin-1. Am J Hypertens, 2002. 15: p. 164-9.

24.   Ogata, T., et al., Stimulation of peroxisome-proliferator-activated receptor alpha (PPAR alpha) attenuates cardiac fibrosis and endothelin-1 production in pressure-overloaded rat hearts. Clin Sci, 2002. 103: p. 284S-288S.

25.   Rhaleb, N.E., et al., Effect of N-acetyl-seryl-aspartyl-lysyl-proline on DNA and collagen synthesis in rat cardiac fibroblasts. Hypertension, 2001. 37: p. 827-32.

26.   Fraccarollo, D., et al., Collagen accumulation after myocardial infarction: effects of ETA receptor blockade and implications for early remodeling. Cardiovasc Res, 2002. 54: p. 559-67.

27.   Huse, M., et al., The TGF beta receptor activation process: an inhibitor- to substrate-binding switch. Mol Cell, 2001. 8: p. 671-82.

28.   Hao, J., et al., Elevation of expression of Smads 2, 3, and 4, decorin and TGF-beta in the chronic phase of myocardial infarct scar healing. J Mol Cell Cardiol, 1999. 31: p. 667-78.

29.   Laping, N.J., et al., Inhibition of transforming growth factor (TGF)-beta1-induced extracellular matrix with a novel inhibitor of the TGF-beta type I receptor kinase activity: SB-431542. Mol Pharmacol, 2002. 62: p. 58-64.

30.   Rodriguez-Barbero, A., et al., Transforming growth factor-beta1 induces collagen synthesis and accumulation via p38 mitogen-activated protein kinase (MAPK) pathway in cultured L(6)E(9) myoblasts. FEBS Lett, 2002. 513: p. 282-8.

31.   McGaha, T.L., et al., Halofuginone, an inhibitor of type-I collagen synthesis and skin sclerosis, blocks transforming-growth-factor-beta-mediated Smad3 activation in fibroblasts. J Invest Dermatol, 2002. 118: p. 461-70.

32.   Dhalla, A.K., et al., chk-YB-1b, a Y-box binding protein activates transcription from rat alpha1(I) procollagen gene promoter. Biochem J, 1998. 336 : p. 373-9.

33.   Norman, J.T., et al., The Y-box binding protein YB-1 suppresses collagen alpha 1(I) gene transcription via an evolutionarily conserved regulatory element in the proximal promoter. J Biol Chem, 2001. 276: p. 29880-90.

34.   Agrawal, S., Importance of nucleotide sequence and chemical modifications of antisense oligonucleotides. Biochim Biophys Acta, 1999. 1489: p. 53-68.

35.   Opalinska, J.B. and A.M. Gewirtz, Nucleic-acid therapeutics: basic principles and recent applications. Nat Rev Drug Discov, 2002. 1: p. 503-14.

36.   Giovannangeli, C. and C. Helene, Triplex-forming molecules for modulation of DNA information processing. Curr Opin Mol Ther, 2000. 2: p. 288-96.

37.   Knauert, M.P. and P.M. Glazer, Triplex forming oligonucleotides: sequence-specific tools for gene targeting. Hum Mol Genet, 2001. 10: p. 2243-51.

38.   Guntaka, R.V., B.R. Varma, and K.T. Weber, Triplex-forming oligonucleotides as modulators of gene expression. Int J Biochem Cell Biol, 2003. 35: p. 22-31.

39.   Roque, F., et al., Safety of intracoronary administration of c-myc antisense oligomers after percutaneous transluminal coronary angioplasty (PTCA). Antisense Nucleic Acid Drug Dev, 2001. 11: p. 99-106.

40.   Kutryk, M.J., et al., Local intracoronary administration of antisense oligonucleotide against c-myc for the prevention of in-stent restenosis: results of the randomized investigation by

the Thoraxcenter of antisense DNA using local delivery and IVUS after coronary stenting trial. J Am Coll Cardiol, 2002. 39: p. 281-7.

41.   Swamynathan, S.K., et al., Targeted disruption of one allele of the Y-box protein gene, Chk-YB-1b, in DT40 cells results in major defects in cell cycle. Biochem Biophys Res Commun, 2002. 296: p. 451-7.

42.   Weiss, R.H. and C.J. Randour, Attenuation of matrix protein secretion by antisense oligodeoxynucleotides to the cyclin kinase inhibitor p21(Waf1/Cip1). Atherosclerosis, 2002. 161: p. 105-12.

43.   Kovacs, A., et al., Triple helix-forming oligonucleotide corresponding to the polypyrimidine sequence in the rat alpha 1(I) collagen promoter specifically inhibits factor binding and transcription. J Biol Chem, 1996. 271: p. 1805-12.

44.   Joseph, J., et al., Antiparallel polypurine phosphorothioate oligonucleotides form stable triplexes with the rat alpha1(I) collagen gene promoter and inhibit transcription in cultured rat fibroblasts. Nucleic Acids Res, 1997. 25: p. 2182-8.

45.   Nakanishi, M., K.T. Weber, and R.V. Guntaka, Triple helix formation with the promoter of human alpha1(I) procollagen gene by an antiparallel triplex-forming oligodeoxyribonucleotide. Nucleic Acids Res, 1998. 26: p. 5218-22.

46.   Franklin, T.J., et al., Inhibition of prolyl 4-hydroxylase in vitro and in vivo by members of a novel series of phenanthrolinones. Biochem J, 2001. 353: p. 333-8.

47.   Nwogu, J.I., et al., Inhibition of collagen synthesis with prolyl 4-hydroxylase inhibitor improves left ventricular function and alters the pattern of left ventricular dilatation after myocardial infarction. Circulation, 2001. 104: p. 2216-21.

48.   Siddiqui-Jain, A., et al., Direct evidence for a G-quadruplex in a promoter region and its targeting with a small molecule to repress c-MYC transcription. Proc Natl Acad Sci U S A, 2002. 99: p. 11593-8.

Chapter 17

Prospects for Gene Therapy for the Fibrosed Heart: Targeting Regulators of Extracellular Matrix Turnover

Hiroshi Ashikaga and Francisco J. Villarreal
University of California, San Diego, California, U.S.A.

1. Introduction

Over the past three decades, molecular and cellular biology has elucidated fundamental biological processes for a number of human ailments. An enormous amount of information on genes and proteins has come to revolutionize therapeutic strategies, and accelerate the evolution of a series of biologically targeted therapies using recombinant DNA technology. The first generation of such therapy includes recombinant forms of naturally occurring proteins in humans. The second-generation drugs are engineered recombinant proteins with application-specific properties, including fusion proteins and monoclonal antibody drugs. The third-generation represents gene therapy in which therapeutic proteins are produced by the patient's cells after transfer of target genes [1]. Whereas first- and second-generation protein drugs have already become a part of standard clinical practice [2], gene therapy has yet to be a clinical reality, despite a substantial technological leap in the last decade. The main stage of gene therapy has shifted from bench to bedside, mainly because of sophistication of recombinant DNA technology and the development of various vector systems. A number of gene therapy clinical trials have been completed to prove its efficacy and safety in various diseases: cystic fibrosis [3-14], malignancy [15-20], AIDS [21, 22], Parkinson's disease [23], and muscular dystrophy [24, 25]. Gene therapy for the cardiovascular system is also in the initial stages of development. The excess deposition of extracellular matrix proteins (i.e., fibrosis) that occurs with cardiac remodeling should be considered as a possible suitable target for gene therapy and is the focus of this chapter.

2. Cardiovascular Gene therapy

Early in the investigation of cardiovascular gene therapy, it became apparent that mere injection of gene transfer vectors into the coronary arteries, which potentially offers therapeutic effects throughout the heart, does not result in widespread transfection of the myocardium *in vivo*. Investigations have identified several physical factors that favor transcoronary gene transfer: high vector concentrations, long vector exposure times, physiological temperatures [26], high coronary flow rates [27], high perfusion pressure [28, 29], and catheter/vector compatibility [30]. In addition, the presence of specific chemical agents to increase coronary vascular permeability turned out to be crucial. These permeability agents include histamine [31], bradykinin, sodium nitroprusside [32], vascular endothelial growth factor (VEGF) [33, 34], substance P [35], and thrombin [36]. Based on these results, a number of preclinical studies succeeded in achieving high transfection rates, using temporary occlusion of great vessels [37-39], total body hypothermia [40, 41], cardiopulmonay bypass [42, 43], or coronary catheters [44-53].

The discoveries at the bench combined with clinical experience in the cardiac catheterization laboratories have led to an acceptance of catheter-based cardiac gene transfer methods, which have proven to be effective in cardiovascular gene therapy trials in human patients [54-56]. However, the success of these clinical trials may, at least partially, be attributable to the nature of the target genes, which are angiogenic growth factors to promote the growth of collateral blood vessels. These growth factors may exert measurable biological effects without transfecting a large number of cells in the heart, provided that each transfected cell secretes substantial amounts of target proteins to stimulate the downstream angiogenic cascade. In contrast, genes coding for proteins that must remain associated to the cell to exert a biological effect (e.g., sarcoplasmic reticulum Ca^{2+}-ATPase) would need to be delivered and expressed in a large number of cells.

3. Targeting the Cardiac ECM

Pathological fibrosis in the myocardium, characterized by excessive collagen deposition in the extracellular matrix (ECM), contributes to left ventricular (LV) dysfunction and poor clinical outcome in myocardial infarction, uncontrolled hypertension, diabetic cardiomyopathy and heart failure [57]. Therefore, therapeutic strategies to reduce collagen deposition may be beneficial for these disease entities. For that purpose, as extensively discussed in other chapters found in this book, one could either inhibit the collagen synthesis in ECM, or enhance collagen degradation by stimulating the production and secretion of matrix metalloproteinases (MMPs).

Gene therapy could enhance the sensitivity of cardiac fibroblasts to humoral factors, such as bradykinin, nitric oxide and adenosine, which are known to physiologically inhibit the production of ECM. Although fibroblasts represent the most abundant cell type in the myocardium, their potential as a target for myocardial gene therapy has been neglected. Fibroblasts undergo DNA replication and therefore are amenable to vectors which rely on cell division, such as retroviruses. For example, gene therapy vectors that overexpress B_2 bradykinin receptors or A_2 adenosine receptors, both of which are responsible for the inhibition of ECM production, may enhance the inhibition of ECM production in fibroblasts.

An alternative approach to diminish cardiac fibrosis is to degrade collagen fibers by overexpressing serine proteases upstream of MMP activation, such as tissue or urokinase plasminogen activator (tPA, uPA, respectively), and thus enhance the production of plasmin (Figure 1). Overexpression of such proteases would trigger an initial catalytic event, which would be ultimately followed by the activation of MMP's and collagen degradation. Thus, the advantage of this approach is that high transfection rates may not be required to achieve a tangible effect on ECM architecture. In support of this approach, genetic ablation of uPA in mice confers protection against post-infarction cardiac rupture but impairs scar formation and infarct revascularization, suggesting the critical role of uPA in the regulation of MMPs [58]. In addition, adenovirus-mediated gene transfer of uPA significantly reduces collagen content in mice with bleomycin-induced pulmonary fibrosis [59].

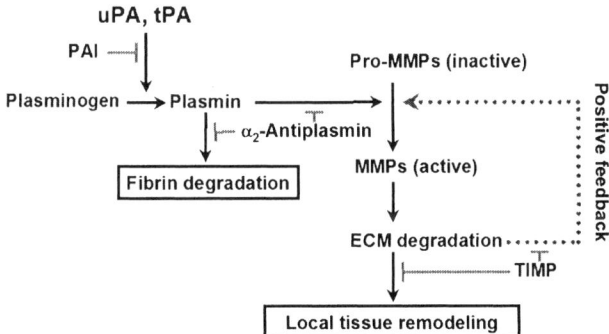

Figure 1. uPA-plasmin cascade. uPA, urokinase plasminogen activator; tPA, tissue plasminogen activator; PAI, plasminogen activator inhibitor; MMPs, matrix metalloproteinases; ECM, extracellular matrix; TIMP, tissue inhibitor of matrix proteinases.

4. Role of Plasmin in ECM Degradation and Remodeling

The study described below [60] was designed to test the hypothesis that overexpression of uPA through transcoronary adenoviral gene transfer in swine would enhance the activation of native MMPs, degrade the cardiac ECM, and alter the local tissue architecture. A total of 6 pigs were studied; animals received adenovirus encoding human uPA (adeno-uPA group, n=3) or β-galactosidase (control adenovirus; adeno-βgal group, n=3).

4.1. Materials and Methods

4.1.1. Surgical preparation

Under general anesthesia with endotracheal intubation, six domestic farm pigs (23-31 kg) underwent left thoracotomy. To measure 3-D myocardial deformation in each heart, three transmural columns of four to six 0.8-mm-diameter gold beads and a 1.7-mm-diameter surface gold bead above each column were placed within the anterior wall between the first (D_1) and the second diagonal branches (D_2) of the left anterior descending (LAD) coronary artery (Figure 2). To provide end-points for a LV long axis, 2-mm-diameter gold beads were sutured to the apical dimple (apex bead, Figure 2) and on the epicardium at the bifurcation of the left anterior descending (LAD) and left circumflex (LCx) coronary arteries (base bead, Figure 2). The local epicardial tangent plane defined by the three epicardial surface gold beads and the LV long axis were used to define a local cardiac coordinate system aligned with the circumferential, longitudinal and radial axes of the LV wall [61]. The chest was then closed and the animal was allowed to recover for 7 days.

4.1.2. Transcoronary Gene Transfer and biplane studies

Animals were positive pressure ventilated and anesthetized utilizing inhaled isoflurane. After control biplane X-ray images of the bead set were digitally recorded (125 frames/sec) [62], the animals underwent catheter-based, transcoronary gene transfer. For this purpose, a 6 Fr Swan-Ganz catheter was introduced from a carotid artery to the proximal LAD, and another 6 Fr Swan-Ganz catheter from a jugular vein to the great cardiac vein. After pretreatment with sodium nitroprusside (100μg/2minutes), adenovirus vector (1.5-2.0×10^{12} vp/animal) encoding either human single chain uPA (n=3, generously provided by T. Sisson and R. Simon, University of Michigan) or β-galactosidase (n=3) was injected into the LAD while both the LAD and the great cardiac vein were occluded simultaneously with a balloon (Figure 2). Biplane X-ray studies were repeated at 3 days and 6 days post gene transfer with the left ventricular end-diastolic pressure matched to that in the control study. End-diastole was defined at the time of the peak of the ECG R-wave at each study. At the end of the final biplane study, the animal was euthanized with an overdose of pentobarbital sodium.

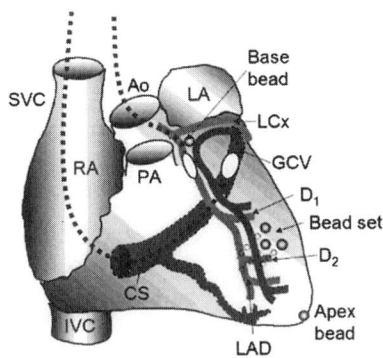

Figure 2. Transcoronary gene transfer. SVC, superior vena cava; IVC, inferior vena cava; RA, right atrium; Ao, aorta; PA, pulmonary artery; CS, coronary sinus; LA, left atrium; LAD, left anterior descending, LCx, left circumflex; GCV, great cardiac vein; $D_{1,2}$, first and second diagonal branch of LAD, respectively.

4.1.3. Histology

Hearts were excised, and rinsed in PBS. A transmural rectangular block of tissue (1.5 cm x 1.5 cm x wall thickness) in the implanted bead set was carefully removed from the ventricular wall. Another block of tissue of the same size was also excised from the posterior wall of the LV at the same level for a control. These transmural rectangular tissues were embedded with O.C.T. compound, quickly frozen in a 2-methybutane bath cooled with dry ice, and stored at –80°C. Specimens were mounted on a freezing microtome, and 5-μm sections were transferred to glass slides. X-gal staining was performed to confirm gene transfection in the adeno-β gal group (n=3). Sections were fixed in buffered 0.1% glutaraldehyde for 5 minutes at room temperature and rinsed briefly in phosphate-buffered saline (PBS). β-Galactosidase staining was performed in 5 mM $K_4Fe(CN)_6$, 5 mM $K_3Fe(CN)_6$, 1 mM $MgCl_2$, and 1 mg/mL X-Gal (5-bromo-4-chloro-3-indoyl-β-D-galactopyranoside) in PBS (pH 7.4) for 3 hours at 37°C. The sections were then fixed again in 4% paraformaldehyde for 10 minutes, rinsed in PBS for 5 minutes, and counterstained with 1% Neutral red for 3 minutes at room temperature. To assess local areas of inflammation, hematoxylin and eosin (H&E) staining was performed. Immunohistochemistry was performed to confirm the presence of human uPA (adeno-uPA group) and to assess collagen degradation by MMP (both groups) using rabbit polyclonal antibodies for human high molecular weight (HMW) uPA (Molecular Innovations, Inc, Southfield, MI) and type I collagen telopeptide (ICTP), respectively (Vectastain ABC system, Vector Laboratories, Burlingame, CA). Tissue MMP activity was measured using an OmniMMP fluorescent peptide substrate (BIOMOL Research lab, Inc.).

4.1.4. Finite strain analysis

The digital images obtained from the biplane X-ray intensifiers were spherically corrected [62] to reconstruct the 3-D coordinates [63] of gold bead markers at end-diastole for each study. End-diastolic transmural distributions of 3-D remodeling strains were calculated as a deformed configuration with the end-diastolic configuration at the control study as the reference state [62]. Six independent finite strains (E_{11}, circumferential; E_{22}, longitudinal; E_{33}, radial E_{12}; E_{23} and E_{13}) were computed in the local cardiac coordinate system [61]. In each set of finite strains, three normal strain components (E_{ii}) reflect myocardial stretch or shortening along the X_i axis, and three shear strains (E_{ij}) represent angle changes between pairs of the initially orthogonal coordinate axes (X_i and X_j).

4.1.5. Statistical analysis

Values are means ± SE unless otherwise specified. *t*-test was used to compare MMP activities in different regions. Statistics were performed using SigmaStat 3.0 (SPSS, Inc. Chicago, IL). Statistical significance was considered as $p < 0.05$.

4.2. Results

In the adeno-βgal group, X-gal staining confirmed diffuse transfection in the bead set area, particularly adjacent to dilated venules (Figure 3).

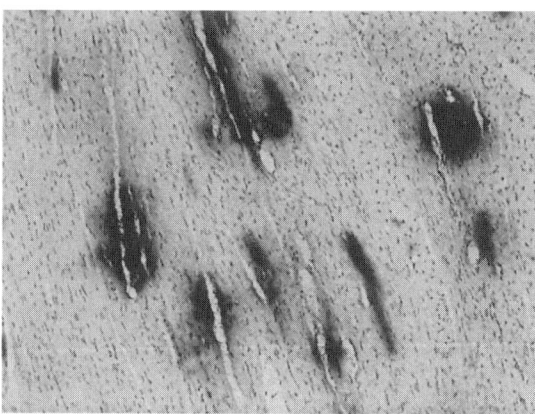

Figure 3. β-galactosidase staining of the gene transfer area (anterior wall, X200).

In the adeno-uPA group, immunohistochemistry confirmed diffuse expression of human uPA in the bead set area (Figure 4).

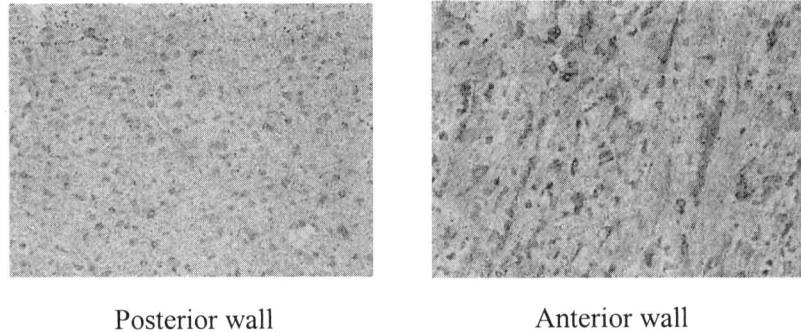

Posterior wall   Anterior wall

Figure 4. Immunohistochemistry for human uPA (x200) in the control (posterior wall) vs. gene transfer (anterior wall) areas. Note the diffuse presence of human uPA (more intense shade of gray) in the anterior wall compared with posterior wall.

In both the adeno-βgal and adeno-uPA groups, the mean MMP activity was higher in the anterior wall than the posterior wall, however differences were not significant (P = NS, Figure 5). MMP activity was higher in the adeno-uPA group than in the adeno-β gal group in both regions, and the difference was statistically significant in the anterior wall (P = 0.004).

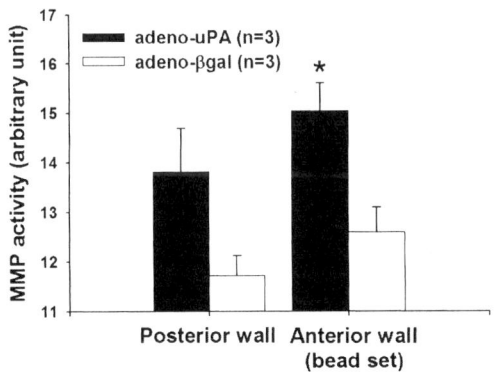

Figure 5. MMP activity in control (posterior wall) vs. gene transfer (anterior wall) areas. *P < 0.05 adeno–uPA vs. adeno-βgal group in the same region.

H&E staining did not show evidence of local inflammation in any of the heart specimens (data not shown). Immunohistochemistry for ICTP detected diffuse presence of ICTP (Figure 6) in both areas of the adeno-βgal group, and in the posterior wall of the adeno-uPA group. A significant reduction of ICTP was observed in the anterior wall of the adeno-uPA group, indicating degradation of type I collagen fibers (upper right hand panel).

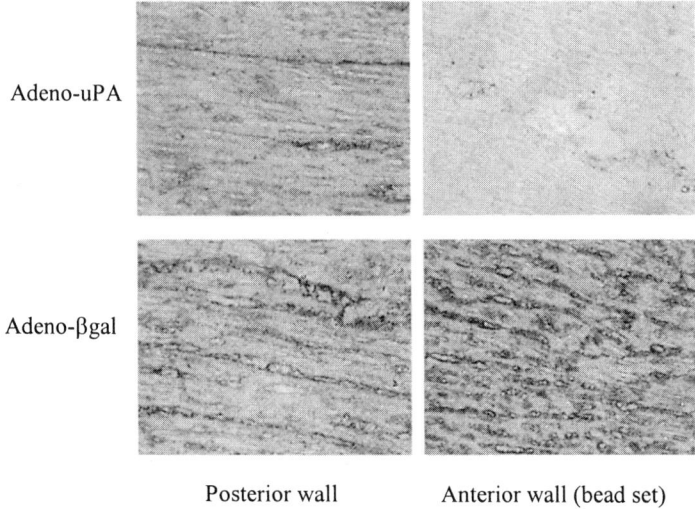

Adeno-uPA

Adeno-βgal

Posterior wall   Anterior wall (bead set)

Figure 6. Immunohistochemistry for ICTP (x200) in control (posterior wall) vs. gene transfer (anterior wall) areas. Note diffuse presence of ICTP (gray intensity) in both areas of the adeno-βgal group, and in the posterior wall of the adeno-uPA group.

Remodeling strain analysis revealed substantial end-diastolic 3-D deformation of the tissue containing the bead set (Figure 7). At 3 days, slight shear between the longitudinal and circumferential axes was oberved without any change in the radial direction. At 6 days, a significant increase in radial wall thickness and circumferential-radial shear were observed.

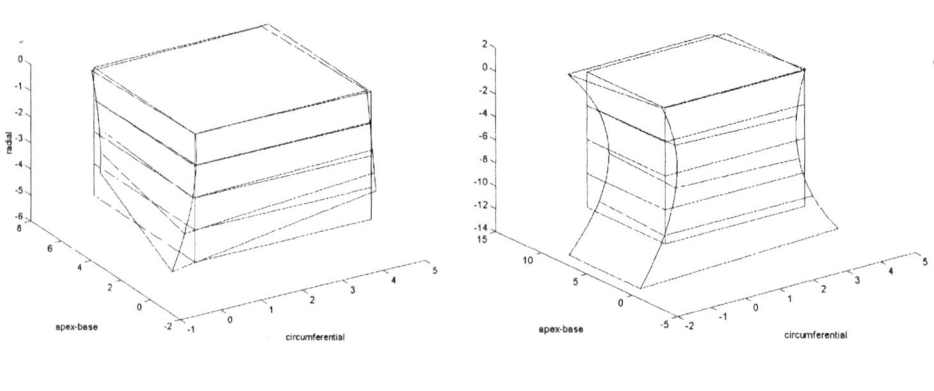

Control vs. 3 days   Control vs. 6 days

Figure 7. Tissue remodeling at the bead set. At each time point, the undeformed cube represents the control state and another cube the deformed state.

4.3 Discussion

On the basis of the limited number of studies performed, our catheter-based transcoronary gene transfer method achieved successful gene delivery in the target area to exert measurable effects. Transcoronary gene delivery is enhanced with coronary sinus occlusion [28]. Figure 3 shows diffuse transfection, particularly near dilated venules, indicating increased transvenular pressures due to coronary sinus occlusion. With successful transfection of human uPA gene, MMP activity was significantly increased in the gene transfer area. MMP activity was also enhanced in the posterior wall, although not significantly, most likely due to the nature of our gene transfer method. Because coronary vasculature forms a continuum from coronary artery to coronary sinus, it is possible that a fraction of the vectors were transfected in the posterior wall via the coronary veins. In fact, Figure 4 shows slight positive uPA staining in the posterior wall. Transfection of β-galactosidase also increased MMP activity in the anterior wall compared with the posterior wall, but the difference was not statistically significant ($P = NS$). Immunohistochemistry with ICTP clearly demonstrated loss of ICTP in the gene transfer area, indicating that increased MMP activity is associated with collagen degradation. Assessment of 3-D remodeling strains revealed substantial alteration of the regional architecture in the gene transfer area.

5. Conclusions and Future Directions

In this study, adenoviral gene transfer of human uPA induced activation of MMPs, degradation of ECM, and alteration of local tissue architecture in the ventricle. Gene transfer of uPA can potentially be a novel therapeutic strategy to reduce pathologic fibrosis in the heart. A larger-scale study will be needed to confirm these results.

Acknowledgments

We thank James W. Covell, John Ross, Jr., and Kenneth R. Chien for intellectual contribution. We also thank Rish Pavelec, Rachel Alexander, Michael Barr, Leslie Hsu and Shirley Reynolds for technical assistance. This study was supported by NIH Grant R01HL43716 (F.V.). H.A. is a recipient of the American Heart Association Postdoctoral Fellowship (Western States Affiliate).

References

1. Buckel, P., Recombinant proteins for therapy. Trends Pharmacol Sci, 1996. 17: p. 450-6.
2. Ashikaga, H., Ben-Yehuda, O., and Chien,K.R., Biotechnology and Cardiovascular Medicine: Recombinant Protein Therapy, in Molecular Basis of Cardiovascular Disease, Chien,K.R., Editor. WB Saunders., 2003, p. Chapter 1.

3.   Alton, E.W. and D.M. Geddes, Gene therapy for cystic fibrosis: a clinical perspective. Gene Ther, 1995. 2: p. 88-95.

4.   Alton, E.W., et al., Cationic lipid-mediated CFTR gene transfer to the lungs and nose of patients with cystic fibrosis: a double-blind placebo-controlled trial. Lancet, 1999. 353: p. 947-54.

5.   Flotte, T., et al., A phase I study of an adeno-associated virus-CFTR gene vector in adult CF patients with mild lung disease. Hum Gene Ther, 1996. 7: p. 1145-59.

6.   Hyde, S.C., et al., Repeat administration of DNA/liposomes to the nasal epithelium of patients with cystic fibrosis. Gene Ther, 2000. 7: p. 1156-65.

7.   Knowles, M.R., et al., A controlled study of adenoviral-vector-mediated gene transfer in the nasal epithelium of patients with cystic fibrosis. N Engl J Med, 1995. 333: p. 823-31.

8.   Knowles, M.R., et al., A double-blind, placebo controlled, dose ranging study to evaluate the safety and biological efficacy of the lipid-DNA complex GR213487B in the nasal epithelium of adult patients with cystic fibrosis. Hum Gene Ther, 1998. 9: p. 249-69.

9.   Porteous, D.J., et al., Evidence for safety and efficacy of DOTAP cationic liposome mediated CFTR gene transfer to the nasal epithelium of patients with cystic fibrosis. Gene Ther, 1997. 4: p. 210-8.

10.   Wagner, J.A., et al., A phase I/II study of tgAAV-CF for the treatment of chronic sinusitis in patients with cystic fibrosis. Hum Gene Ther, 1998. 9: p. 889-909.

11.   Wagner, J.A., et al., Safety and biological efficacy of an adeno-associated virus vector-cystic fibrosis transmembrane regulator (AAV-CFTR) in the cystic fibrosis maxillary sinus. Laryngoscope, 1999. 109: p. 266-74.

12.   Wagner, J.A., et al., Maxillary sinusitis as a surrogate model for CF gene therapy clinical trials in patients with antrostomies. J Gene Med, 1999. 1: p. 13-21.

13.   Wagner, J.A., et al., A phase II, double-blind, randomized, placebo-controlled clinical trial of tgAAVCF using maxillary sinus delivery in patients with cystic fibrosis with antrostomies. Hum Gene Ther, 2002. 13: p. 1349-59.

14.   Zabner, J., et al., Repeat administration of an adenovirus vector encoding cystic fibrosis transmembrane conductance regulator to the nasal epithelium of patients with cystic fibrosis. J Clin Invest, 1996. 97: p. 1504-11.

15.   Dalgleish, A.G., Cancer vaccines. Br J Cancer, 2000. 82: p. 1619-24.

16.   Habib, N., et al., Clinical trial of E1B-deleted adenovirus (dl1520) gene therapy for hepatocellular carcinoma. Cancer Gene Ther, 2002. 9: p. 254-9.

17.   Hasenburg, A., et al., Thymidine kinase gene therapy with concomitant topotecan chemotherapy for recurrent ovarian cancer. Cancer Gene Ther, 2000. 7: p. 839-44.

18.   Jaffee, E.M., et al., A phase I clinical trial of lethally irradiated allogeneic pancreatic tumor cells transfected with the GM-CSF gene for the treatment of pancreatic adenocarcinoma. Hum Gene Ther, 1998. 9: p. 1951-71.

19.   Kauczor, H.U., et al., CT-guided intratumoral gene therapy in non-small-cell lung cancer. Eur Radiol, 1999. 9: p. 292-6.

20.   Li, D., et al., Combination nonviral interleukin 2 and interleukin 12 gene therapy for head and neck squamous cell carcinoma. Arch Otolaryngol Head Neck Surg, 2001. 127: p. 1319-24.

21.   Deeks, S.G., et al., A phase II randomized study of HIV-specific T-cell gene therapy in subjects with undetectable plasma viremia on combination antiretroviral therapy. Mol Ther, 2002. 5: p. 788-97.

22.   Mitsuyasu, R.T., et al., Prolonged survival and tissue trafficking following adoptive transfer of CD4zeta gene-modified autologous CD4(+) and CD8(+) T cells in human immunodeficiency virus-infected subjects. Blood, 2000. 96: p. 785-93.

23.   During, M.J., et al., Subthalamic GAD gene transfer in Parkinson disease patients who are candidates for deep brain stimulation. Hum Gene Ther, 2001. 12: p. 1589-91.

24.   Law, P.K., et al., Human gene therapy with myoblast transfer. Transplant Proc, 1997. 29: p. 2234-7.

25. Mendell, J.R., et al., Myoblast transfer in the treatment of Duchenne's muscular dystrophy. N Engl J Med, 1995. 333: p. 832-8.

26. Donahue, J.K., et al., Ultrarapid, highly efficient viral gene transfer to the heart. Proc Natl Acad Sci U S A, 1997. 94: p. 4664-8.

27. Yuan, Y., et al., Flow modulates coronary venular permeability by a nitric oxide-related mechanism. Am J Physiol, 1992. 263: p. H641-6.

28. Logeart, D., et al., How to optimize in vivo gene transfer to cardiac myocytes: mechanical or pharmacological procedures? Hum Gene Ther, 2001. 12: p. 1601-10.

29. Wright, M.J., et al., In vivo myocardial gene transfer: optimization and evaluation of intracoronary gene delivery in vivo. Gene Ther, 2001. 8: p. 1833-9.

30. Marshall, D.J., et al., Biocompatibility of cardiovascular gene delivery catheters with adenovirus vectors: an important determinant of the efficiency of cardiovascular gene transfer. Mol Ther, 2000. 1: p. 423-9.

31. Pilati, C.F. and M.B. Maron, Effect of histamine on coronary microvascular permeability. Am J Physiol, 1984. 247: p. H1-7.

32. Yuan, Y., et al., Histamine increases venular permeability via a phospholipase C-NO synthase-guanylate cyclase cascade. Am J Physiol, 1993. 264: p. H1734-9.

33. Nagata, K., et al., Phosphodiesterase inhibitor-mediated potentiation of adenovirus delivery to myocardium. J Mol Cell Cardiol, 2001. 33: p. 575-80.

34. Wu, H.M., et al., VEGF induces NO-dependent hyperpermeability in coronary venules. Am J Physiol, 1996. 271: p. H2735-9.

35. Baluk, P., et al., Endothelial gaps: time course of formation and closure in inflamed venules of rats. Am J Physiol, 1997. 272: p. L155-70.

36. van Nieuw Amerongen, G.P., et al., Transient and prolonged increase in endothelial permeability induced by histamine and thrombin: role of protein kinases, calcium, and RhoA. Circ Res, 1998. 83: p. 1115-23.

37. Maurice, J.P., et al., Enhancement of cardiac function after adenoviral-mediated in vivo intracoronary beta2-adrenergic receptor gene delivery. J Clin Invest, 1999. 104: p. 21-9.

38. Miyamoto, M.I., et al., Adenoviral gene transfer of SERCA2a improves left-ventricular function in aortic-banded rats in transition to heart failure. Proc Natl Acad Sci U S A, 2000. 97: p. 793-8.

39. Schmidt, U., et al., Restoration of diastolic function in senescent rat hearts through adenoviral gene transfer of sarcoplasmic reticulum Ca(2+)-ATPase. Circulation, 2000. 101: p. 790-6.

40. Ikeda, Y., et al., Restoration of deficient membrane proteins in the cardiomyopathic hamster by in vivo cardiac gene transfer. Circulation, 2002. 105: p. 502-8.

41. Hoshijima, M., et al., Chronic suppression of heart-failure progression by a pseudophosphorylated mutant of phospholamban via in vivo cardiac rAAV gene delivery. Nat Med, 2002. 8: p. 864-71.

42. Davidson, M.J., et al., Cardiac gene delivery with cardiopulmonary bypass. Circulation, 2001. 104: p. 131-3.

43. Jones, J.M., et al., Adenoviral gene transfer to the heart during cardiopulmonary bypass: effect of myocardial protection technique on transgene expression. Eur J Cardiothorac Surg, 2002. 21: p. 847-52.

44. Barr, E., et al., Efficient catheter-mediated gene transfer into the heart using replication-defective adenovirus. Gene Ther, 1994. 1: p. 51-8.

45. Kaplitt, M.G., et al., Long-term gene transfer in porcine myocardium after coronary infusion of an adeno-associated virus vector. Ann Thorac Surg, 1996. 62(6): p. 1669-76.

46. Giordano, F.J., et al., Intracoronary gene transfer of fibroblast growth factor-5 increases blood flow and contractile function in an ischemic region of the heart. Nat Med, 1996. 2: p. 534-9.

47. Lai, N.C., et al., Intracoronary delivery of adenovirus encoding adenylyl cyclase VI increases left ventricular function and cAMP-generating capacity. Circulation, 2000. 102:

p. 2396-401.

48. Shah, A.S., et al., Intracoronary adenovirus-mediated delivery and overexpression of the beta(2)-adrenergic receptor in the heart : prospects for molecular ventricular assistance. Circulation, 2000. 101: p. 408-14.

49. Shah, A.S., et al., In vivo ventricular gene delivery of a beta-adrenergic receptor kinase inhibitor to the failing heart reverses cardiac dysfunction. Circulation, 2001. 103: p. 1311-6.

50. Donahue, J.K., et al., Focal modification of electrical conduction in the heart by viral gene transfer. Nat Med, 2000. 6: p. 1395-8.

51. Boekstegers, P., et al., Myocardial gene transfer by selective pressure-regulated retroinfusion of coronary veins. Gene Ther, 2000. 7: p. 232-40.

52. Kupatt, C., et al., Retroinfusion of NFkappaB decoy oligonucleotide extends cardioprotection achieved by CD18 inhibition in a preclinical study of myocardial ischemia and retroinfusion in pigs. Gene Ther, 2002. 9: p. 518-26.

53. Kupatt, C., et al., VEGF165 transfection decreases postischemic NF-kappa B-dependent myocardial reperfusion injury in vivo: role of eNOS phosphorylation. FASEB J, 2003. 17: p. 705-7.

54. Laitinen, M., et al., Catheter-mediated vascular endothelial growth factor gene transfer to human coronary arteries after angioplasty. Hum Gene Ther, 2000. 11: p. 263-70.

55. Grines, C.L., et al., Angiogenic Gene Therapy (AGENT) trial in patients with stable angina pectoris. Circulation, 2002. 105: p. 1291-7.

56. Losordo, D.W., et al., Phase 1/2 placebo-controlled, double-blind, dose-escalating trial of myocardial vascular endothelial growth factor 2 gene transfer by catheter delivery in patients with chronic myocardial ischemia. Circulation, 2002. 105: p. 2012-8.

57. Jugdutt, B.I., Remodeling of the myocardium and potential targets in the collagen degradation and synthesis pathways. Curr Drug Targets Cardiovasc Haematol Disord, 2003. 3: p. 1-30.

58. Heymans, S., et al., Inhibition of plasminogen activators or matrix metalloproteinases prevents cardiac rupture but impairs therapeutic angiogenesis and causes cardiac failure. Nat Med, 1999. 5: p. 1135-42.

59. Sisson, T.H., et al., Treatment of bleomycin-induced pulmonary fibrosis by transfer of urokinase-type plasminogen activator genes. Hum Gene Ther, 1999. 10: p. 2315-23.

60. Ashikaga, H., et al., Effects of gene transfer of urokinase-type plasminogen activator on extracellular matrix and ventricular remodelling in swine. J Am Coll Cardiol, 2003. 41: p. A864-2.

61. Meier, G.D., et al., Kinematics of the beating heart. IEEE Trans Biomed Eng, 1980. 27: p. 319-29.

62. Ashikaga, H., et al., Transmural mechanics underlying left ventricular torsional recoil during isovolumic relaxation. Am J Physiol, 2003.

63. MacKay, S.A., M.J. Potel, and J.M. Rubin, Graphics methods for tracking three-dimensional heart wall motion. Comput Biomed Res, 1982. 15: p. 455-73.

Chapter 18

Therapeutic Potential of TIMPs in Heart Failure

Suresh C. Tyagi, Ph.D.
University of Louisville School of Medicine, Louisville, Kentucky, U.S.A.

1. Introduction

The accumulation of oxidized extracellular matrix (ECM) between the endothelium and cardiac muscle, and endocardial endothelial dysfunction, are hallmarks of congestive heart failure. The induction of oxidative stress, decrease in endothelial cell density, activation of matrix and disintegrin metalloproteinases, collagenolysis, and repression of cardiac inhibitor of metalloproteinase (CIMP) are associated with deposition of oxidized matrix. As elaborated in the chapter below, studies that employ CIMP (also known as tissue inhibitor of metalloproteinase-4) as genetic material or a proteomic therapeutic based agent may improve the heart's response to oxidative stresses.

2. Structure and function of MMPs and their inhibitors.

Currently, there are more than 40 known metalloproteinases (MMPs) including disintegrin metalloproteinases (DMPs). Based on their substrate specificity MMPs are grouped into four major classes [1-3]. Interstitial collagenases, gelatinases, stromelysins and other such as membrane bound (MT-MMPs) and DMPs. MMPs can be inhibited in tissues by four tissue inhibitor of metalloproteinase (TIMPs 1-4) [4, 5]. Mammalian TIMPs are two-domain molecules [5]. The high level of sequence divergence between the four TIMPs suggests that they probably possess different functional properties. TIMP-1 is known to inhibit MMP-1 and -2. TIMP-2 inhibits MMP-2. TIMP-3 is expressed in vascular tissues and can inhibit MMP-3, and -7. TIMP-4 (also known as CIMP) is expressed at high levels in the heart and inhibits MMP-9 and DMPs [4, 5]. Most of the TIMPs have molecular weight in the range of 25 kDa. TIMPs have 60-70% sequence homologies in their primary structure. The tertiary structure of TIMP molecules reveal MMP binding domains and a membrane binding domain, suggesting other role of TIMPs in ECM remodeling (figure 1).

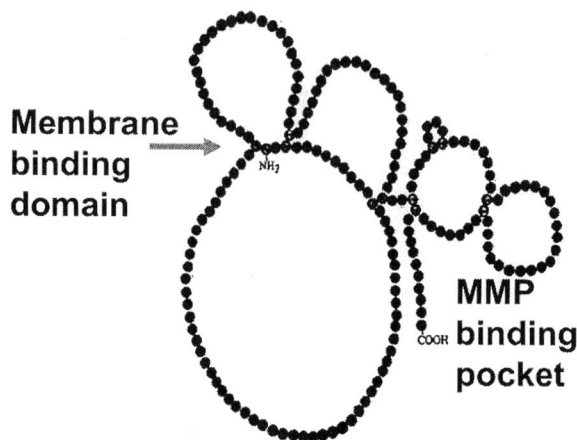

Figure 1. Tertiary structure of TIMP: There are four TIMPs. All TIMPs have ~60-70% sequence homology. In TIMP structure there are four kinger regions and a global domain. The C-terminal is an MMP binding domain. The N-terminal has receptor binding domains.

3. Redox and proteolytic stresses

The balance of activity between MMPs and TIMPs can regulate ECM remodeling [1-3, 6]. As illustrated in figure 2, in the scenario of increased MMP-1, -2, -9 and DMP, if TIMP-1 and -2 are induced, this would lessen collagen degradation and lead to the development of tissue fibrosis. At the same time if CIMP production is decreased this can lead to the degradation of ultra structural collagen, elastin, fibronectin and integrins. The disruption of the elastin, fibronectin and integrin complex via signaling through the focal adhesion complex can potentially yield increases in the cellular redox stress. As described in figure 3, we postulate that increased levels of CIMP will "shed"(i.e. protect) the cell from redox and proteolytic stresses. The induction of redox stress which generates superoxide is associated with increased NADPH oxidase and decreased catalase and superoxide dismutase activities (SOD). The nitric oxide (NO) generated by eNOS, l-arginine and BH4 reacts with superoxide yielding peroxynitrite. Peroxynitrite generated from oxidative stress can activate latent resident MMPs [7] and oxidize CIMP and other matrix molecules. In this regard, the administration of CIMP by protein transfer or by gene transfer can thus, protect cells from redox and proteolytic stresses.

4. Congestive heart failure (CHF)

CHF is associated with endocardial endothelial (EE) dysfunction and

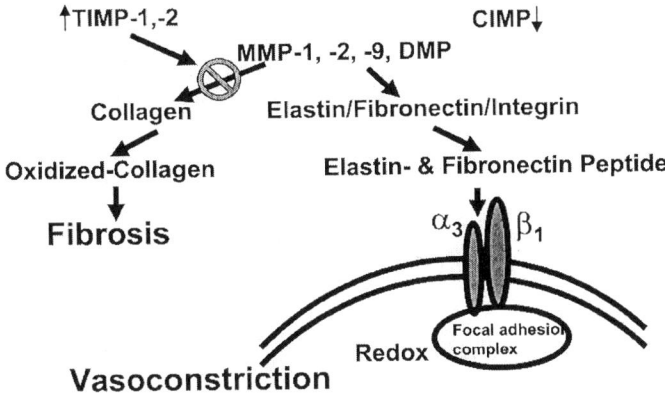

Figure 2. A balance between proteinase and antiproteinase activities is required for physiological remodeling. In cardiac pathologies MMP-1, -2, -9 and DMP are induced. TIMP-1 and -2 are also induced but CIMP is decreased. TIMP-1 and -2 inhibit MMP-1 and -2 and yield collagen accumulation. MMP-9 and DMP degrade collagen, elastin, fibronectin and integrins. CIMP inhibits MMP-9 and DMP. Elastin and fibronectin peptides induce redox state and vascular contractions through integrins.

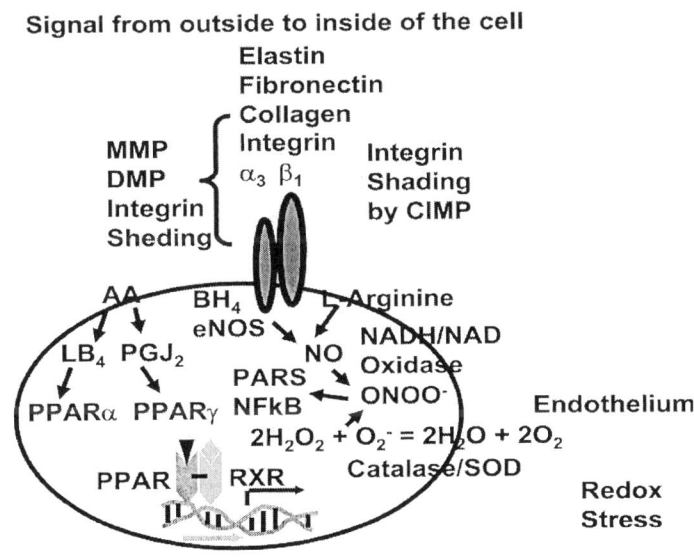

Figure 3. Extracellular matrix degradation leads to signals from outside to inside of cells. Normally CIMP protects (sheds) the integrin from redox and proteolytic stresses. Redox stress increases NAPDH oxidase and decreases catalase/SOD. To compensate AA generates LB4 and PGJ2 agonist for PPAR. The PPAR in coordinance with RXR induces catalase/SOD and decreases NADPH oxidase. The eNOS, BH4 and l-arginine generate NO. NO and superoxide induces PARS and NFκB.

increased endothelial to myocyte distance. This distance is primarily due to increased accumulation of interstitial matrix. Our laboratory has identified a link between increased interstitial matrix accumulation and EE dysfunction [8]. Although chronic exposure of ROS leads to capillary cell apoptosis, interstitial fibrosis, and oxidized-matrix accumulation between endothelium and myocyte [8-11], the mechanism by which oxidized matrix is accumulated is unclear. The efficiency of cardiac filling is inversely proportional to extracellular matrix content [12]. Moreover, oxidative-modification of collagen enhances tissue stiffness [13, 14]. Although interstitial collagen I is the most abundant ECM in the myocardium, there is substantial elastin in the basement membrane of capillary endothelium. Perivascular as well as interstitial fibrosis are prominent in CHF. MMPs degrades interstitial collagen however, emerging data suggests that the more collagen is degraded the more is synthesized [15]. Ultimately, synthesis can exceed degradation and ultimately lead to the accumulation of oxidized-collagen. MMP-2 and -9 also degrade elastin [16]. In the presence of reduced eNOs activity to ameliorate left ventricular load, latent MMPs are activated; this in turn favors the degradation of ultra structural collagen and elastin. Because the turnover of ultra structure collagen and elastin is lower than oxidized-collagen [17], the cardiac ECM is replaced by oxidatively-modified stiffer collagen. This enhances accumulation of oxidized-matrix between endothelium and myocyte, decreases eeNO diffusion, and impairs EE function.

5. TIMPs in oxidative stress and signal transduction

Under normal physiological conditions, a critical balance between MMPs/DMPs and TIMPs is maintained [1-3, 6]. However, TIMPs can also modulate cell functions. Several lines of evidence suggest that TIMP-1 can yield antimitogenic activity [18, 19]. However, we have shown that TIMP-1 has proliferative activity on endothelial cells [20]. TIMP-2 has also been shown to be a growth stimulatory protein for transformed fibroblasts [21]. Baker et al [22] demonstrated that TIMP-3 induces apoptosis in vascular smooth muscle cells and regresses neointimal growth. Results from our laboratory have shown that CIMP (TIMP-4) induces apoptosis in transformed cardiac fibroblasts and has no effect in normal fibroblast cells [23].

MMPs are regulated primarily at three levels: 1) at the transcriptional level by cytokines, growth factors, neurohormones, and oxidative stress; 2) by their inhibitors TIMP/CIMP; 3) by the activation of zymogen by oxidative stress and/or by proteolytic cleavage [1-3, 6]. Alterations at any of these stages can lead to an imbalance in the concentration and activity of MMPs and TIMPs and may induce the development of excess matrix degradation and/or production. Reactive oxygen species (ROS) stimulate MMPs [24]. The *in vivo* inhibition of NO can also increase MMP activity [25]. The inactivation of MMP by NO may be multi factorial: 1) because NO is an antioxidant, NO may inhibit redox-sensitive MMP gene activation [26]; 2) independent of MMP, NO may directly induce CIMP; 3) NO may block the metal ion active site in the latent form of MMP. This suggests that the rate of MMP activation is correlated to the rate of NO oxidation. Also TIMPs are sensitive to oxidative inactivation [26, 27], and during inflammation, oxidants disrupt the balance between proteinases and antiproteinases [28, 29]. The nitration of TIMP-1 has been shown to decrease its inhibitory activity [27]. Serine proteinases can activate latent myocardial MMP [1, 6, 24]. Serine proteinases also degrade inhibitors of MMPs. MMPs degrade inhibitors of serine proteinases [1, 6]. Therefore, a vicious cascade of activation/inactivation of proteinase and antiproteinase can be initiated in which accumulation of oxidized matrix is favored.

It has been shown that synthetic inhibitors (i.e. drugs) of MMPs improve cardiac dysfunction post myocardial infarction [30] and in pacing induced models of heart failure [31]. However, the specificity of these inhibitors to myocardial MMP is unclear. Furthermore, these synthetic inhibitors may not perturb the oxidative stress. On the other hand CIMP has residues (tyrosine and cysteine) which can sequester, in part, ROS, impairing their ability to activate MMPs. CIMP may thus, play a central role in NO generation and redox signaling in EE cells (figure 4). Therefore, the administration of CIMP may ultimately decrease the accumulation of oxidized-matrix via increasing NO.

Sixteen percent of the myocardium is composed of capillaries, including lumen and the endothelium [32]. Capillary endothelium, strategically located between the superfusing luminal blood and the underlying cardiac muscle, plays an important role in controlling myocardial performance [33]. A gradient of NO concentration (i.e. high in EE and low in midmyocardium) has been demonstrated, suggesting a role of EE NO in beating hearts [34]. During protracted cycles of silent ischemia/reperfusion by increased pulse pressure and increased chronic preload [33], ROS are generated: $2O_2 + 2H_2O = 2H_2O_2 + O_2^-$ (super oxide). The production of ROS can be dependent or independent of inflammatory and/or mitochondrial NADH/NAD oxidase [35]. ROS masks the activity of superoxide dismutase and catalase [36], causing a decrease in endothelial NO availability and activates latent resident myocardial MMPs [24]. In addition, decreased SOD and catalase activity can stimulate increased levels of cytokines, growth factors, and neurohormones [37-39] which initiate a vicious cycle of oxidative stress. For example, neurohormones such as angiotensin II [40] increase oxidative stress by lowering the levels of bradykinin and prostaglandins [41, 42], and also induce NADH/NAD oxidase [43]. The consequence of this intricate sequence of events is decreased NO. Therefore, increased ROS is one of the primary inciting stimuli of decreased endothelial NO.

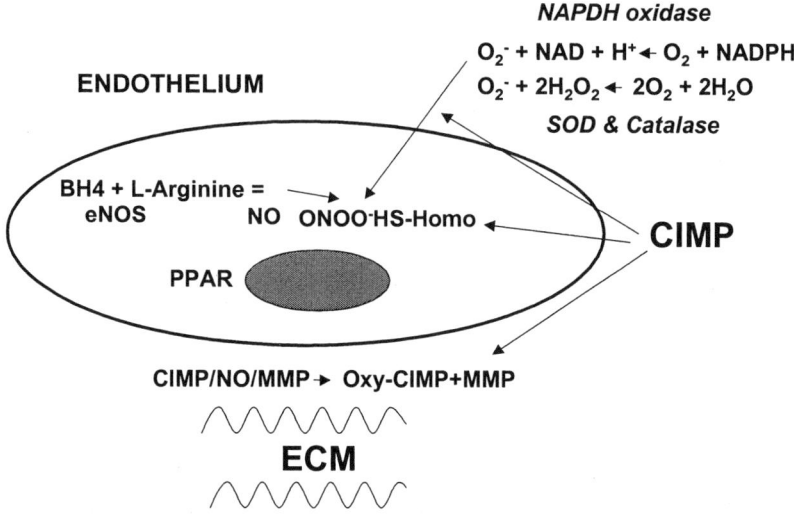

Figure 4. CIMP has the potential to inhibit MMPs and can behave as a redox sink during oxidative stress, and since it possesses a receptor binding domain it can act directly on endothelial cells. This leads to decrease redox and proteolytic stresses, and restores endothelial cell integrity.

6. CIMP gene transfer

Numerous studies have indicated that a chronically elevated work load is an important cause of heart failure. As mentioned above, endothelial dysfunction is a hallmark of CHF. The contribution of increased oxidized-matrix accumulation and decreased efficiency of eeNO in CHF is unclear. This process is enhanced by increased MMP activity, unabated proteolysis, and fibrosis. Thus, it is important to develop agents that can increase the levels of eeNO and enhance cardiac muscle relaxation. To this end, CIMP protein transfer to hearts with chronic increases in work load should benefit in that CIMP can diminish oxidative and proteolytic stresses. It is thus, currently of great interest to generate CIMP transgenic mice and measure cardiac function during chronic heart failure.

We performed a recent study [44] in which an arteriovenous fistula (AVF) was created in C57BL/J6 mice. CIMP was administered to AVF and sham mice using protein transfer into the peritoneal cavity via miniosmotic pumps for 4 weeks. Mice were grouped as follows: sham; sham + CIMP; AVF; and AVF + CIMP (n=6). In vivo left ventricular (LV) pressure was measured in all animals. We measured plasma and LV tissue levels of CIMP. LV levels of NADPH oxidase activity, were increased in AVF mice and decreased in AVF mice treated with CIMP. Compared with sham animals, CIMP was decreased in AVF mice, and CIMP protein transfer increased plasma and LV tissue levels of

CIMP in AVF mice. In situ zymography demonstrated robust increases in MMP activity in the hearts from AVF mice compared with shams, and treatment with CIMP decreased myocardial MMP activity. In AVF mice, the cardiac pressure-length relationship was similar to that observed in sham mice after administration of CIMP. Contractile responses of normal LV rings were measured in the presence and absence of CIMP. CIMP shifted the pressure-length relationship to the left, attenuated LV dilatation, and had no effect on calcium chloride-mediated contraction. Treatment of AVF mice with CIMP significantly abrogated the contractile dysfunction and decreased the oxidative stress in volume overload-induced heart failure. These and similar studies demonstrate that the mechanism of reverse remodeling by MMP inhibition specifically targeting the heart is an important and emerging area. MMP inhibition using pharmacological compounds in myocardial infarction patients currently is in clinical trials, and there is great anticipation as to determining the outcome of these studies. CIMP delivery should also be considered in the future as a possible option for treatment of heart disease.

7. Acknowledgments

This work was supported in part by NIH grants HL-71010 and HL-74185

References

1. Tyagi, S.C., Proteinases and myocardial extracellular matrix turnover. Mol Cell Biochem, 1997. 168: p. 1-12.
2. Loechel, F., et al., Human ADAM 12 (meltrin alpha) is an active metalloprotease. J Biol Chem, 1998. 273: p. 16993-7.
3. Hayden, M.R. and S.C. Tyagi, Arteriogenesis: Angiogenesis within Unstable Atherosclerotic Plaque-- Interactions with Extracellular Matrix. Curr Interv Cardiol Rep, 2000. 2: p. 218-227.
4. Tyagi, S., Dynamic role of extracellular matrix metalloproteinases in heart failure. Cardiovasc Pathol, 1998. 7: p. 153-159.
5. Brew, K., D. Dinakarpandian, and H. Nagase, Tissue inhibitors of metalloproteinases: evolution, structure and function. Biochim Biophys Acta, 2000. 1477: p. 267-83.
6. Nagase, H. and W. JF, Matrix metalloproteinase. J Biol Chem, 1999. 274: p. 21491-21494.
7. Mujumdar, V.S., G.M. Aru, and S.C. Tyagi, Induction of oxidative stress by homocyst(e)ine impairs endothelial function. J Cell Biochem, 2001. 82: p. 491-500.
8. Miller, A., et al., Reversal of endocardial endothelial dysfunction by folic acid in homocysteinemic hypertensive rats. Am J Hypertens, 2002. 15: p. 157-63.
9. Michel, J.B., et al., Morphometric analysis of collagen network and plasma perfused capillary bed in the myocardium of rats during evolution of cardiac hypertrophy. Basic Res Cardiol, 1986. 81: p. 142-54.
10. Amann, K., et al., Myocyte/capillary mismatch in the heart of uremic patients. J Am Soc Nephrol, 1998. 9: p. 1018-22.

11. Miller, A., et al., Hyperhomocysteinemia induces multiorgan damage. Heart & Vessels, 2000. 15: p. 135-143.

12. Patel, R., et al., Simavastatin induces regression of cardiac hypertrophy and fibrosis and improves cardiac function in a transgenic rabbit model of human hypertrophic cardiomyopathy. Circulation, 2001. 104: p. 317-324.

13. Capasso, J.M., T.F. Robinson, and P. Anversa, Alterations in collagen cross-linking impair myocardial contractility in the mouse heart. Circ Res, 1989. 65: p. 1657-64.

14. Matsubara, L.S., et al., Alterations in myocardial collagen content affect rat papillary muscle function. Am J Physiol, 2000. 279: p. H1534-9.

15. Aimes, R.T. and J.P. Quigley, Matrix metalloproteinase-2 is an interstitial collagenase. Inhibitor-free enzyme catalyzes the cleavage of collagen fibrils and soluble native type I collagen generating the specific 3/4- and 1/4-length fragments. J Biol Chem, 1995. 270: p. 5872-6.

16. Senior, R.M., et al., Human 92- and 72-kilodalton type IV collagenases are elastases. J Biol Chem, 1991. 266: p. 7870-5.

17. Rucklidge, G.J., et al., Turnover rates of different collagen types measured by isotope ratio mass spectrometry. Biochim Biophys Acta, 1992. 1156: p. 57-61.

18. Moses, M.A. and R. Langer, A metalloproteinase inhibitor as an inhibitor of neovascularization. J Cell Biochem, 1991. 47: p. 230-5.

19. Hayakawa, T., et al., Growth-promoting activity of tissue inhibitor of metalloproteinases-1 (TIMP-1) for a wide range of cells. A possible new growth factor in serum. FEBS Lett, 1992. 298: p. 29-32.

20. Tyagi, S.C., et al., Induction of tissue inhibitor of metalloproteinase and its mitogenic response to endothelial cells in human atherosclerotic and restenotic lesions. Can J Cardiol, 1996. 12: p. 353-62.

21. Nemeth, J.A. and C.L. Goolsby, TIMP-2, a growth-stimulatory protein from SV40-transformed human fibroblasts. Exp Cell Res, 1993. 207: p. 376-82.

22. Baker, A.H., et al., Divergent effects of tissue inhibitor of metalloproteinase-1, -2, or -3 overexpression on rat vascular smooth muscle cell invasion, proliferation, and death in vitro. TIMP-3 promotes apoptosis. J Clin Invest, 1998. 101: p. 1478-87.

23. Tummalapalli, C.M., B.J. Heath, and S.C. Tyagi, Tissue inhibitor of metalloproteinase-4 instigates apoptosis in transformed cardiac fibroblasts. J Cell Biochem, 2001. 80: p. 512-21.

24. Tyagi, S.C., A. Ratajska, and K.T. Weber, Myocardial matrix metalloproteinase(s): localization and activation. Mol Cell Biochem, 1993. 126: p. 49-59.

25. Radomski, A., et al., The role of nitric oxide and metalloproteinases in the pathogenesis of hyperoxia-induced lung injury in newborn rats. Br J Pharmacol, 1998. 125: p. 1455-62.

26. Tyagi, S.C., S. Kumar, and S. Borders, Reduction-oxidation (redox) state regulation of extracellular matrix metalloproteinases and tissue inhibitors in cardiac normal and transformed fibroblast cells. J Cell Biochem, 1996. 61: p. 139-51.

27. Frears, E.R., et al., Inactivation of tissue inhibitor of metalloproteinase-1 by peroxynitrite. FEBS Lett, 1996. 381: p. 21-4.

28. Stricklin, G.P. and J.R. Hoidal, Oxidant-mediated inactivation of TIMP. Matrix Suppl, 1992. 1: p. 325.

29. Shabani, F., J. McNeil, and L. Tippett, The oxidative inactivation of tissue inhibitor of metalloproteinase-1 (TIMP-1) by hypochlorous acid (HOCI) is suppressed by anti-rheumatic drugs. Free Radic Res, 1998. 28: p. 115-23.

30. Rohde, L.E., et al., Matrix metalloproteinase inhibition attenuates early left ventricular enlargement after experimental myocardial infarction in mice. Circulation, 1999. 99: p. 3063-70.

31.   Spinale, F.G., et al., Matrix metalloproteinase inhibition during the development of congestive heart failure : effects on left ventricular dimensions and function. Circ Res, 1999. 85: p. 364-76.

32.   Hoppeler, H. and S. Kayar, Capillary and oxidative capacity of muscles. News in physiol Sci, 1988. 3: p. 113-116.

33.   Cox, M.J., et al., Apoptosis in the left ventricle of chronic volume overload causes endocardial endothelial dysfunction in rats. Am J Physiol, 2002. 282: p. H1197-205.

34.   Pinsky, D.J., et al., Mechanical transduction of nitric oxide synthesis in the beating heart. Circ Res, 1997. 81: p. 372-9.

35.   Babior, B.M., NADPH oxidase: an update. Blood, 1999. 93: p. 1464-76.

36.   Roos, D., et al., Protection of human neutrophils by endogenous catalase: studies with cells from catalase-deficient individuals. J Clin Invest, 1980. 65: p. 1515-22.

37.   Laycock, S.K., et al., Effects of chronic norepinephrine administration on cardiac function in rats. J Cardiovasc Pharmacol, 1995. 26: p. 584-9.

38.   Givertz, M.M. and W.S. Colucci, New targets for heart-failure therapy: endothelin, inflammatory cytokines, and oxidative stress. Lancet, 1998. 352: p. SI34-8.

39.   Chen, C.Y., Y.L. Huang, and T.H. Lin, Association between oxidative stress and cytokine production in nickel-treated rats. Arch Biochem Biophys, 1998. 356: p. 127-32.

40.   Tyagi, S., M. Hayden, and J. Hall, Role of angiotensin in angiogenesis and cardiac fibrosis in heart failure, Angiotensin II Receptor Blockade: Physiological and clinical implications. Prog Exp Cardiol, 1998. 2: p. 537-549.

41.   Varin, R., et al., Improvement of endothelial function by chronic angiotensin-converting enzyme inhibition in heart failure : role of nitric oxide, prostanoids, oxidant stress, and bradykinin. Circulation, 2000. 102: p. 351-6.

42.   Li, P., et al., Angiotensin-(1-7) augments bradykinin-induced vasodilation by competing with ACE and releasing nitric oxide. Hypertension, 1997. 29: p. 394-400.

43.   Zhang, H., et al., Angiotensin II-induced superoxide anion generation in human vascular endothelial cells: role of membrane-bound NADH-/NADPH-oxidases. Cardiovasc Res, 1999. 44: p. 215-22.

44.   Cox, M.J., et al., Attenuation of oxidative stress and remodeling by cardiac inhibitor of metalloproteinase protein transfer. Circulation, 2004. 109: p. 2123-2128.

INDEX